THE MIDDLE ATMOSPHERE AS OBSERVED FROM BALLOONS, ROCKETS AND SATELLITES

THE MIDDLE ATMOSPHERE AS OBSERVED FROM BALLOONS, ROCKETS AND SATELLITES

A ROYAL SOCIETY DISCUSSION

ARRANGED BY THE BRITISH NATIONAL COMMITTEE ON SPACE RESEARCH AND SOLAR–TERRESTRIAL PHYSICS, UNDER THE LEADERSHIP OF

SIR HARRIE MASSEY, F.R.S., SIR GRANVILLE BEYNON, F.R.S., J. T. HOUGHTON, F.R.S., AND L. THOMAS

HELD ON 12 AND 13 DECEMBER 1978

LONDON
THE ROYAL SOCIETY
1980

Printed in Great Britain for the Royal Society
at the
University Press, Cambridge

ISBN 0 85403 137 5

First published in *Philosophical Transactions of the Royal Society of London*,
series A, volume 296 (no. 1418), pages 1–268

Published by the Royal Society
6 Carlton House Terrace, London SW1Y 5AG

PREFACE

The aim of atmospheric research is to describe and understand the three-dimensional fields of structure, composition and dynamics. Many complications arise from the interactions between these fields and this is particularly true for the middle atmosphere, the region between altitudes of about 10 and 100 km. In the past decade, particular attention has been paid to the composition of the stratosphere, because of concern for possible biological and climatic effects associated with reductions in the ozone layer. It has been suggested that such reductions could be caused by the injection of nitric oxide from supersonic aircraft or the release of other chemicals, such as chlorofluoromethanes, at ground level. In the studies of the associated chemistry, extensive use has been made of balloon and aircraft-borne measurements of constituents and of laboratory data on reaction rate coefficients; similarly, rocket-borne measurements have formed the basis of investigations of both neutral and ion chemistry at greater heights. Thermal emission measurements from satellites have provided global information on temperature structure and enabled deductions to be made concerning the large-scale motion field. The corresponding global data on constituents from satellite instruments are beginning to be provided; they are urgently required for the verification of theoretical models.

This volume contains papers presented at a two-day Discussion Meeting held on 12 and 13 December 1978 at the Royal Society. These papers are divided almost equally between studies of middle atmosphere structure and dynamics, and observations of middle atmosphere composition and chemical reactions. Almost half of the papers are based on satellite-borne remote-sensing measurements since a wide coverage in space and time of observations is required. In this respect, the volume points the way for future developments as well as providing a review of the present knowledge and understanding of the middle atmosphere.

J. T. Houghton

November 1979

CONTENTS

Phil. Trans. R. Soc. Lond. A **296**, 3–5 (1980)
Printed in Great Britain

Introduction

By J. T. Houghton, F.R.S.†
Department of Atmospheric Physics, Clarendon Laboratory, Oxford, OX1 3PU, U.K.

In this introduction at the beginning of a meeting devoted to the presentation of observations regarding the middle atmosphere, it is appropriate first to define the region of the atmosphere under consideration, then briefly to survey the history of observation of the middle atmosphere, and finally to list a few of the major problem areas at the present time.

The middle atmosphere is a rather recent name given to the region of the atmosphere between altitudes of 10–15 km and about 100 km (figure 1). Its lower boundary is the tropopause, the level at which temperature ceases to fall with height and which defines the upper boundary of the troposphere (or turning sphere), a region dominated by convection. Between the tropopause and the high temperature peak around 50 km (the stratopause) is the stratosphere, so called because it is a region where temperature increases with height and where, therefore, vertical eddy motions are strongly inhibited. Above the stratosphere is the mesosphere (or middle-sphere) up to the mesopause, the temperature minimum at around 85 km altitude, above which is the thermosphere. At some level, commonly between 100 and 120 km, is the turbopause, below which the atmosphere is reasonably well mixed, the mixing being dominated by eddy processes and above which molecular diffusion dominates so that molecular species are separated out according to their respective masses. Above these levels, also, the action of electric and magnetic fields on the ionized component of the atmosphere is important in determining the motion. Methods of analysing atmospheric structure and dynamics in the higher atmosphere are, therefore, very different from those in the middle atmosphere and it is convenient to define the top of the middle atmosphere at around 100 km.

The history of observation in the middle atmosphere goes back to the years around 1900 when Teisserenc de Bort, in investigations of the atmosphere's temperature structure from balloons, reached the stratosphere on a number of occasions. Table 1 lists some of the milestones in middle atmosphere observation between then and 1967 when the first satellite observations of stratospheric temperature structure were made from the Tiros 7 spacecraft by Kennedy and Nordberg.

The major features of middle atmosphere temperature structure are controlled by the absorption and emission of radiation. Absorption of ultraviolet solar radiation by ozone leads to the high temperature around 50 km; emission from the 15 μm infrared band of carbon dioxide is the main cause leading to the cold temperatures at the mesopause. Solar radiation absorbed by atomic and molecular oxygen leads to the rapid increase of temperature above the mesopause. Even a cursory inspection of figure 1, however, will demonstrate that radiation alone cannot explain the latitudinal distribution of temperature. For instance, the polar mesopause in the summer hemisphere is much colder than in the winter hemisphere despite substantial heating by absorption of solar radiation on the summer side. Atmospheric motions are responsible for transporting large quantities of heat and momentum; in the process they

† Present address: Science Research Council, Appleton Laboratory, Ditton Park, Slough SL3 9JX, U.K.

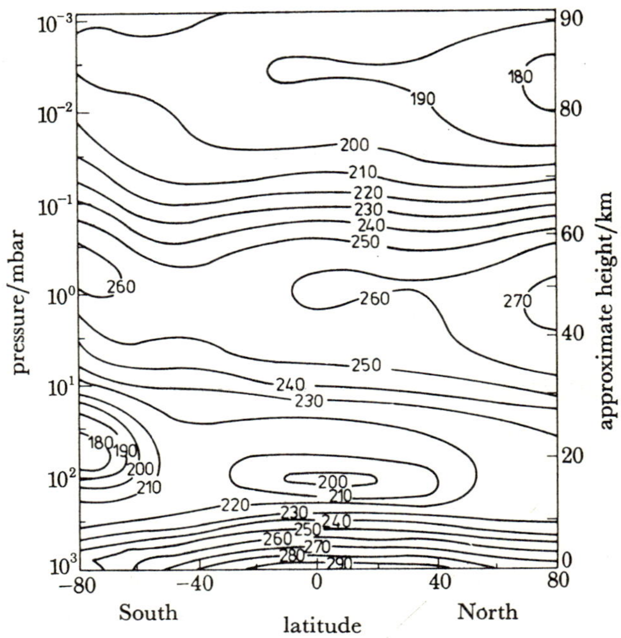

FIGURE 1. Temperature (K) cross section of the atmosphere from 80° N to 80° S as deduced from radiance measurements from the selective chopper radiometer on Nimbus 5 and the pressure modulator radiometer on Nimbus 6 for 4 August 1975. After Houghton (1978).

TABLE 1. MILESTONES IN MIDDLE ATMOSPHERE RESEARCH

1900	discovery of the stratosphere: Teisserenc de Bort
1922	discovery of high temperature around 50 km: Lindemann & Dobson
1927	ozone network set up: Dobson
1930	theory of ozone layer: Chapman
1943	dryness of stratosphere discovered: Brewer & Dobson
1949	Dobson–Brewer circulation postulated
1952	stratospheric warming discovered: Scherhag
1958	radiative sources and sinks computed: Murgatroyd & Goody
1959	meteorological rocket network set up in U.S.
1961	theory of planetary-wave propagation: Charney & Drazin
1961	discovery of quasi-biennial oscillation: Veryard, Ebdon & Reed
1964	theory of mesospheric circulation: Leovy
1967	remote sounding of stratospheric temperature from Tiros 7: Kennedy & Nordberg

TABLE 2. MAJOR TOPICS OF MIDDLE ATMOSPHERE RESEARCH

radiation:	detailed radiative budget of lower stratosphere; non-L.t.e. (local thermodynamic equilibrium) in mesosphere; effect of variable solar input (in ultraviolet and in particle flux)
chemistry:	detailed reactions (especially involving ozone); detailed distribution of constituents (e.g. hydrogen, nitrogen, chlorine compounds); effect of artificially produced constituents (e.g. chlorofluoromethanes); neutral and ion chemistry of D-layer
dynamics:	propagation of planetary waves; mechanism of stratospheric warmings; interaction with troposphere; wave–mean flow interactions (e.g. mechanism of quasi-biennial oscillation); propagation and absorption of gravity waves; control of mean mesospheric circulation

also transport minor atmospheric constituents, namely water vapour, ozone, atomic oxygen and other compounds such as the oxides of nitrogen which are involved in the very complex chemistry of atmospheric ozone production and destruction.

The complex interplay which occurs in the middle atmosphere between radiation, dynamics and chemistry is what gives the region much of its fascination as a scientific discipline. Some of the major topics of middle atmosphere research being tackled at the present time are listed in table 2. To further our understanding regarding these questions will require a large increase in observations of structure and composition, especially in their coverage in space and time. Associated with the observational programme, further development will be needed of a variety of models which include the important processes in the three areas of chemistry, radiation and dynamics and which also include details of the interaction with the troposphere below where much of the dynamical forcing originates, and where many of the minor constituents have their origin.

Because of the possibilities presented by new observational methods and by the availability of satellites, a Middle Atmosphere Programme has been formulated under the Special Committee on Solar–Terrestrial Physics (Scostep) of the International Council of Scientific Unions, whose purpose is to encourage and coordinate research in the middle atmosphere in the early 1980s. That such a programme should be set up is also appropriate in the light of the concern which exists regarding the possibility that substantial changes in the ozone layer may occur because of the release into the atmosphere of chemicals (such as chlorofluoromethanes) arising as a result of man's activities. We can, therefore, look forward to a considerable increase in research activity in the middle atmosphere during the next decade. The purpose of this meeting is to summarize our present knowledge and to provide a perspective against which future plans can be formulated.

References and bibliography (Houghton)

Brewer, A. W. 1949 *Q. Jl R. met. Soc.* **75**, 351–363.
Chapman, S. 1930 *Mem. R. met. Soc.* **3**, 103–125.
Charney, J. G. & Drazin, P. G. 1961 *J. geophys. Res.* **66**, 83–109.
Dobson, G. M. B. 1930 *Proc. R. Soc. Lond.* A **129**, 411–433.
Dobson, G. M. B., Brewer, A. W. & Cwilong, B. M. 1946 *Proc. R. Soc. Lond.* A **185**, 144–175.
Dobson, G. M. B., Harrison, D. N. & Lawrence, J. 1927 *Proc. R. Soc. Lond.* A **114**, 521–541.
Houghton, J. T. 1978 *Q. Jl R. met. Soc.* **104**, 1–29.
Houghton, J. T. 1977 *The physics of atmospheres.* Cambridge University Press.
Kennedy, J. S. & Nordberg, W. 1967 *J. atmos. Sci.* **24**, 711–719.
Leovy, C. 1964 *J. atmos. Sci.* **21**, 327–341.
Lindemann, F. A. & Dobson, G. M. B. 1922 *Proc. R. Soc. Lond.* A **102**, 411–437.
Murgatroyd, R. J. & Goody, R. M. 1958 *Q. Jl R. met. Soc.* **83**, 225–234.
Murgatroyd, R. J. & Singleton, F. 1961 *Q. Jl R. met. Soc.* **87**, 125–135.
Reed, R. J. 1965 *J. atmos. Sci.* **22**, 331–333.
Scherhag, R. 1952 *Ber. dt. Wetterd.* **6**, 51–63.
Veryard, R. G. & Ebdon, R. A. 1961 *Met. Mag., Lond.* **90**, 125–143.

Phil. Trans. R. Soc. Lond. A **296**, 7–18 (1980)
Printed in Great Britain

Climatology of the stratosphere and mesosphere

BY KARIN LABITZKE
Meteorologisches Institut der Freien Universität Berlin, Federal Republic of Germany

The use of all available radiosonde, rocketsonde and satellite data, has elucidated the climatology of the atmosphere up to the mesopause.

The seasonal variation of the temperature in the stratosphere and mesosphere will be described, mainly from monthly mean maps. Emphasis will be given to longitudinal and hemispheric differences.

During the winter season when, owing to the development of the cold stratospheric polar vortex, the winds in the stratosphere are mainly from the west, the planetary waves of the troposphere penetrate into the stratosphere, and even into the mesosphere during the northern winter. This results in longitudinal differences over both hemispheres, though the effect is stronger over the Northern Hemisphere.

The hemispheric differences are particularly large between the winters, as the stratospheric midwinter warmings are different over both hemispheres. As a consequence, the transition from the winter to the summer circulation also evolves differently. During summer, the southern stratosphere is warmer than the northern because of the difference in solar heating of ozone due to the ellipticity of the Earth's orbit.

1. INTRODUCTION

For the investigation of the lower and middle stratosphere radiosonde data have now been available for about 20 years (since 1958), i.e. since the International Geophysical Year. Above this level rocketsondes have been an important data source, especially for the fine structure of the atmosphere. Since 1969, remote temperature soundings from satellites provide data on a global scale to study the structure of the upper stratosphere and recently also of the mesosphere.

The information obtained from these data is very large and several climatologies and model atmospheres have been derived (e.g. *C.I.R.A.* 1972; Labitzke *et al.* 1972; Newell *et al.* 1972; van Loon *et al.* 1972; Newell *et al.* 1974; Knittel 1976). In the following only a few aspects of the climatology of the stratosphere and mesosphere will be pointed out.

2. ZONAL MEAN TEMPERATURES

Zonal mean temperatures describe the gross structure of the temperature distribution, omitting the zonal asymmetries. In figure 1 the zonal mean temperatures (based on the radiosonde data of several years) are shown in a vertical meridional section for the January of the Northern Hemisphere. Typical features of the middle stratosphere are: the cold centre over polar latitudes, the warm region over middle latitudes and the cold tropical tropopause. In the upper stratosphere, near 10 mbar,† the pattern becomes simpler, as the temperature increases regularly from polar to low latitudes.

It must be kept in mind, however, that the stratospheric winter is a highly disturbed season and disturbances such as stratospheric warmings can destroy the polar vortex and alter the

† 1 bar = 10^5 Pa.

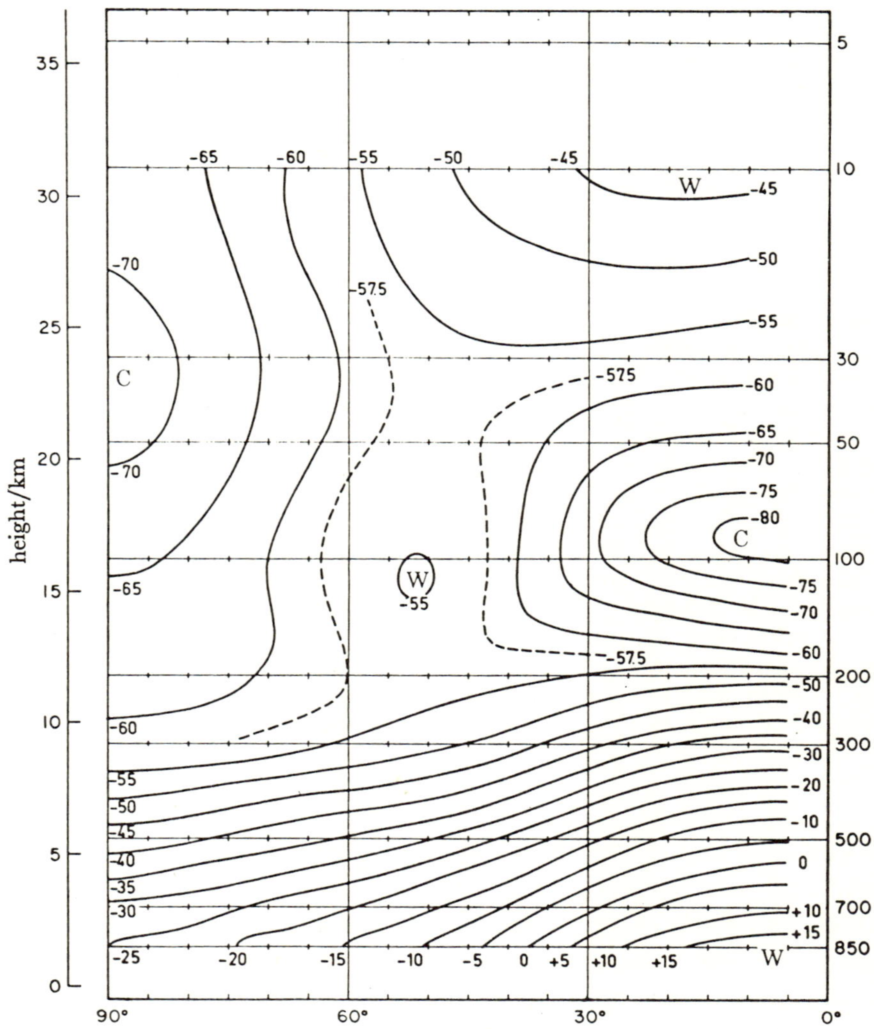

FIGURE 1. Vertical meridional section of zonal mean temperature (°C), for January, N.H. (from Labitzke *et al.* 1972).

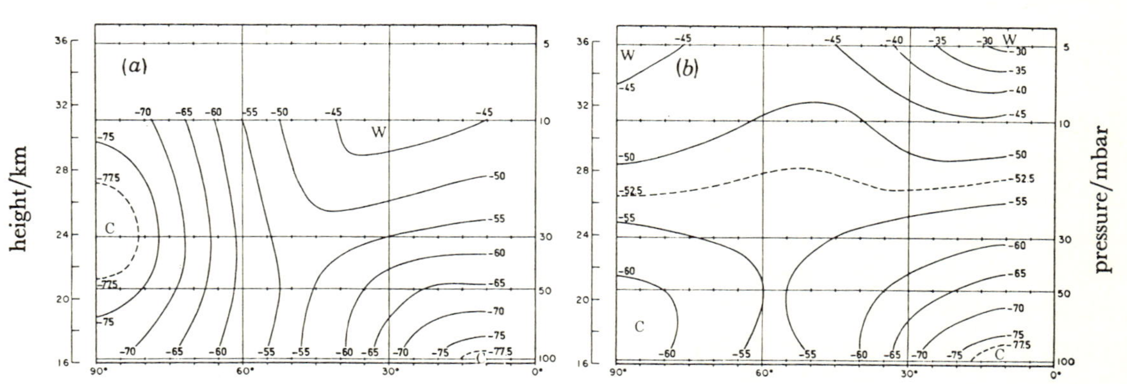

FIGURE 2. Vertical meridional sections of zonal mean temperatures (°C) (*a*) for a cold January and (*b*) for a warm January (from van Loon *et al.* 1972).

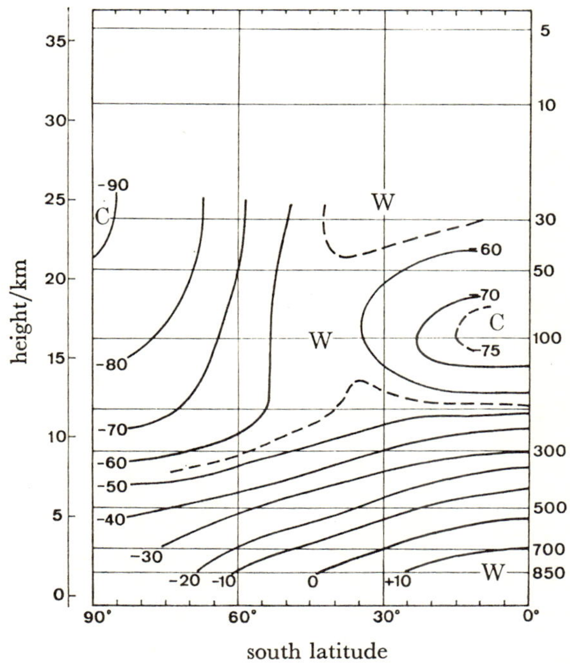

FIGURE 3. Vertical meridional section of zonal mean temperatures (°C), for July, S.H. (Data for the stratosphere from Knittel (1976); for the troposphere from Taljaard *et al.* (1969)).

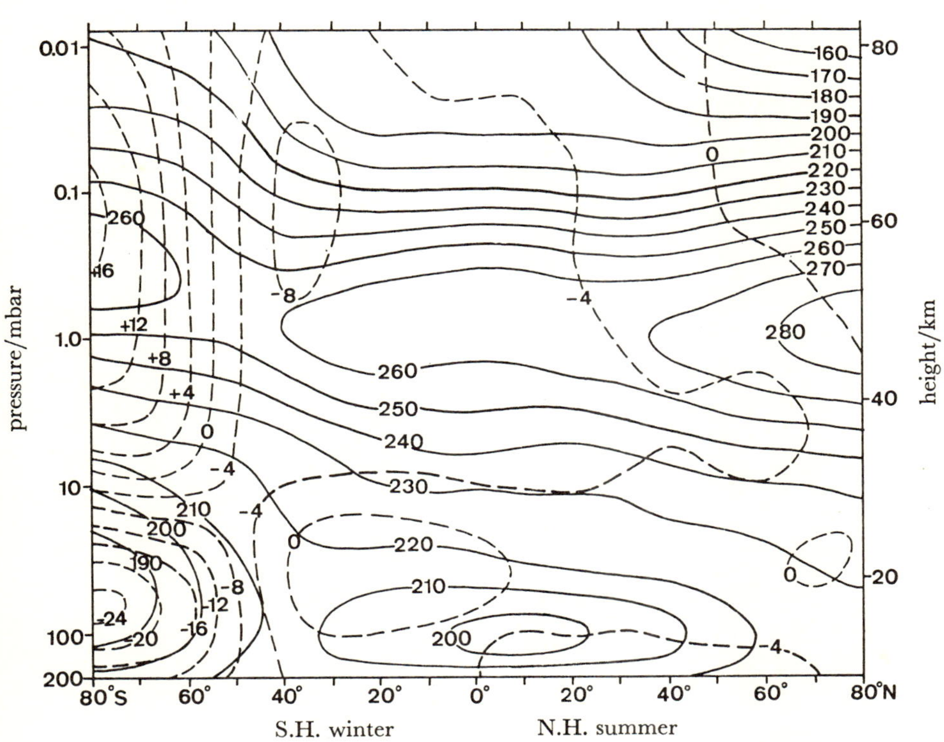

FIGURE 4. Vertical meridional section of zonal mean temperatures (K) for July. Dashed lines indicate deviations from *C.I.R.A.* (1972) (from Labitzke & Barnett 1978).

temperature distribution of a whole month such that the temperature gradient is reversed over high latitudes. This is demonstrated in Figure 2, where the zonal mean temperatures of the 'cold' January 1967 are compared with those of the 'warm' January 1968.

The temperature structure of the middle stratosphere of the Southern Hemisphere winter, namely of July, is given in figure 3. Typically, the antarctic polar vortex is much colder than

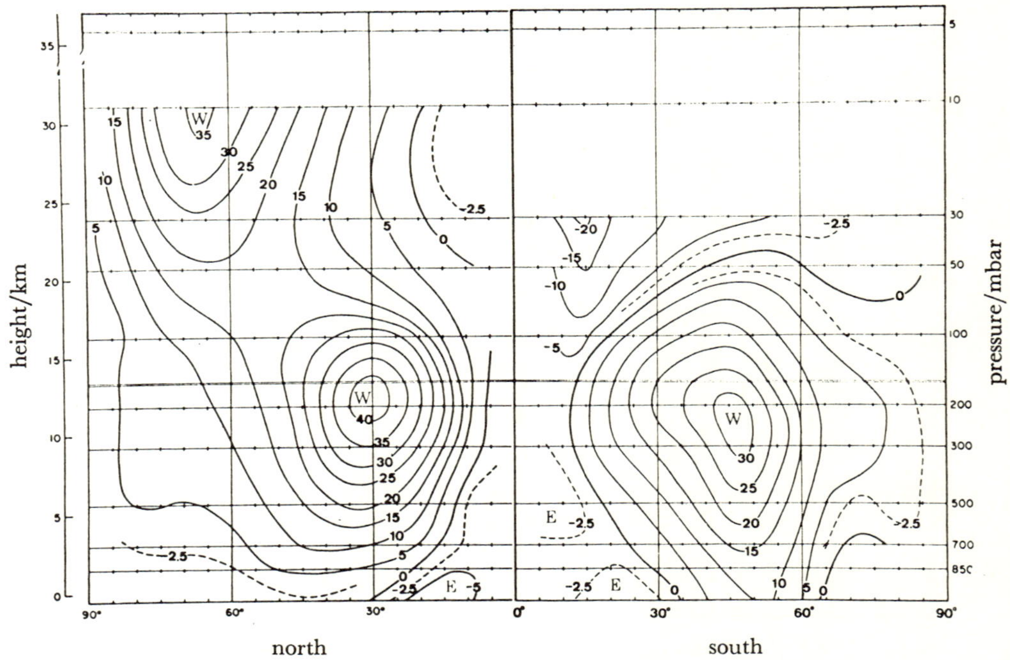

FIGURE 5. Vertical meridional section of zonal mean winds (in m/s) for January (from Knittel 1976).

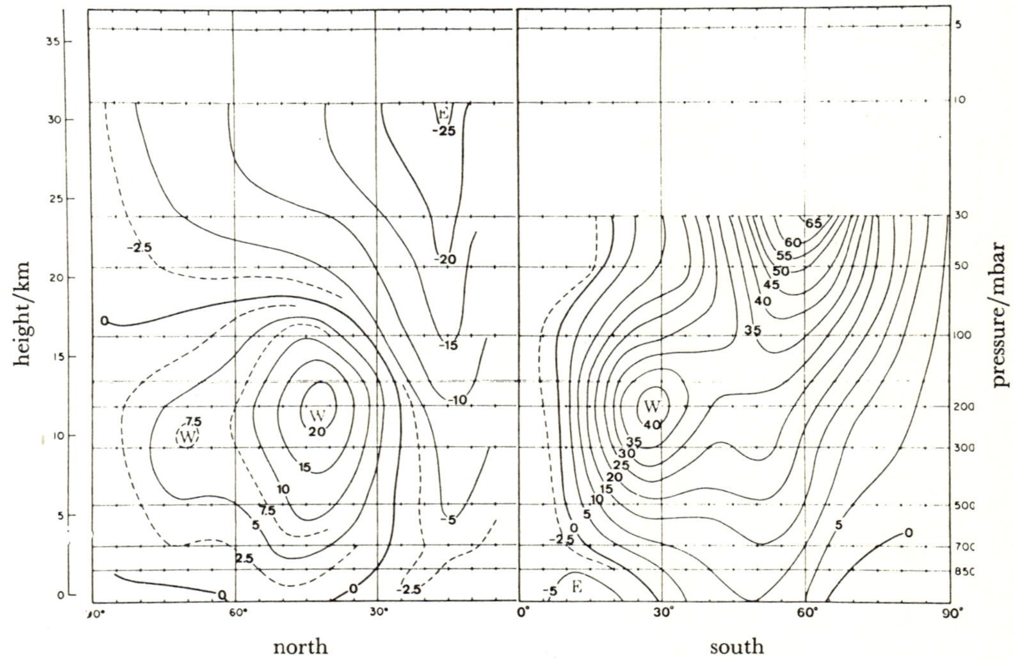

FIGURE 6. Vertical meridional section of zonal mean winds (in m/s) for July (from Knittel 1976).

the arctic, and the warm region over middle latitudes is displaced about 10° latitude towards the equator.

Until recently the only way of studying the temperature structure of the upper stratosphere and of the mesosphere on a large scale was by using rocketsonde data. The distribution in time and space of these soundings was inadequate to establish a complete climatology. Today, different experiments on board satellites are capable of measuring the emitted thermal radiation of the Earth's atmosphere up to the mesopause region, daily on a global scale. With these new data it will be possible within the next years to get a complete analysis of the temperature structure of the atmosphere and of its changes up to the mesopause.

The zonal mean temperature cross section for July, figure 4, is based on retrieved temperatures from satellite measured radiances. Several years of data were available for the stratosphere, but only July 1975 for the mesosphere. The main features of the mesosphere are, however, well reflected: the very strong temperature decrease over high latitudes from the stratopause to the mesopause during summer and the reversed horizontal temperature gradient with temperatures increasing from the summer pole to the winter pole. Over the Northern Hemisphere the differences from the *C.I.R.A.* (1972) are below 4 K during this time of the year.

3. Zonal mean winds

In meridional sections from pole to pole the distribution of the mean zonal winds is shown in figure 5 for January and in figure 6 for July. In the stratosphere the polar night jet between 60 and 70° latitude is the main feature of the winter and this jet is more than twice as strong during the southern winter. During summer, winds from the east accompany the summer anticyclone. They reach further down over the Northern Hemisphere.

Over the tropics the direction of the winds depends on the phase of the quasi-biennial oscillation (review by Wallace 1973) and cannot be well reflected in a mean over several years.

4. Monthly mean maps

Monthly mean maps of temperatures or satellite measured radiances (which can be interpreted as mean temperatures over a certain layer) provide an excellent tool for showing longitudinal and hemispheric differences.

At first, the temperatures during early winter over both hemispheres (i.e. December, N.H. and June, S.H.) are compared. In the middle stratosphere (30 mbar) the antarctic region is much colder than the arctic (figure 7). This is valid for the whole winter. The well known warm region north of Japan, which is part of the standing wave 1 of the N.H., has only a weakly developed counterpart in the southern subtropics between the longitudes 170 W and 130 E. This warm region develops into an active region only in late spring (October) when the transition into summer conditions takes place. In the upper stratosphere (2 mbar), figure 8, the pattern of the temperature distribution is rather different over the two hemispheres: over the N.H. the strongly developed standing wave 1 slopes westward with height and the maximum of the temperature wave is found about 50° longitude further west in the 2 mbar level compared with that of the 30 mbar level; over middle and high southern latitudes the amplitude of the temperature wave 1 is again much smaller than over the N.H., and typically the temperature gradient is rather weak over the antarctic polar region.

The radiance data of channel 3000 of the p.m.r., figure 9, reflect the temperature pattern of the upper mesosphere with the warm polar region. The planetary wave number 1 is still evident during the northern winter but barely noticeable during the southern winter. These differences will be discussed further in § 5.

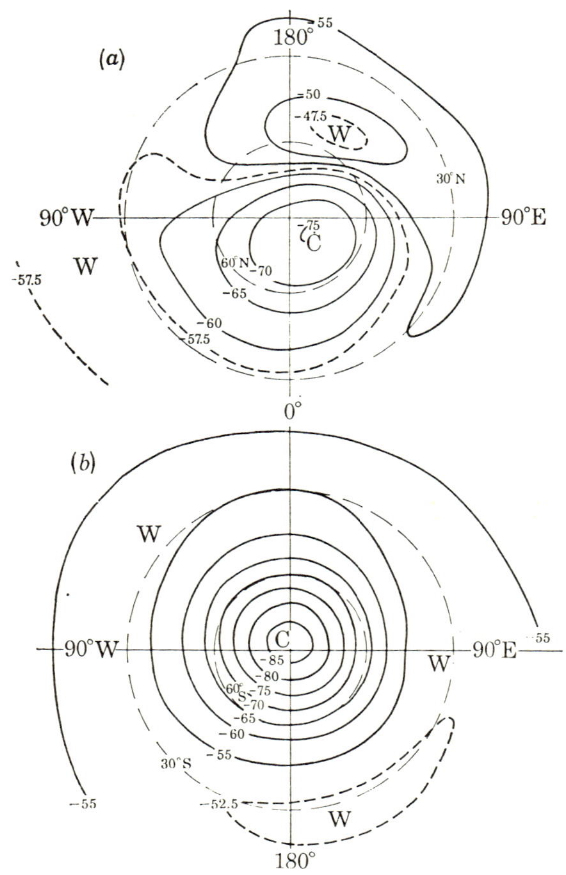

FIGURE 7. Monthly mean 30 mbar temperatures (°C), (*a*) December, N.H., and (*b*) June, S.H. (from Labitzke 1977).

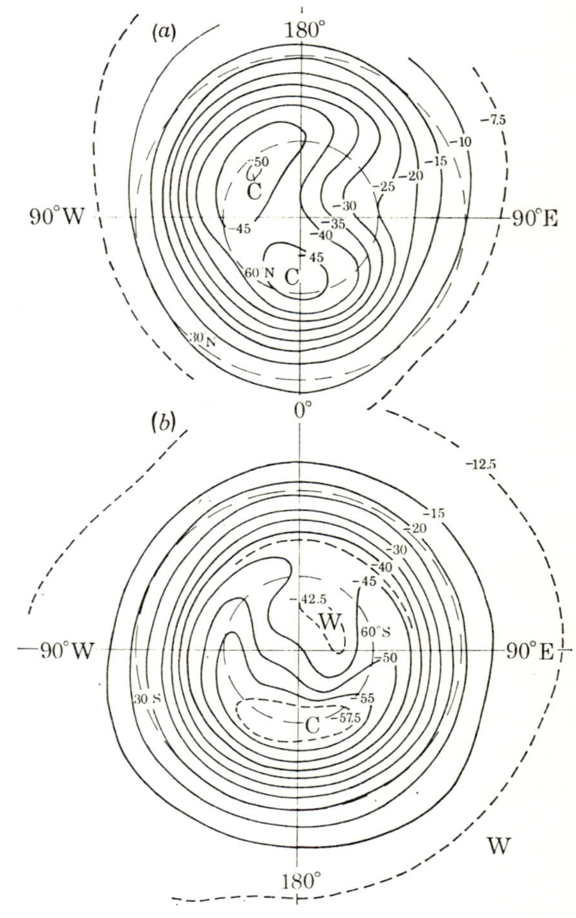

FIGURE 8. Monthly mean 2 mbar temperatures (°C), (*a*) December, N.H., and (*b*) June, S.H. (from Labitzke 1977).

In the upper stratosphere the transition from winter to summer has already started over Antarctica in early July, while a late winter cooling in high northern latitudes, caused by dynamic processes in connection with stratospheric warmings and typical of the northern winter, delays the transition to a summer pattern for about 2 months. Figure 10 shows mean maps of 2 years of radiances of the s.c.r. (A) for the spring season of both hemispheres. While the transition to summer is well in progress in September over the antarctic, no indication of a transition is noticeable in March over the arctic.

During summer the warm polar anticyclone is at its peak in the stratosphere. The isotherms of the 10 mbar level of the northern summer, figure 11*a*, indicate only a weak diurnal wave. As the analysis is based on radiosonde data of 00 U.T. the minimum is found during night over the Atlantic and the maximum during day over the Pacific. Together with the warm anti-cyclone, east winds prevail in the summer stratosphere (cf. figures 5 and 6) and lower mesosphere and prohibit the propagation of the planetary waves. The northern summer stratosphere is colder than the southern (Barnett 1974) because of the difference in solar heating of ozone. This is due to the ellipticity of the Earth's orbit, which produces a 6 % modulation of the solar input.

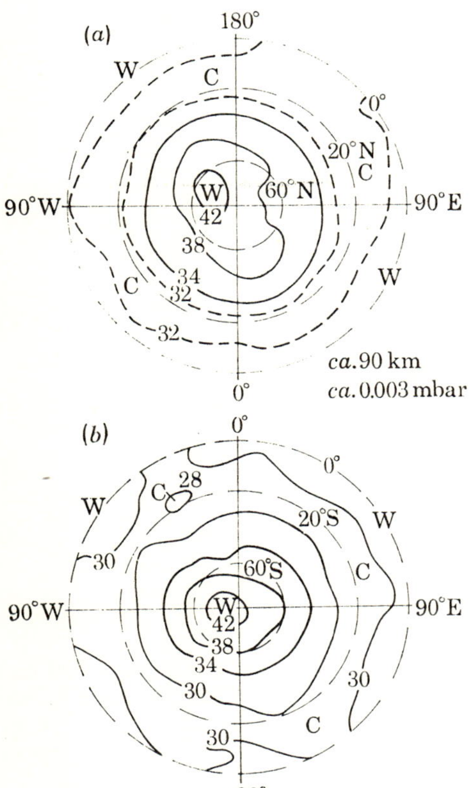

FIGURE 9. Monthly mean radiances [in mW/(m² sr (cm⁻¹))] of the p.m.r. Channel 3000, (*a*) January 1976, N.H., and (*b*) July 1975, S.H. (from Labitzke & Barnett 1978).

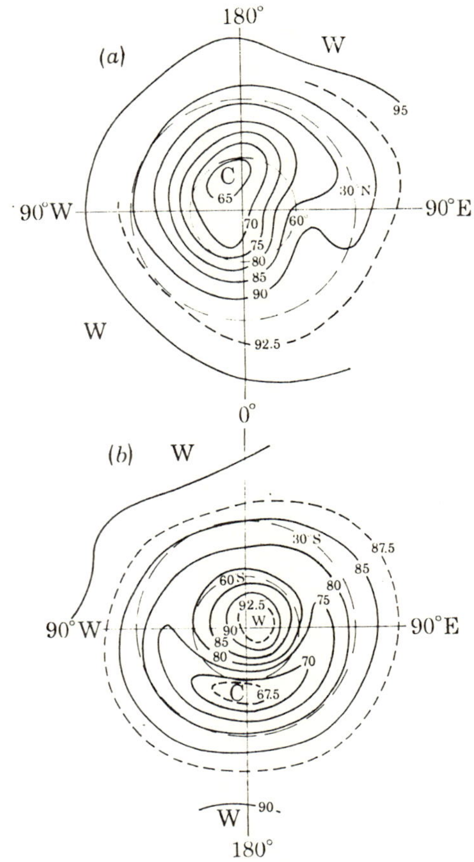

FIGURE 10. Monthly mean radiances [in mW/(m² sr (cm⁻¹))] of the s.c.r. channel A, (*a*) March, N.H., and (*b*) September, S.H. (from Labitzke 1974).

The polar mesosphere is cold during summer, figure 11*b*, and the radiance data for the southern summer show no longitudinal differences because tidal oscillations cannot be observed as the data shown here are based on daily mean values of a 24 h period.

In figure 12, the temperature differences between the hemispheres are summarized for the lower and upper stratosphere and for high and middle latitudes. The timescale is arranged so that the same seasons in the two hemispheres can be compared directly. The most obvious feature is that in winter the difference over high latitudes, which amounts to 18 °C at the 100 mbar level vanishes in the upper stratosphere and reverses sign during spring, as discussed above. Also, the warmer southern summer values are clearly noticeable in all four sets of curves. Another interesting feature is the opposite behaviour between middle and high latitudes in the lower stratosphere during the winter half-year.

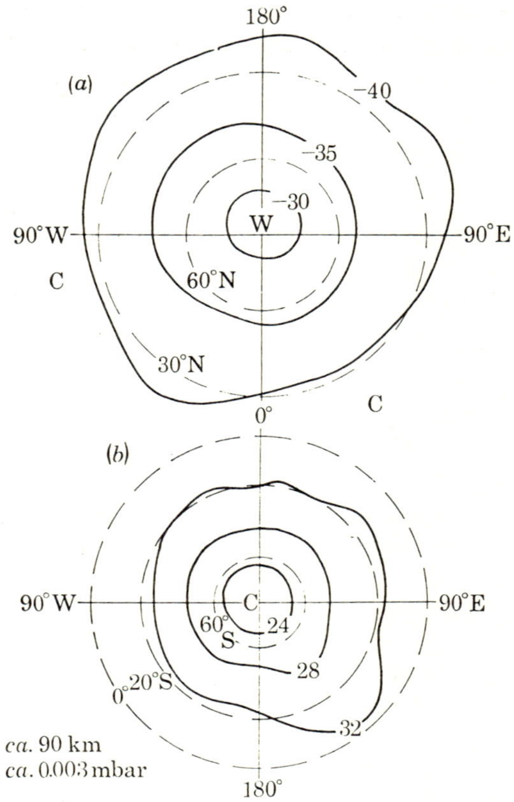

FIGURE 11. (*a*) Monthly mean 10 mbar temperatures (°C), July, N.H. (from Labitzke *et al.* 1972); (*b*) monthly mean radiances [in mW/(m² sr cm⁻¹)] of the p.m.r. channel 3000, January 1976, S.H. (from Labitzke & Barnett 1978).

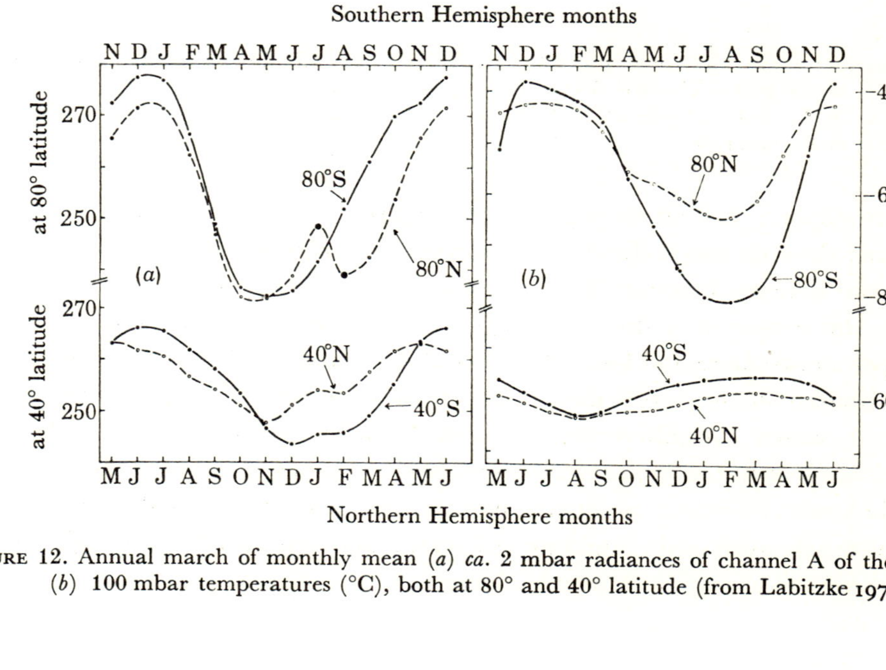

FIGURE 12. Annual march of monthly mean (*a*) *ca.* 2 mbar radiances of channel A of the s.c.r. (K), (*b*) 100 mbar temperatures (°C), both at 80° and 40° latitude (from Labitzke 1974).

5. Planetary waves

As mentioned before, planetary waves, especially of wave numbers 1 and 2, play an important role in the general circulation of the stratosphere during winter, when the polar night jet is well developed. The quasi-stationary waves have their maximum amplitudes between 70 and 60° latitude, along the axis of the polar night jet.

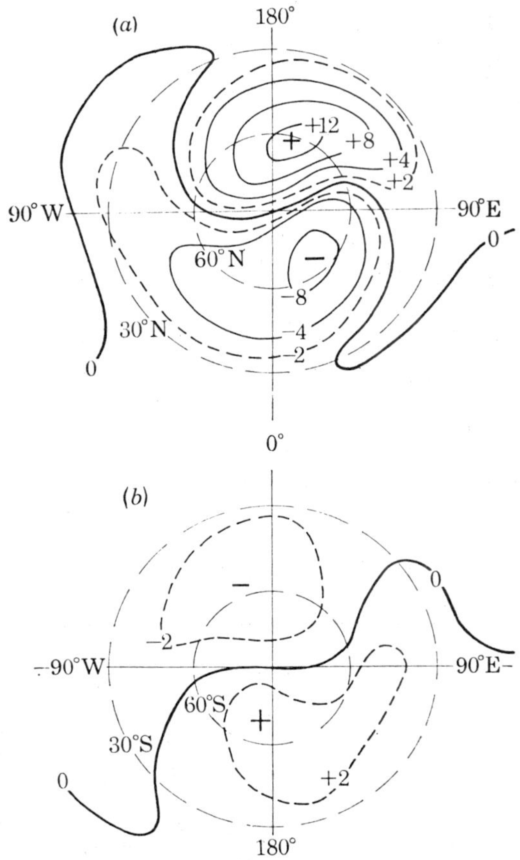

Figure 13. Latitude anomalies of the 30 mbar temperatures (°C) for (*a*) December, N.H., and (*b*) June, S.H. (data from Knittel 1976).

During winter the quasi-stationary waves are much stronger developed over the N.H. than over the S.H., as demonstrated in figure 13, where the deviations of the 30 mbar temperatures from the zonal means are analysed. A Fourier analysis for latitude 60° gives the following results: the amplitude of wave 1 is 10 °C at 60° N and 3 °C at 60° S, and the amplitude of wave 2 is 3 °C at 60° N and 0.1 °C at 60° S.

To give an indication of the slope of the quasi-stationary waves with height, figure 14 was constructed. It indicates at 60° latitude the deviations from the zonal mean temperatures in the middle stratosphere and the deviations from the zonal mean radiance data in the upper stratosphere and mesosphere (only one year of data available here). There appears to be not much difference in the vertical slope of the standing waves, but the amplitude of wave 1 remains two times larger also in the upper stratosphere and mesosphere during the northern winter.

The transition to a summer circulation takes place during October in the middle southern stratosphere, in spite of the fact that the upper stratosphere has already warmed. This transition is brought about by an intense development of the quasi-stationary planetary waves which during this time of the year reach even larger amplitudes than they do during the northern winter (figure 15).

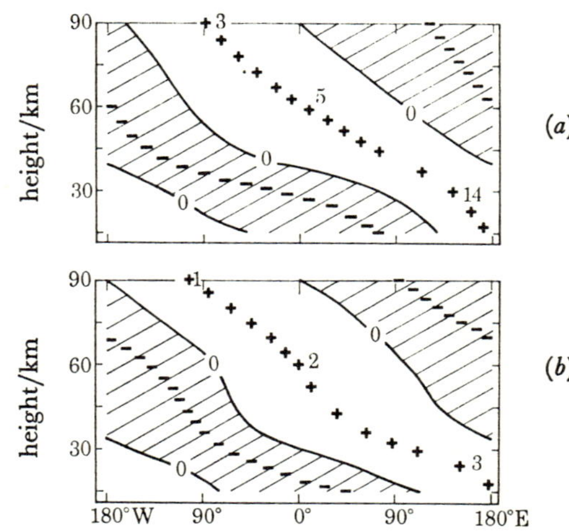

FIGURE 14. Vertical section along 60° latitude of the deviations of the temperatures (lower stratosphere) or of the radiances (upper stratosphere and mesosphere) from the zonal mean (*a*) December, (*b*) June. For selected levels the value of maximum positive deviation of radiances or temperatures is given. These values should be compared only between respective levels.

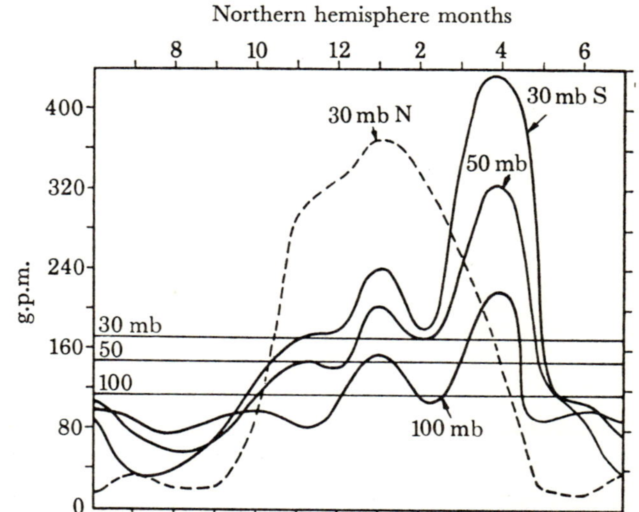

FIGURE 15. Annual march of amplitudes (g.p.m.) of planetary height – wave 1 at 60° S for the 100 mbar, 50 mbar and 30 mbar level, and at 60° N for the 30 mbar level (from Knittel 1976).

6. VARIABILITY

After a review has been given of the 'mean' state of the stratosphere and mesosphere, a short remark must be made on the natural variability of the stratospheric climate. In figure 16, monthly mean temperatures for the North Pole are shown from different sources. They indicate

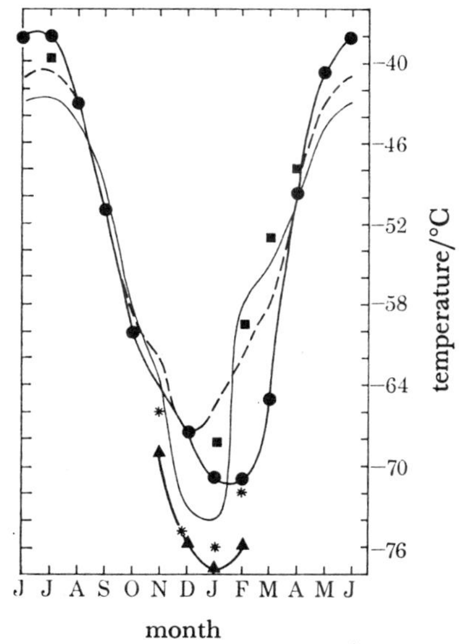

FIGURE 16. Monthly mean 50 mbar temperatures (°C) for the North Pole. Averages over different periods by different authors (partly by Labitzke *et al.* 1972). ▲, Berlin 1971/72 + 1973/74; – –, Wege 1949–53; ●, Berlin 1964/7–1969/6; ——, Muench 1955/7–1959/6; *, Muench 1955/56; ■, Ebdon 1957–61.

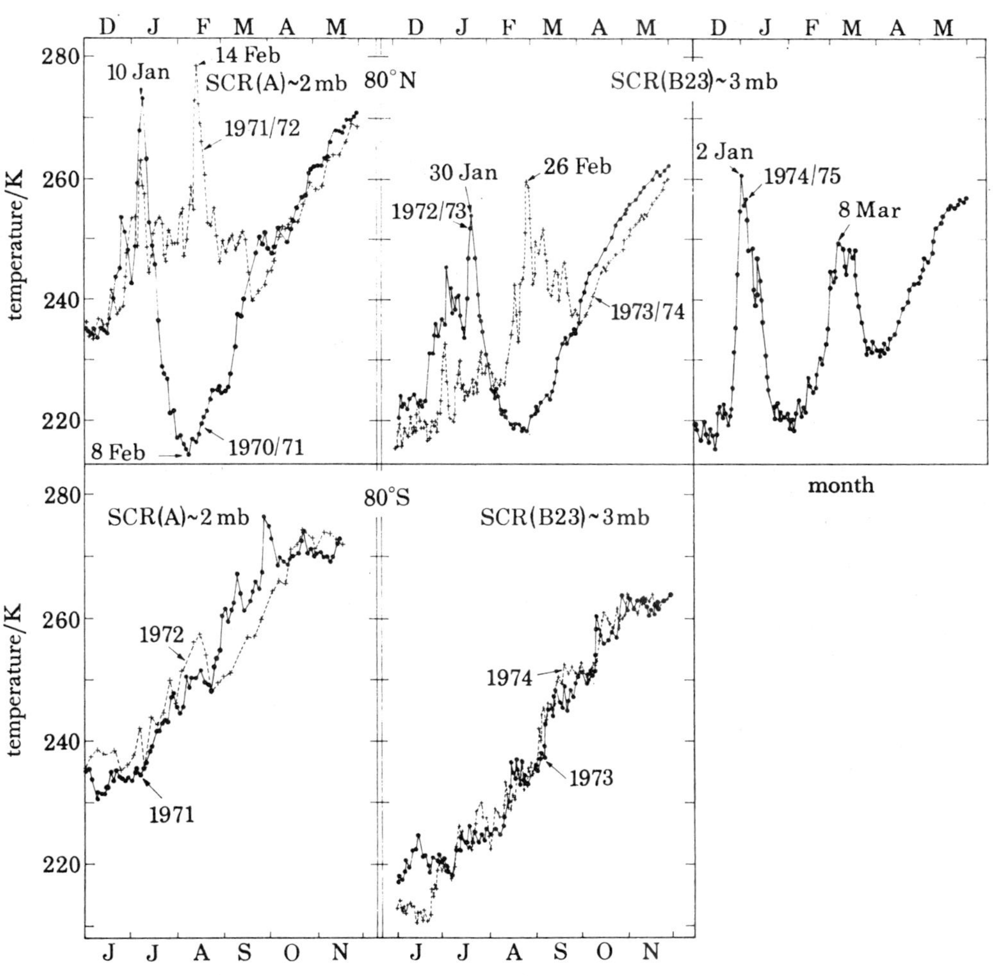

FIGURE 17. Daily zonal means at 80° latitude of radiances (converted into equivalent blackbody temperatures) of upper stratospheric channels of the s.c.r., flown on Nimbus 4 and 5 (from Labitzke 1977).

clearly that cold winters without major warmings result in much lower mean temperatures and, if an average is made during a period when several major warmings took place (such as during the period used by Ebdon), quite another value might be established as a mean. A connection with the solar cycle has been pointed out before (Kriester 1971; Kriester-Naujokat 1978; Labitzke *et al.* 1972). For a better comparison of the data of the middle stratosphere it has been suggested to derive mean data for the winter, using only the undisturbed winter months (Labitzke *et al.* 1972).

The satellite data have already clearly indicated (figure 17) that the upper stratosphere is highly disturbed during each northern winter, where a distinction can be made between winters with a late winter cooling period (usually following major warmings) and without. However, during the southern winters the temperature minimum is usually reached already during winter solstice and a rather regular warming trend towards summer values is obvious.

7. Conclusion

The material presented here showed clearly that the recent satellite experiments have largely increased our knowledge about the climatology of the stratosphere and mesosphere, though more years of data are needed to gain a representative climatology of the upper stratosphere and of the mesosphere. The discussion of the data pointed (*a*) to the effects of the planetary waves of the troposphere which penetrate into the stratosphere and even into the mesosphere, except during summer; these waves create longitudinal differences which should be treated in future model atmospheres; (*b*) to hemispheric differences which in winter have their origin in differences between the general circulation of the troposphere; and (*c*) to the interannual variability and possible variations within the solar cycle.

References (Labitzke)

Barnett, J. J. 1974 *Q. Jl R. met. Soc.* **100**, 505–530.
C.I.R.A. 1972 Cospar *International Reference Atmosphere*. Berlin: Akademie-Verlag.
Knittel, J. 1976 *Met. Abh. Inst. Met. Berl.* A **2** (no. 1).
Kriester, B. 1971 *Met. Rdsch.* **24**, 71–73.
Labitzke, K. *et al.* 1972 *Meteor. Abh. F. U. Berlin* **100** (no. 4).
Labitzke, K. 1974 *J. geophys. Res.* **79**, 2171–2175.
Labitzke, K. 1977 Cospar *Space Resarch*, **XVII**, 159–165, Oxford and New York: Pergamon Press.
Labitzke, K. & Barnett, J. J. 1978 Cospar *Space Research*, **XVIII**. (In the press.)
Naujokat, B. 1978 Paper presented at the S.T.P.-Symposium, Innsbruck, Austria.
Newell, R. E., Kidson, J. W., Vincent, D. G. & Boer, G. J. 1972 *The general circulation of the tropical atmosphere and interactions with extratropical latitudes*, vol. 1. Cambridge, Mass., and London: The M.I.T. Press.
Newell, R. E., Kidson, J. W., Vincent, D. G. & Boer, G. J. 1974 *The general circulation of the tropical atmosphere and interactions with extratropical latitudes*, vol. 2. Cambridge, Mass., and London: The M.I.T. press.
Taljaard, J. J., van Loon, H., Crutcher, H. L. & Jenne, R. L. 1969 *Climate of the upper air: Southern Hemisphere*, part 1. NAVAIR 50-1C-55, Chief of Nav. Oper., Washington, D.C.
van Loon, H., Jenne, R. L. & Labitzke, K. 1972 *Met. Abh. Inst. Met. Berl.* **100** (no. 5).
Wallace, J. M. 1973 *Rev. Geophys. Space Phys.* **11**, 191–222.

Phil. Trans. R. Soc. Lond. A **296**, 19–40 (1980) [19]
Printed in Great Britain

Seasonal and diurnal variations of middle atmosphere winds

BY G. V. GROVES
Department of Physics and Astronomy, University College London,
Gower Street, London WC1E 6BT

Previously published meridional cross sections of monthly mean W–E and S–N wind components from 60 to 130 km are reviewed in the light of wind data obtained by ground-based radio methods over various timescales including that of the solar cycle. A mean meridional circulation in the winter hemisphere is implied in which a dominant mesospheric cell between low and high latitudes is confined to lower heights at higher latitudes: a second meridional circulation is tentatively located above the former in the lower thermosphere extending downwards to lower heights at higher latitudes and introducing the possibility of polar and equatorial air mixing at mid-latitudes at heights close to 90 km. Meridional circulation is discussed in relation to the winter anomaly in D-region radio-wave absorption.

Diurnal and semidiurnal wind oscillations at heights between 25 and 60 km are presented which are the results of a comprehensive analysis of data from short series of rocket launchings held at various Western Hemisphere sites between 1965 and 1974. Phase and amplitude profiles at different times and locations rarely show similarities, but at latitudes greater than about 30°, S–N components are consistently in quadrature with and similar to the W–E components. An interpretation of the oscillations at these latitudes in terms of a superposition of tidal modes indicates the presence of quasi-steady high-order modes excited at tropospheric heights by sources of limited ($\sim 10^3$ km) extent. At latitudes less than about 30°, steady or quasi-steady diurnal and semi-diurnal components are not necessarily the dominant components of daily variation. At high latitudes diurnal phases generally show little change with height in comparison with observations at lower latitudes in accord with the latitudinal properties of positive and negative diurnal modes.

1. INTRODUCTION

Since the early 1950s, and particularly during the 1960s, rocket techniques have provided a major source of new information on upper atmosphere structure. The role of the rocket in this type of experimentation has usually been that of a vehicle of transport from which to eject one or more types of experimental package in flight. Data have been acquired for different height regions by means of parachutes or similar forms of decelerating device (often with a temperature sensing package) as well as chaff, falling spheres, grenades and chemical releases. Atmospheric temperatures, pressures and densities have been investigated by these techniques, but the parameter most frequently measured over different height intervals extending into the thermosphere has been the wind velocity.

At heights close to 90 km, ground-based radio techniques have also been available for wind determination. Although their observational height range is of limited extent compared with that of rocket techniques, they are particularly suitable for the study of time dependences which, in upper atmosphere winds, are traditionally classified in terms of a prevailing component, being that part of the wind velocity that does not change in the course of a day, tidal motions, turbulence and gravity waves of a few hours periodicity or less. Seasonal and longer-term effects contribute to the prevailing component but in addition day-to-day variations (of about 4 days periodicity) have also been found to contribute significant energy peaks to the spectra of meteor wind data (Teptin 1972).

In the first part of this paper (§§ 2–6), seasonal wind variations at heights above 60 km are reviewed, particular attention being given to S–N components and meridional circulation. The second part of the paper deals with diurnal wind variations as investigated by short series of rocket launchings extending over 1–3 days at heights of 25–60 km (or to 90 km on a few occasions). On 19–20 March 1974, 70 rockets were launched at eight Western Hemisphere sites; the results from these launchings are presented and discussed in §§ 7–13 in conjunction with those of earlier launchings.

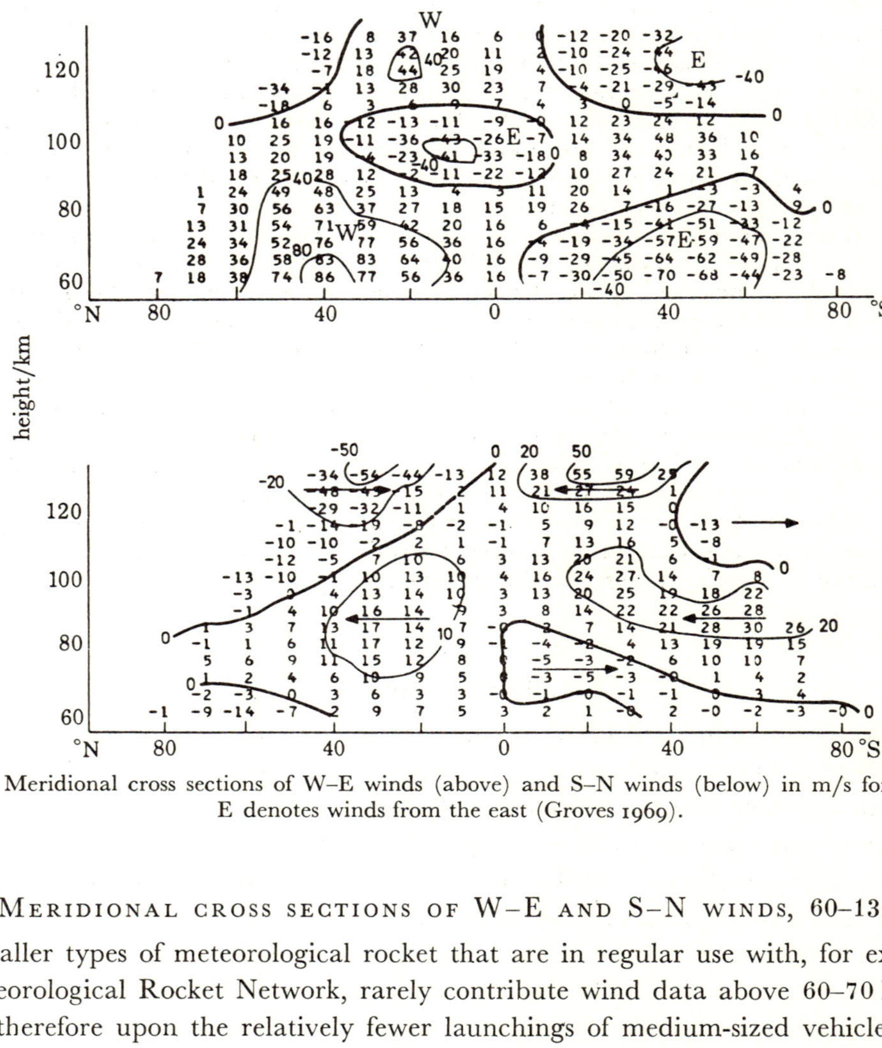

FIGURE 1. Meridional cross sections of W–E winds (above) and S–N winds (below) in m/s for 1 January. E denotes winds from the east (Groves 1969).

2. MERIDIONAL CROSS SECTIONS OF W–E AND S–N WINDS, 60–130 km

The smaller types of meteorological rocket that are in regular use with, for example, the U.S. Meteorological Rocket Network, rarely contribute wind data above 60–70 km altitude and it is therefore upon the relatively fewer launchings of medium-sized vehicles that data acquisition at these heights has depended. In relation to the complex nature of wind variations in space and time, coverage has been extremely sparse. Nevertheless, it has been possible by combining rocket data (and gun-probe data at certain sites) with radio wind data to construct meridional cross sections of mean W–E and S–N wind components at monthly intervals (Groves 1969). The rocket and gun-probe data available at that time were from grenade experiments at Woomera, Wallops Island and Fort Churchill and chemical releases at Bardados,

Eglin, Woomera and other mid-latitude sites. Data were provided by ground-based radio techniques sited in Europe and the Southern Hemisphere at Adelaide and at Mawson.

Figure 1 shows the models for 1 January. In constructing these cross sections seasonal symmetry was assumed between Northern and Southern Hemispheres in an attempt to overcome the shortage of data, i.e. Southern Hemisphere data were combined with Northern Hemisphere data after a 6-month change of date and a change of sign of S–N components. On account of the poor longitudinal distribution of data and unknown magnitude of standing waves, the extent to which the models represented zonal mean values was uncertain; values of 20 m/s for S–N wind components at 90 km being considered to be unlikely to be globally representative. As cross sections of observed averages, however, the models provided a convenient reference against which to compare other individual measurements that might become available. The W–E wind model in a slightly revised form was included in *C.I.R.A.* (1972, part 2).

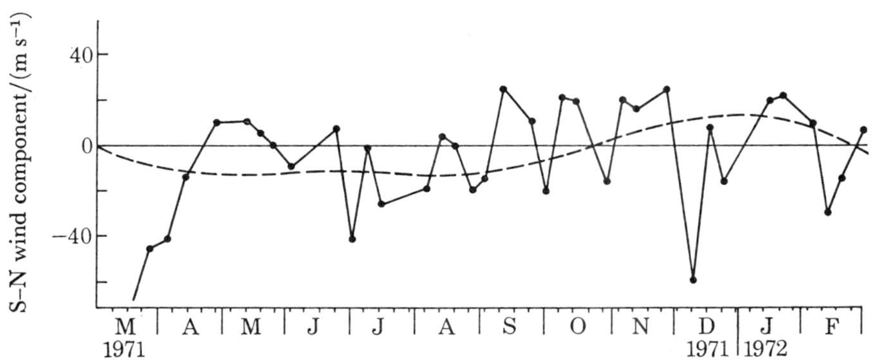

FIGURE 2. The 24 h averages of S–N wind components taken at weekly intervals by the Jamaica meteor-radar technique. The dashed curve is the 1969 S–N wind model for 15° N (Scholefield & Alleyne 1975).

3. Comparisons of recent radio wind data with S–N and W–E models

In recent years a number of comparisons have been made between radio wind observations and the S–N and W–E models of § 2. Many of these observations are from longitudes where data were not previously available, notably in the Western Hemisphere, and offer the opportunity of checking the main features of the models and their seasonal trends for a range of heights centred close to 90 km.

The first measurements by partial radio-wave reflexions at Saskatoon (52° N, 107° W) (Gregory & Rees 1971) showed satisfactory agreement with the 1969 models, but more extensive comparisons up to 110 km (Gregory & Mason 1975) have shown satisfactory agreement to 85 km only and differences in the annual variation of zonal flow above 100 km.

Mean meteor winds measured at Durham, New Hampshire (43° N, 71° W) from May to December 1970 were in general agreement with the models, the meridional winds showing flow of 10 m/s towards the equator during summer and a similar flow towards the poles in winter (Clark 1975). A similar magnitude of *ca.* 10 m/s towards the equator in summer (9 August–4 September 1974) was found for 90–100 km at Atlanta (34° N, 84° W) (Roper 1975).

In 1970, meteor wind data were obtained for the first time in Jamaica (18° N, 77° W) and results from March 1971–February 1972 (Scholefield & Alleyne 1975) showed a seasonal variation in good agreement with the 1969 models. The general trend of the S–N component (figure 2) was towards the equator in summer and towards the pole in winter, although data points deviated appreciably from the model at times: the particularly large deviation in March–April 1971 corresponded to replacement of easterly by northerly flow.

Meridional flow from November 1970 to November 1971 in the meteor region at Christchurch, New Zealand (43° S, 72° E) was towards the equator in local summer and towards the pole in local winter, being not in conflict with the pattern of flow suggested by figure 1 (Wilkinson & Baggaley 1975).

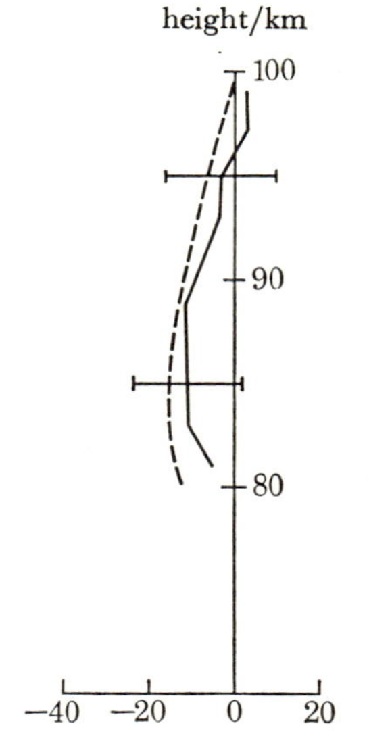

FIGURE 3. S–N prevailing wind components at Adelaide, S. Australia. Key: ——, observed ionospheric drift by partial reflexion technique 15–21 June 1973; ---, the 1969 S–N wind model for the same latitude and time of year (Vincent & Stubbs 1977).

Monthly mean ionospheric drifts from 80 to 100 km for 1972 have been obtained at Adelaide (35° S, 139° E) by the partial reflexion technique and compared with the 1969 models (Stubbs 1976). Overall, the models provided a better picture of the 1972 winter than the 1972 summer for which they fairly consistently exceeded observed S–N drift velocities and underestimated W–E magnitudes. For 15–21 June 1973 prevailing components have been derived with estimates of error, and S–N components compare well with model values as shown in figure 3 (Vincent & Stubbs 1977).

4. Day-to-day and long-term wind components

The spread of data points in figure 2 is evidence of the presence of variations on timescales of greater than one day and less than a few weeks. Such variations are now recognized as a general feature of radio wind observations with various possible dominating periodicities according to location, altitude and time of year. Spectral analysis of meteor radar winds at Obninsk (55° N, 36° E) and LF radio drift measurements between Kühlungsborn (54° N, 12° E) and Collm (51° N, 13° E) during the period November 1972 to March 1973 show peak intensities at periods of 10–16 days for the zonal, 16–20 days for the meridional component and of 2.5–6 days in both components (Lysenko *et al.* 1975). The results of other investigators tend to fall in the same ranges. Periods between 2 and 5 days, and amplitudes between 10 and 20 m/s were found by Muller & Kingsley (1974) at Sheffield (53° N, 1° W), the absence of a marked seasonal effect being taken to indicate an indirect process of coupling with tropospheric variations of similar periodicities and not direct planetary wave propagation. Simultaneous wind variations at Garchy (47° N, 3° E) and Obninsk have been identified as a planetary wave of a 2-day period and longitudinal wavenumber around 3 (Glass *et al.* 1975). At Adelaide (35° S, 139° E) the variation is again found at all times of the year and is comparable in magnitude to that of the mean flow, being quasi-periodic with *ca.* 10–15 day periods (Elford 1976). Long-period waves also appear in spectral analyses of data from Durham, New Hampshire (Clark 1975) and Saskatoon (Belmont *et al.* 1977). The quasi-periodic character of the variations is well illustrated by figure 4, which shows the S–N prevailing wind for winter 1972–3 at Obninsk, the results having been subjected to harmonic analysis with a sliding 24 h interval shifted in steps of 1 h, and at Kühlungsborn–Collm, where values refer to midnight (Lysenko *et al.* 1975).

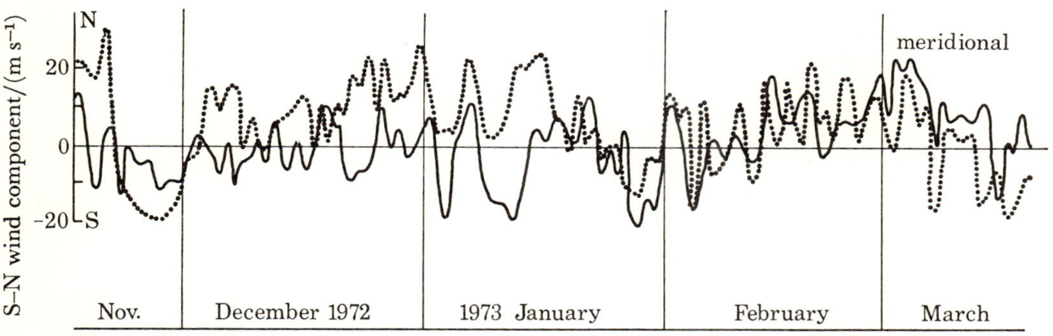

FIGURE 4. S–N prevailing wind components for winter 1972–3 at 95 km. Key: ——, Obninsk; ...., Kühlungsborn/Collm (Lysenko *et al.* 1975).

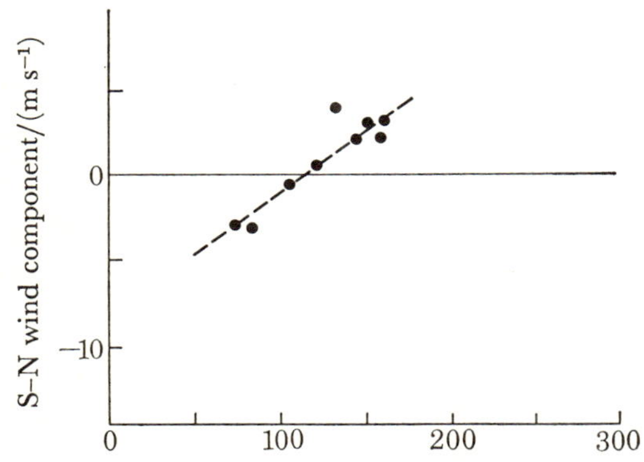

FIGURE 5. Prevailing S–N wind components at 95 km for the winter months November–February plotted against solar 10.7 cm radio flux. Data are from Kühlungsborn/Collm (1957/58–1974/75) and Obninsk (1964/65–1972/73) (Portnyagin *et al.* 1977).

Long term variations can be studied at only a few sites which have an adequate time record. During the years 1966–72, meteor winds at Adelaide showed a sudden change in early 1969 of the vertical gradient of the meridional component at 90 km, from slightly positive (*ca.* $+1\ \mathrm{m\,s^{-1}\,km^{-1}}$) to slightly negative (*ca.* $-1\ \mathrm{m\,s^{-1}\,km^{-1}}$), without evidence of any substantial complementary change in the zonal flow (Elford 1976). From the 1956–73 run of ionospheric drift measurements at Kühlungsborn–Collm, an oscillation of about 22-month period in the zonal prevailing wind has been reported (Sprenger *et al.* 1975). A dependence on solar activity of the prevailing wintertime (November to February) winds was shown some years ago for both W–E and S–N components by the Kühlungsborn–Collm observations (Sprenger & Schminder 1969). This dependence has been confirmed by subsequent data up to 1975 and by the radar meteor wind measurements at Obninsk, as shown for the S–N components in figure 5 (Portnyagin *et al.* 1977). No well defined correlation between wind speeds and solar activity has been found for summer months.

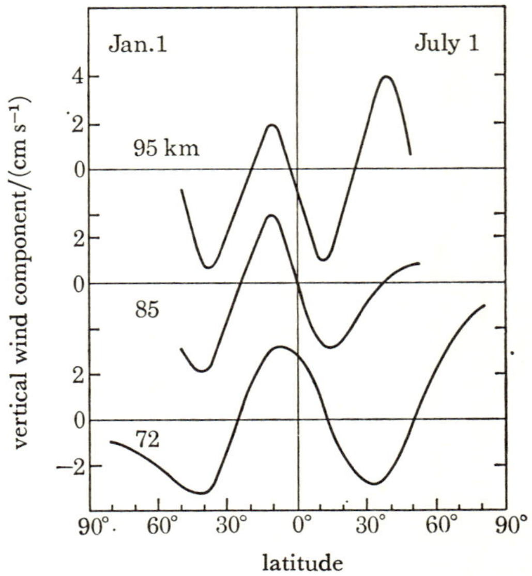

FIGURE 6. Variation with latitude of the mean velocity component normal to the isobaric surfaces at heights of about 72, 85 and 95 km in the N hemisphere for January 1 and July 1. Positive values represent upward motion (Ebel 1974).

5. AN INFERRED MERIDIONAL CIRCULATION

Thermally driven circulations between summer and winter hemispheres were depicted in the two dimensional theoretical 20–80 km models of Leovy (1964), with meridional winds increasing to 1.8 m/s at 70 km. This pattern of interhemispheric flow differs from that shown in figure 1 where maximum velocities reach much higher values of 10–20 (or more) m/s close to 95 km, and the latitudinal dependence is not that of a simple interhemispheric circulation cell. Subsequent radio wind measurements (§ 3) have provided further evidence of the higher values of S–N velocities shown by the model in the region of 90 km at the solstices; and, since many of these observations are at new longitudes, model values now appear unlikely to be greatly in excess of zonally averaged means. If such is the case, a mean meridional circulation may be derived on the basis of the S–N and W–E monthly wind models.

The problem has already been investigated at heights between 70 and 100 km by Ebel (1974) who used the steady state equations and the 1969 models to calculate the corresponding mean vertical motion (along with the required heat and momentum sources). The results obtained for equinox and solstice conditions show that the strength and direction of the circulation cells undergo a marked annual change. During the equinoxes the circulation is predominantly symmetric with respect to the equator, having four cells pole-to-pole. At the solstices it is strongly asymmetric with five cells pole-to-pole (reducing to three at the lower heights considered) and this circulation appears to be superimposed on a single-cell pole-to-pole circulation of the type proposed by Kellogg (1961). For winter the calculations depict upward motion between 5° and 20° latitude at all heights considered (figure 6): downward motion is obtained between 30° and a higher latitude which is about 50° for the upper half of the 70–100 km altitude range increasing to close to 90° latitude at 70 km. In other words, the mesospheric circulation cell over the winter hemisphere involves air rising at low latitudes, merging with a small interhemispheric flow and descending as it moves towards the winter pole.

The flow towards the equator at 60–65 km in the S–N model of figure 1 is indicative of a closed or nearly closed circulation in which descending polar air turns equatorwards. The upper boundary of this circulation cell is the zero S–N velocity line in figure 1, which passes through 80 km altitude at 80° latitude and slopes upwards to 111 km at 20° latitude: the gradient of the line is about 1:200. On its upper side the flow reverses, reaching high velocities in the direction of the equator at 120 km (figure 1). No calculations of vertical motion are available at these heights and nothing definite can be said about the lower thermosphere circulation, but it may be noted that the flow towards the equator in the model would be consistent with a circulation in which air descends over the winter polar region and ascends at low latitudes. A tentative picture that may therefore be presented of the winter hemisphere meridional circulation is that of two separate but likewise rotating circulations, one in the mesosphere and one in the lower thermosphere. Their common boundary on average is represented by the sloping zero S–N line of figure 1, which is the dividing line between two contrary moving air streams, one originating at high latitudes and one at low latitudes. At 90–95 km altitude and nearer the equator than mid-latitudes, the circulation is within the uppermost part of the mesospheric cell, whereas, nearer the pole than mid-latitudes at these heights, the circulation is within the lowermost part of the thermospheric cell.

The proximity of two contrary flow regions invites consideration of the extent to which they mix and of possible aeronomic consequences. While these matters lie beyond the scope of this paper, attention is drawn to the possible contribution of horizontal as well as vertical turbulent mixing. The presence of small-scale turbulent motions is well established for the 85–105 km region, but the day-to-day variations, such as those apparent in figure 4, would effect horizontal transport of air with path lengths of the order of 1000 km, and may be responsible for large scale eddy mixing. On such timescales, horizontal displacements across a zero S–N velocity line in figure 1 are of course possible since figure 1 represents the monthly mean situation.

6. Meridional circulation and the winter anomaly

The striking correlations that have been observed between the winter anomaly in D-region radio-wave absorption and meteorological parameters such as stratospheric temperatures point to atmospheric circulation as being relevant to the ionospheric phenomenon. As the latter in its most conspicuous form consists of days or groups of days of excessive absorption any relationship with meteorological parameters would more reasonably be strongest with departures from the average than with the average itself. A number of physical processes are likely to be involved in establishing the internal atmospheric conditions appropriate to anomalous absorption and here we consider only the meridional circulation as a contributing effect. In this case absorption would be expected to correlate significantly with S–N wind components in any set of observational data.

The winter observations at Arenosillo (37° N, 7° W) of 1969–70 showed that absorption increased with increasing winds from southern directions at 85 km and with increasing winds from northern directions between 90 and 93 km (Rose *et al.* 1972*a*). Considered in conjunction with figure 1, these results would indicate an increase in mass flow above and below the zero S–N velocity line at times of greater absorption, with the upper cell extending downwards to possibly 10 km below its normal position.

The absence of the winter anomaly south of about 30° latitude directs attention to the latitudinal form of the meridional circulation at the altitude of the anomalous absorption, which is typically close to 84–85 km (Laštovička 1972). On moving from mid to low latitudes in the winter hemisphere a number of changes in the meridional circulation occur: S–N wind velocities decrease and the zero S–N velocity line (figure 1) continues to increase in altitude and becomes ill defined. Ebel (1974) sees the 30° latitude limit as coinciding with the reversal from downward motion at mid-latitudes to upward motion at low latitudes, as shown in figure 6.

Although no definite conclusions can be reached at the present stage about the role of the meridional circulation in contributing to the internal atmospheric conditions appropriate to anomalous absorption, it appears worthy of consideration in future investigations.

Table 1. Diurnal launch series 1965–74

site	latitude	longitude W	date	number of successful launchings
Thule	76° 33′ N	68° 49′	24–26 Oct. 1968	14
Fort Churchill	58° 44′	93° 49′	6–8 Sept. 1966	10
			8–9 Sept. 1966	10
			4–5 Jan. 1968	12
			23–25 Oct. 1968	18
			19–20 Mar. 1974	8
Wallops Island	37° 50′	75° 29′	19–20 Mar. 1974	13
Arenosillo	37° 06′	06° 44′	24–28 Feb. 1970	27
White Sands	32° 23′	106° 29′	30 June–2 July 1965	17
			9–11 Oct. 1965	16
Cape Kennedy	28° 27′	80° 32′	13–15 Dec. 1967	25†
			23–25 Oct. 1968	17
Antigua	17° 09′	61° 47′	19–20 Mar. 1974	8
Fort Sherman	09° 20′	79° 59′	19–20 Mar. 1974	8
Kourou	05° 08′	52° 37′	19–22 Sept. 1971	13
			19–20 Mar. 1974	10
Natal	05° 55′ S	35° 10′	1966–68	24
			19–20 Mar. 1974	8
Ascension Island	07° 59′	14° 25′	11–12 Apr. 1966	13
			12–13 Apr. 1966	13
			24–26 Oct. 1968	14
			19–20 Mar. 1974	8
Mar Chiquita	37° 45′	57° 25′	19–20 Mar. 1974	7

† Also analysed as 13–14 Dec. 1967 (12 launchings) and 14–15 Dec. 1967 (14 launchings).

7. Diurnal launch series 1965–74

Table 1 summarizes the sets of data from short series of rocket launchings at different sites whose analysis for diurnal and semidiurnal wind components is presented here. On 19–20 March 1974, 70 rockets equipped with standard Datasonde instrumentation were launched at eight Western Hemisphere sites by N.A.S.A. Wallops Flight Center in cooperation with other agencies, to study atmospheric tides and their latitudinal variation (Schmidlin *et al.* 1975). The locations of the eight sites are shown in figure 7. Most of the other series of launchings shown in table 1 were carried out within the U.S. Meteorological Rocket Network during the years 1965–8, the height range of investigation being approximately 25–60 km. The 23–25 October 1968 launchings took place simultaneously at Ascension Island, Cape Kennedy, Fort Churchill and Thule, whereas other launchings were conducted on different dates at various

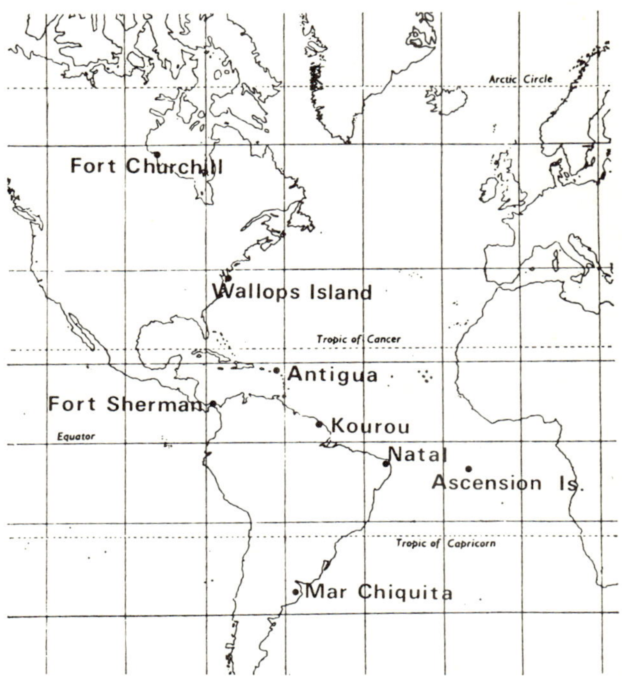

FIGURE 7. The eight sites from which diurnal series of rockets were launched on 19–20 March 1974.

low, mid and high latitude sites. The 24–28 February 1970 launchings were of Skua II rockets carrying a very lightweight chaff to measure winds from about 75 to 95 km altitude over Arenosillo (Rose *et al.* 1972*b*). The acoustic-grenade technique provided results to 90 km over Kourou on 19–22 September 1971 (Smith *et al.* 1974) and the same technique provided results from 24 launchings at Natal in 1966–8 (Smith *et al.* 1968, 1969, 1970). Where no specific reference is quoted in the text, the data source has been the Meteorology Data Reports of World Data Center A.

8. Method of harmonic analysis of data

It has been assumed that a wind component, v, may be represented as a sum of mean, diurnal and semidiurnal components by

$$v = v_0 + A_1 \cos \tfrac{1}{12}\pi t + B_1 \sin \tfrac{1}{12}\pi t + A_2 \cos \tfrac{1}{6}\pi t + B_2 \sin \tfrac{1}{6}\pi t, \qquad (1)$$

where t is local mean time in hours, departures from apparent local solar time being neglected. The symbols v_0, A_1, B_1, A_2, B_2 are five unknowns which may be determined, together with estimates of their standard deviations, by the method of least squares from, in principle, six or more observations. The analysis has been previously described (Groves 1967) and does not require equal time intervals between observations. All observed wind values have been given an equal weight, although the analysis permits the introduction of different weights.

Results are presented below in terms of the amplitudes and phases of the harmonic components, which may be readily calculated from the values determined for the *A*s and *B*s in equation (1): phase is expressed by the local time at which an oscillation maximizes. The horizontal lines in figures 8–18 are centred on the least-squares calculated value of a quantity and are equal in length to two standard deviations. At heights where amplitudes are small, phases become indeterminate as indicated by very large error bars or a data point in brackets. Attention has been given to the removal of a background trend in v by replacing v_0 in (1) by a polynomial in time. When the total observing time exceeded 2 days, a cubic was adopted. Many series of launchings extend over only 1 day and in most cases the analysis of these has been performed with v_0 expressed linearly in time: when there were no more than eight launchings in a series, the background trend was ignored.

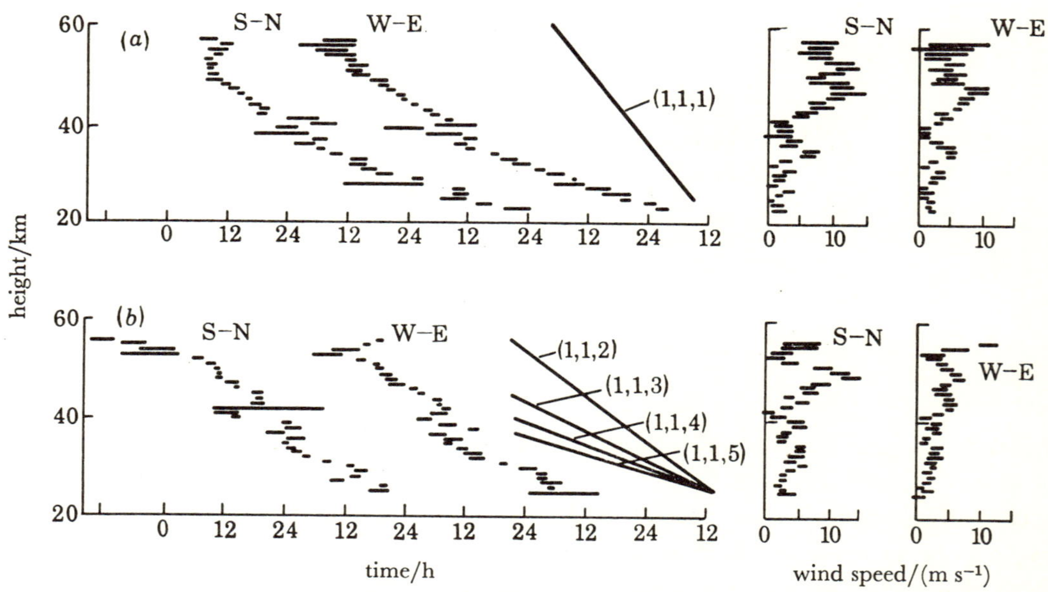

FIGURE 8. Diurnal wind components at Cape Kennedy. The straight lines indicate the approximate theoretical slopes of the phase profiles of the first five positive migrating modes. (*a*) 23–25 October 1968, 17 launchings in 48 h, (*b*) 13–15 December 1967, 25 launchings in 48 h.

9. Diurnal winds at mid-latitudes

The diurnal wind oscillations derived for 23–25 October 1968 and 13–15 December 1967 at Cape Kennedy have previously been compared and discussed (Groves 1976). The results are shown here in figure 8 in comparison with the theoretical phase profiles (the straight lines) for the migrating Hough modes (1, 1, 1) to (1, 1, 5): observed phases are seen to correspond most nearly with modes of a higher order than the leading (1, 1, 1) mode. Phase profiles for 23–25 October 1968 show about 2.25 rotations of the wind vector between 25 and 55 km and are similar to that of the (1, 1, 3) mode. For 13–15 December 1967 the average gradient of the phase profiles corresponds more closely with that of the (1, 1, 2) mode.

A notable property of the Cape Kennedy results of 23–25 October 1968 (figure 8) is that the profiles of S–N and W–E amplitudes show no significant differences and that W–E phases at all heights are later than S–N phases by close to 6 h: in other words, the diurnal wind vector at any given height rotated clockwise with no change of magnitude within the limits of accuracy of the data. If the oscillation may be regarded as a superposition of positive ($n > 0$) Hough modes $(1, s, n)$, then the above property implies approximate equality of W–E and S–N Hough wind functions for each mode that is significantly present. By examination of numerical evaluations of Hough wind functions for various s and n (not demonstrated here) it is found that this apparently stringent condition holds (to within about 20 %) at latitudes between 20° and 40° latitude, provided that eastward travelling modes are generally small compared with westward travelling modes.

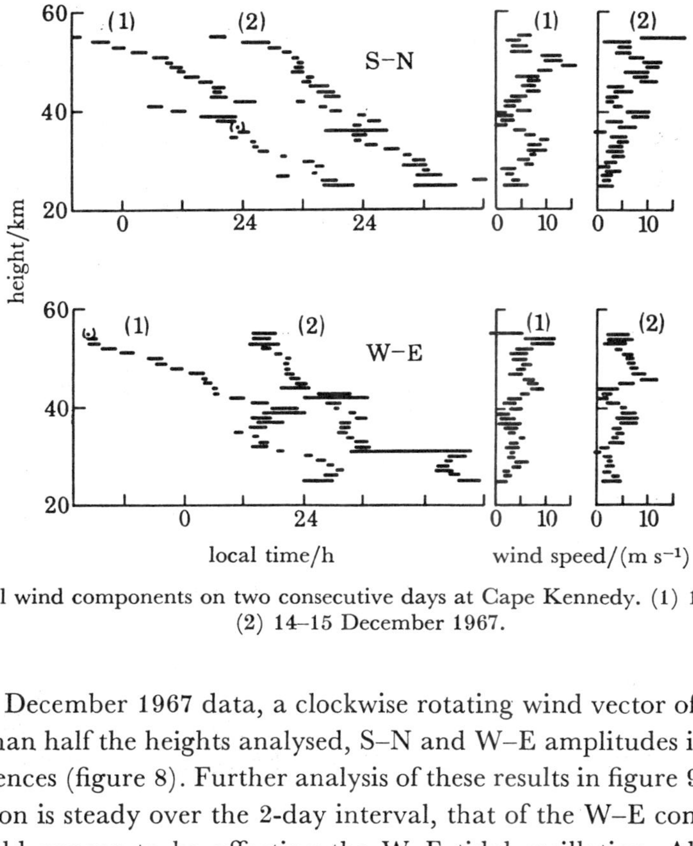

FIGURE 9. Diurnal wind components on two consecutive days at Cape Kennedy. (1) 13–14 December 1967; (2) 14–15 December 1967.

For the 13–15 December 1967 data, a clockwise rotating wind vector of constant magnitude is found at less than half the heights analysed, S–N and W–E amplitudes in particular showing significant differences (figure 8). Further analysis of these results in figure 9 shows that, whereas the S–N oscillation is steady over the 2-day interval, that of the W–E component is not: some disturbance would appear to be affecting the W–E tidal oscillation. Although little can be said about the nature of such a disturbance, the direction of slope of the phase profiles (to earlier times with increasing height) points to a source of excitation, that is at lower heights and probably in the troposphere.

Data have now been analysed for diurnal winds at slightly higher latitudes than Cape Kennedy and the results (figures 10 and 11) are found to be generally consistent with the diurnal rotation of a wind vector of approximately constant magnitude at any given height. The number of launchings in each series is shown in table 1, but data from all launchings may not be available at every height. Reference lines at 12 h and 18 h have been added to the plots of S–N and W–E phases respectively for Northern Hemisphere sites to aid comparisons. The

S–N phase profiles at White Sands are quite well reproduced with a 6 h delay by the W–E phase profiles. So also is that for Arenosillo (figure 11). For the Southern Hemisphere site of Mar Chiquita, observed phases are consistent with an anticlockwise rotation of the wind vector in accord with tidal theory. Mar Chiquita phase profiles have approximately the same average slope as those taken at the same time at Wallops Island, which is at the same latitude (within 0.1°) in the opposite hemisphere. A 12 h shift between S–N phases at the two sites therefore indicates a tidal excitation that is predominantly symmetric with respect to the equator; but amplitude profiles differ between the two sites to an extent that suggests that non-migrating modes, i.e. modes that have both local-time and longitudinal dependences, may be significant even for a separation of only 18.1° in longitude.

Results of the White Sands launchings of 30 June–2 July 1965 have previously been reported in some detail by Beyers *et al.* (1966).

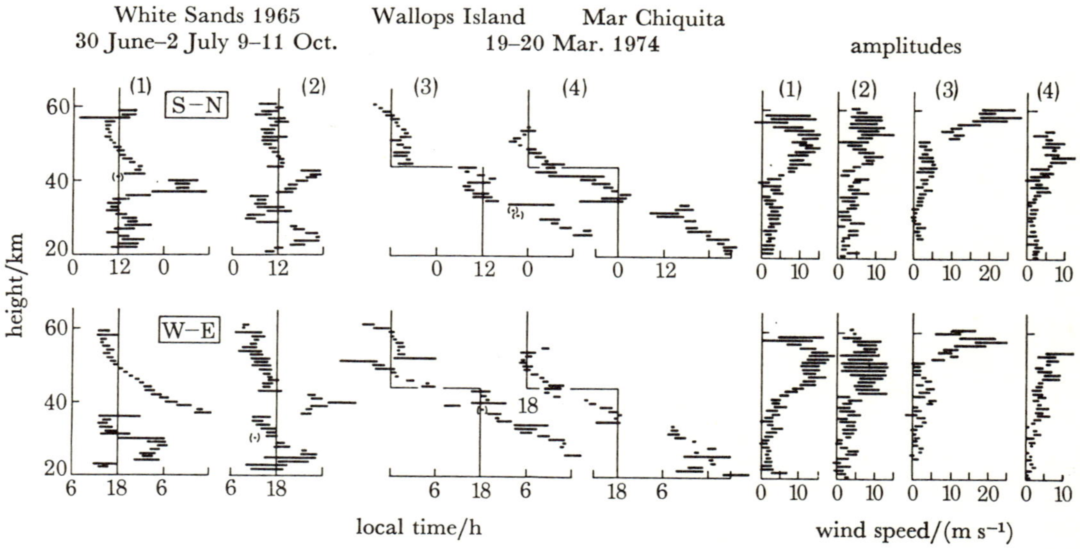

FIGURE 10. Diurnal wind components at White Sands, Wallops Island and Mar Chiquita.

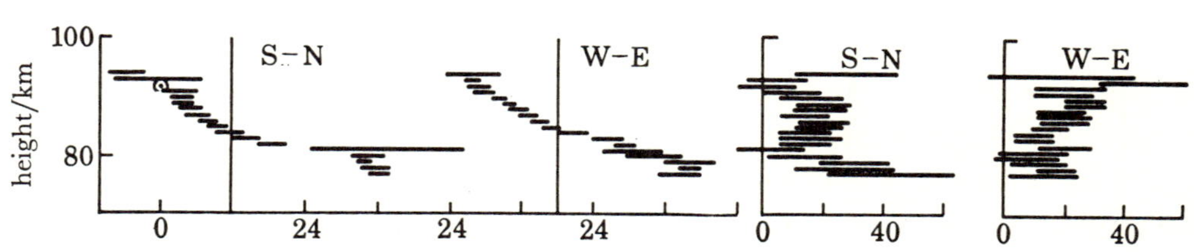

Figure 11. Diurnal wind components at Arenosillo, 24–28 February 1970.

10. Diurnal winds at low latitude

S–N and W–E Hough wind functions for positive diurnal modes (not presented here) are found by inspection to become increasingly different in form at latitudes of less than about 15°; and hence at low latitude sites S–N and W–E amplitude and phase profiles would be expected to differ appreciably: the observational results (figures 12–15) confirm this expectation.

The results for Antigua show that W–E phases are later than S–N phases by 6 h at most heights except for the interval from 36 to 51 km (figure 12): short wavelengths appear to be propagating in the W–E component below 36 km and in the S–N component below 51 km changing to longer wavelengths above these heights.

Four sets of results for Ascension Island are compared in figure 13. The two sets showing greatest similarity are those for April 1966 which are based on the first 13 and the last 13 launchings of a series of 24 over a 2-day interval. A detailed analysis of the April 1966 data

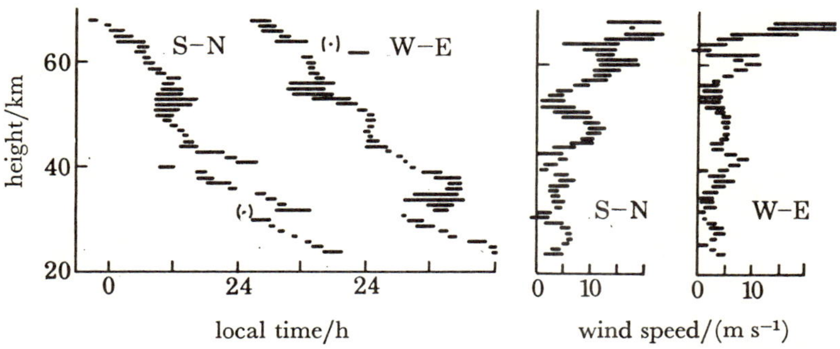

FIGURE 12. Diurnal wind components at Antigua, 19–20 March 1974.

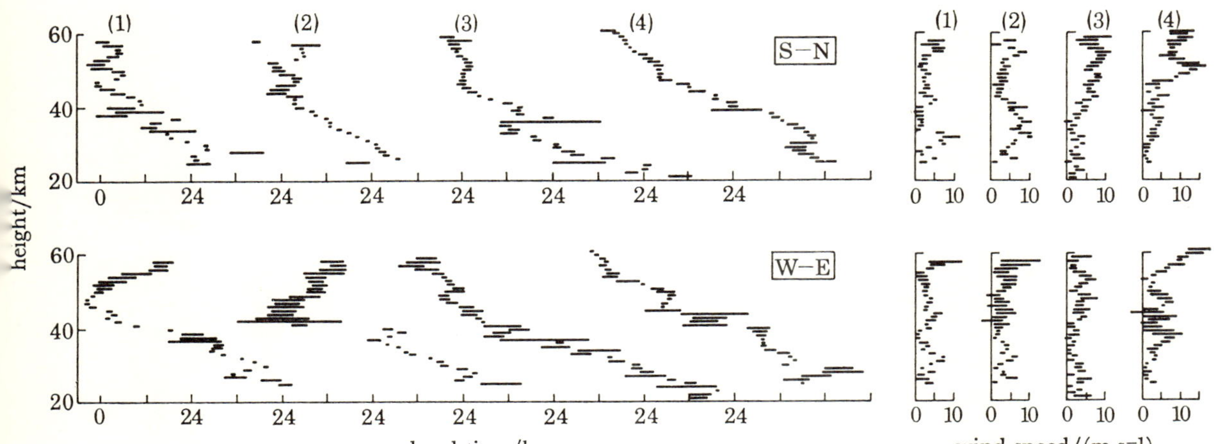

FIGURE 13. Diurnal wind components at Ascension Island. (1) 11–12 April 1966; (2) 12–13 April 1966; (3) 24–26 October 1968; (4) 19–20 March 1974.

has been given by Beyers & Miers (1968), who noted considerable variation in both S–N and W–E components between the two days and concluded that the diurnal variation may have been masked or distorted by some other short term disturbance. On close comparison of the two sets of April 1966 results in figure 13, it is found that 40% of the phases on the second day differ significantly from those on the first. The situation is similar to that reported above for 13–15 December 1967 at Cape Kennedy: a steady or quasi-steady tidal oscillation appears to be replaced or dominated by variations that arise from a non-tidal source, i.e. one that does not correlate with the hour angle of the Sun over several days. For other low latitude analyses reported here (figures 14 and 15) it is not possible, with the number and distribution of available launchings, to compare consecutive cycles and no indication can therefore be given of the presence or otherwise of non-tidal effects.

A comparison between the results for Fort Sherman in figure 14 and those taken on the same date (19–20 March 1974) at Ascension Island is of interest as the two sites are at nearly the same latitude in opposite hemispheres. A similar comparison in § 9 for Wallops Island and Mar Chiquita indicated a predominantly symmetric excitation with respect to the equator, but from figures 13 and 14 the Fort Sherman and Ascension Island oscillations are seen to differ markedly in phase and amplitude. These differences may be attributed to the presence of non-tidal effects as discussed above or to the presence of non-migrating modes as referred to in § 9 in conjunction with the 65.6° change of longitude between the two sites. The 19–20 March 1974 results (figure 15) for Kourou and Natal, which are also sites at approximately equal latitudes in opposite hemispheres, again show very dissimilar phase and amplitude profiles.

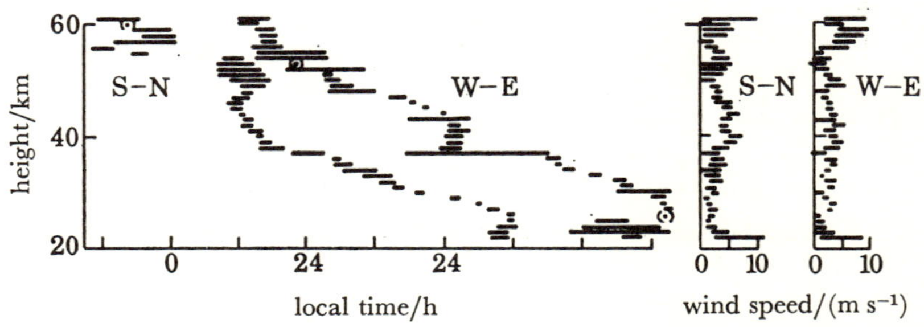

FIGURE 14. Diurnal wind components at Fort Sherman, 19–20 March 1974.

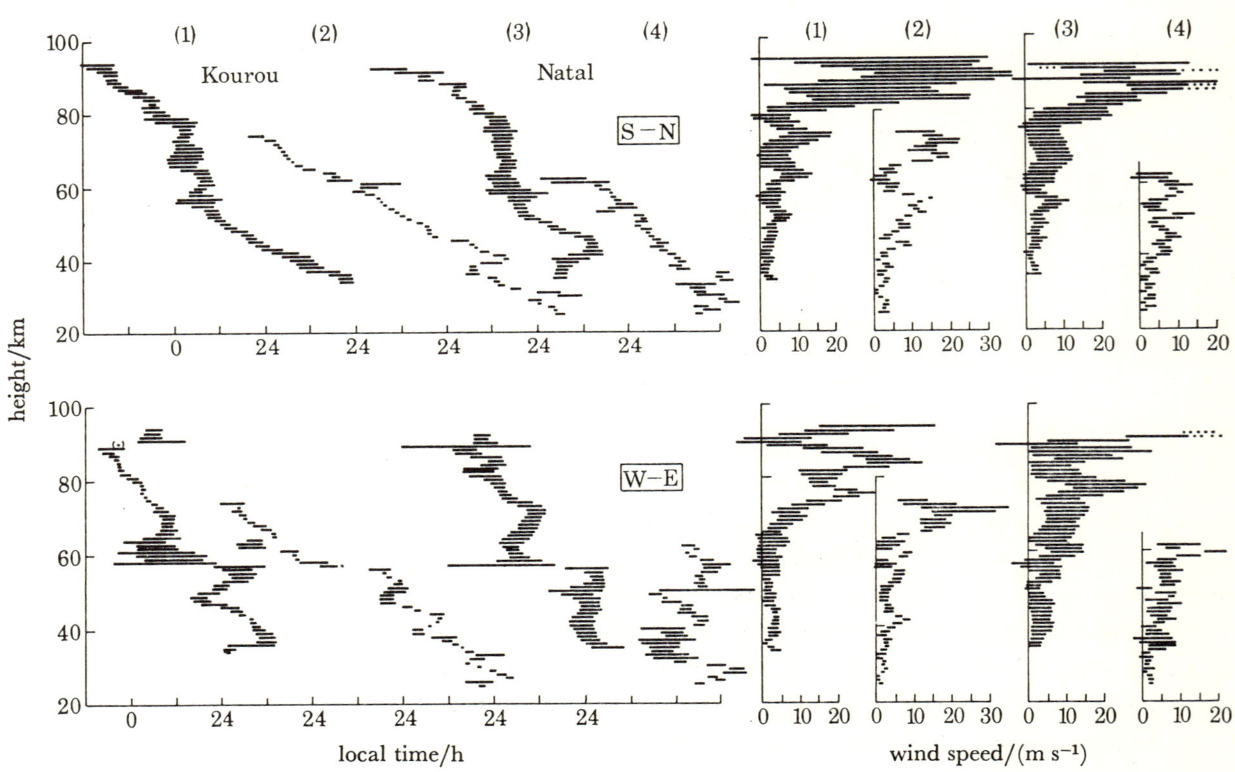

FIGURE 15. Diurnal wind components at Kourou and Natal. (1) Kourou, 19–22 September 1971; (2) Kourou, 19–20 March 1974; (3) Natal, 1966–68; (4) Natal, 19–20 March 1974.

Also shown in figure 15 are results obtained by the acoustic-grenade technique at Natal (Groves 1974) and Kourou (Groves 1975*a*). The Natal launchings differ from the other launch series on account of their distribution over an interval of 22 months; consequently the results from them would be expected to reveal a component of oscillation that is steady for much of the year. The Kourou acoustic-grenade phase profiles (figure 15) match those for Natal of 1966–8 to a remarkable extent except at the lower heights. As reported previously, however (Groves 1975*a*), the Kourou temperature oscillation showed significant differences from that for Natal, indicating some variation in the source of excitation for which the atmosphere response was of observational significance in temperature and not in winds. Further evidence of variation in the source of excitation between different dates is provided by the 19–20 March 1974 wind results at Natal and Kourou which differ significantly from the acoustic-grenade results at the corresponding sites (figure 15).

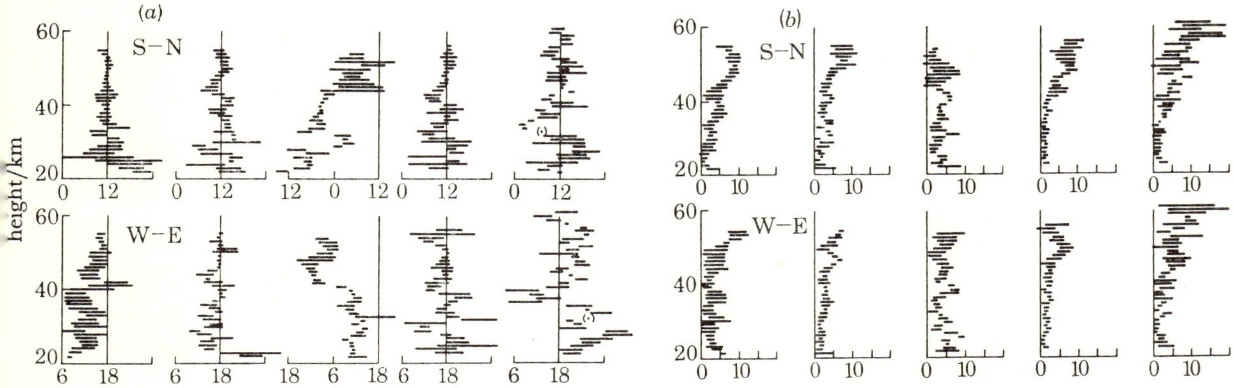

FIGURE 16. Diurnal wind components at Fort Churchill. (*a*) Phases, (*b*) amplitudes.

11. Diurnal winds at Fort Churchill

It is a well known property of diurnal Hough functions (Chapman & Lindzen 1970; Groves 1975*b*) that positive modes are confined to low and mid latitudes, whereas negative modes attain their greatest values at high latitudes. At Fort Churchill tidal oscillations can therefore be expected to be composed of negative modes. Another property of negative modes is that, in contrast to the vertically propagating and oscillatory nature of positive modes, they show no rotation in height and have amplitudes that decay in either vertical direction away from the level of excitation, such that stratospheric oscillations would be excited by stratospheric (ozone) heating. The height independence of phase at high latitude was confirmed by the analysis of Reed *et al.* (1969) of summer data between 30 and 60 km from the combined stations of Fort Churchill and Fort Greely for the diurnal S–N wind oscillation: observed S–N wind phases were very close to 12 noon, the value which accords theoretically with an expected 12 noon phase for the ozone heating rate.

Figure 16 shows S–N and W–E diurnal winds for Fort Churchill obtained from series of launchings conducted on the dates shown. The results are independent of those of Reed *et al.*

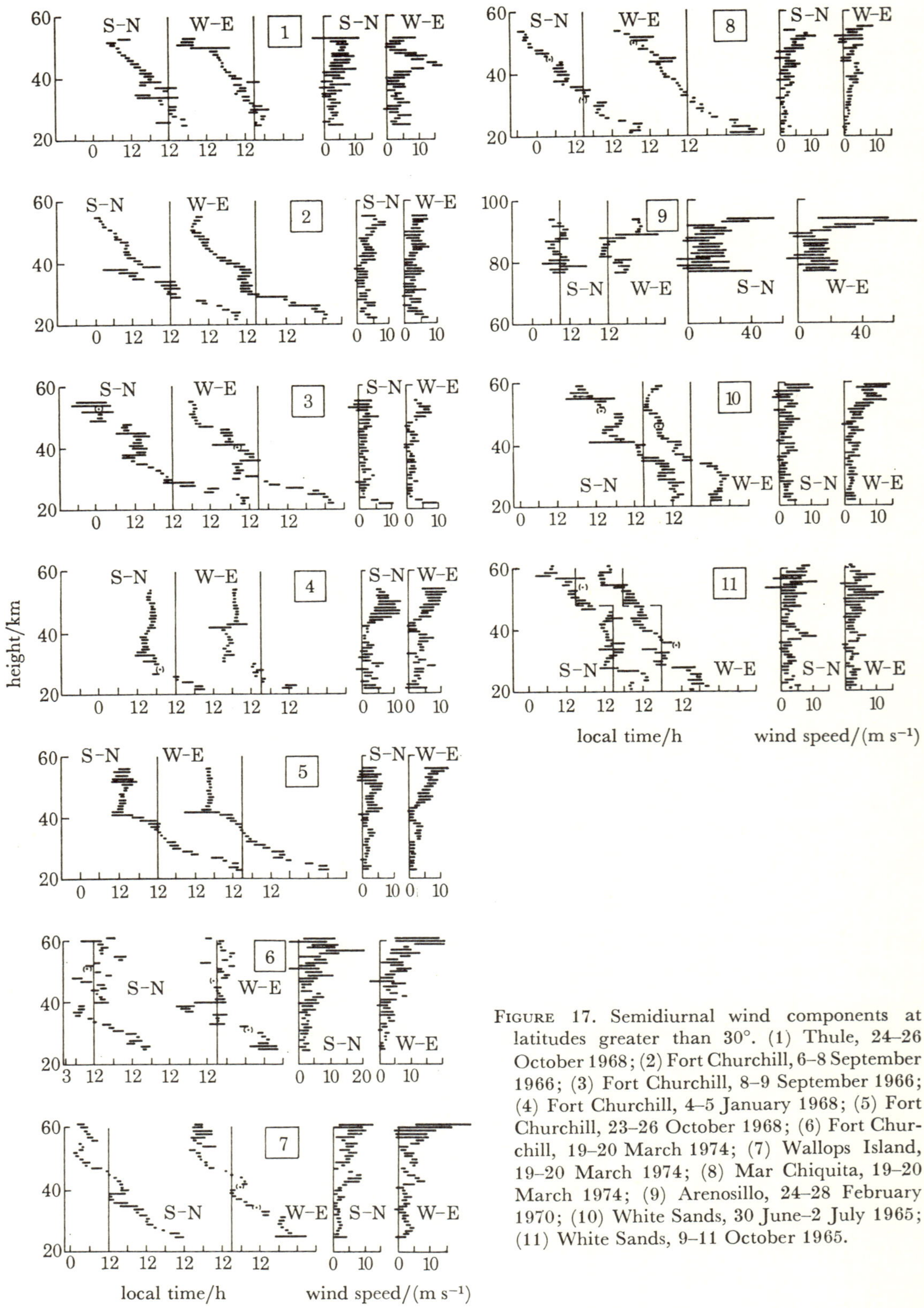

FIGURE 17. Semidiurnal wind components at latitudes greater than 30°. (1) Thule, 24–26 October 1968; (2) Fort Churchill, 6–8 September 1966; (3) Fort Churchill, 8–9 September 1966; (4) Fort Churchill, 4–5 January 1968; (5) Fort Churchill, 23–26 October 1968; (6) Fort Churchill, 19–20 March 1974; (7) Wallops Island, 19–20 March 1974; (8) Mar Chiquita, 19–20 March 1974; (9) Arenosillo, 24–28 February 1970; (10) White Sands, 30 June–2 July 1965; (11) White Sands, 9–11 October 1965.

(1969) which were derived from data widely distributed in time over the summer months of several years up to and including 1966. S–N phases at or close to 12 noon are common in figure 16 with the notable exceptions, to be discussed in the next paragraph, of 4–5 January 1968 and certain heights below 45 km. For the W–E wind oscillation the corresponding theoretical phase is 1800 h, the rotation of the wind vector being clockwise. With the same exceptions this value is fairly well supported observationally. The two launch series in September 1966 comprise the first ten and the last ten launchings respectively of a series of 19 launchings at 4 h intervals over 3 days. The general form of the phase profiles in figure 16 clearly differs in character from that at other latitudes (figures 8–15) where positive modes are able to propagate.

Below 40 km ozone lifetimes increase sufficiently for atmospheric motion to become an effective transport mechanism and longitudinal asymmetries in ozone distribution arise (Heath *et al.* 1973; Krueger *et al.* 1973; Dütsch 1974). Maximum heating rates may then be displaced in longitude from the subsolar longitude, and phases of the S–N wind below about 45 km would not necessarily be at 12 noon. In the absence of information on ozone distributions such relations cannot, however, be examined. The unusual phase profiles of 4–5 January 1968 also direct attention to the possibility of a disturbed ozone distribution, in this case up to 55 km, and considering the time of year this might arise from a stratospheric warming and its effect on temperature-dependent photochemistry (Barnett *et al.* 1975). On examining M.N.R. synoptic charts of North America, it is found that this series of launchings coincided with a major warming which started in December 1967 and persisted into the beginning of 1968: by the end of December, the temperature gradient between polar and mid latitudes had reversed throughout the entire middle and upper stratosphere making the event the earliest major warming ever to be recorded (E.S.S.A. 1970; N.O.A.A. 1971). Of the other above Fort Churchill launchings, none took place at the time of a stratospheric warming.

12. Semidiurnal winds at latitudes greater than 30°

Figure 17 shows semidiurnal wind phases and amplitudes at six different sites, data being available at Fort Churchill for four different dates. The 19 launchings of 6–9 September 1966 at Fort Churchill have been divided into two groups consisting of the first ten and the last ten launchings respectively: each group extends over three consecutive cycles and the results of the first group are quite well reproduced by the second group indicating the presence of quite a steady oscillation over this 3-day interval. Comparisons between other sets of results are, in contrast, notable for conspicuous differences in phase and amplitude profiles.

One feature that stands out in figure 17 is that S–N and W–E profiles show similarities of phase and, to a lesser extent, of amplitude, W–E phases being later than S–N phases by close to 3 h, corresponding to a clockwise rotation of the wind vector. For the Southern Hemisphere site of Mar Chiquita an anticlockwise rotation is indicated.

Another feature of the phase profiles in figure 17 is their failure to comply with previous theoretical results for semidiurnal oscillations, which present a mainly constant phase arising from the assumed predominance of the (2, 2, 2) mode. In the classical treatment (Chapman & Lindzen 1970) a rapid phase reversal is obtained just below 30 km between adjacent regions of almost constant phase extending up to the mesopause and down to the surface. In the more recent calculations of Lindzen & Hong (1974) with a more realistic atmosphere, the chief modification is to the level at which the reversal occurs.

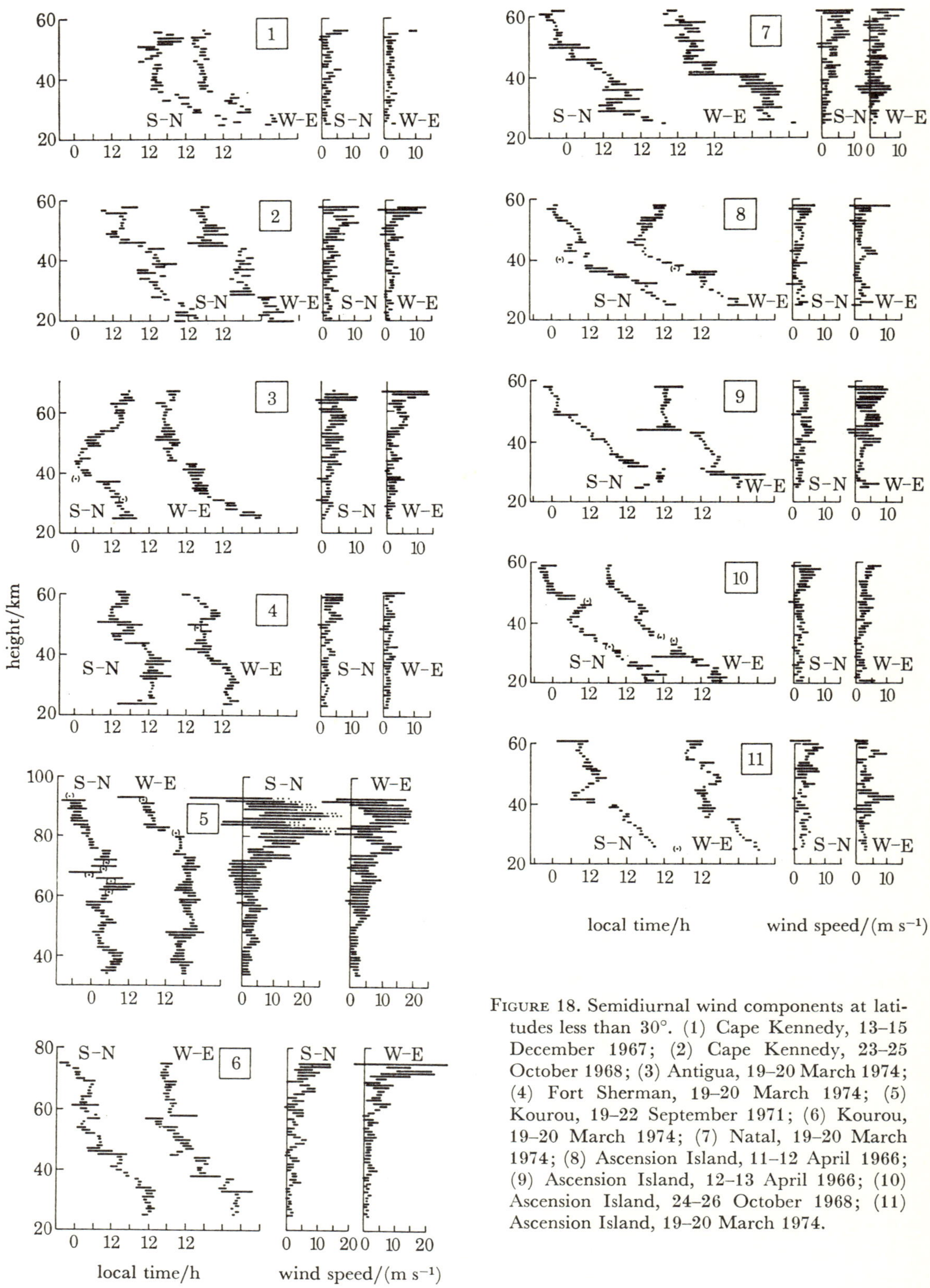

FIGURE 18. Semidiurnal wind components at latitudes less than 30°. (1) Cape Kennedy, 13–15 December 1967; (2) Cape Kennedy, 23–25 October 1968; (3) Antigua, 19–20 March 1974; (4) Fort Sherman, 19–20 March 1974; (5) Kourou, 19–22 September 1971; (6) Kourou, 19–20 March 1974; (7) Natal, 19–20 March 1974; (8) Ascension Island, 11–12 April 1966; (9) Ascension Island, 12–13 April 1966; (10) Ascension Island, 24–26 October 1968; (11) Ascension Island, 19–20 March 1974.

Interpretation of the results in figure 17 in terms of tidal Hough modes requires the introduction of significant contributions from higher-order semidiurnal modes. Some guide to the order of modes present is provided by the phase profiles which, for particular height ranges, complete one rotation (i.e. 12 h change of phase) within 14 km or less of vertical distance. Among the migrating semidiurnal modes, those having such oscillatory structures would be (2, 2, n) where $n \gtrsim 14$. As with the diurnal components (§ 9) the direction of slope of the phase profiles (to earlier times with increasing height) points to a source of excitation at lower heights, most probably in the troposphere, where a common identity would be expected with the source of excitation of the diurnal components.

The above account of the properties of the semidiurnal oscillations closely follows that previously given (Groves 1976) on the basis of the Fort Churchill results of 6–9 September 1966 and 23–25 October 1968 and the Wallops Island and Mar Chiquita results of 19–20 March 1974. The further observations that have since been analysed and now appear in figure 17 support the previous account. The similarity between S–N and W–E semidiurnal profiles at mid and high latitudes, and the clockwise rotation of the wind vector at northern latitudes and anticlockwise rotation at southern latitudes, are in accord with the properties of Hough wind functions as previously discussed (Groves 1976). These properties were previously examined for migrating modes (2, 2, n) and further calculations (not presented here) show that they also hold for both eastward and westward travelling non-migrating modes (2, s, n), $s \neq 2$.

13. Semidiurnal winds at latitudes less than 30°

Figure 18 shows sets of results of semidiurnal oscillations derived at sites of less than 30° latitude. The two sets of results for Cape Kennedy at 28.5° N are consistent with a clockwise rotating wind vector at most observation heights within the limits of error and could be grouped with the results in figure 17. For Antigua at 17° N and also at still lower latitudes, S–N and W–E Hough wind functions become appreciably dissimilar and corresponding dissimilarities are to be expected between S–N and W–E semidiurnal wind components. The observational results for Antigua, Fort Sherman and most other launch sites in figure 18 show conspicuous differences between S–N and W–E components, which would appear to confirm this expectation.

The analyses for Ascension Island on 11–12 and 12–13 April 1966 (figure 18) show that both S–N and W–E semidiurnal components change radically between the two days. As with the diurnal components at low latitude (§ 10), a non-tidal variation would appear at times to replace or dominate a steady or quasi-steady tidal oscillation. Insufficient data are available to examine the frequency of such occurrences.

14. Discussion

14.1. *Seasonal wind variations*

Meridional cross sections of S–N and W–E wind components from 60 to 130 km at monthly intervals were prepared in 1969 and similar W–E wind models appear in *C.I.R.A.* (1972). These cross sections were prepared by averaging and smoothing observed wind components and have provided reference models against which to compared newly acquired data. Recent wind measurements by radio techniques, 80–100 km, tend to confirm the main features of the models and their seasonal trends. Although many of the recent observations are from new longitudes, data coverage is still inadequate to allow reliable zonal mean motions to be obtained. With the new data, however, zonal means would not appear unlikely to differ greatly from model values and in particular from the high (*ca.* 10 m/s) S–N model velocities at heights close to 90 km.

Consideration has been given in § 5 to the mean meridional circulation that may be inferred from the wind models by reference to the 1974 calculations of Ebel for vertical wind velocities and by interpretation of the sloping zero S–N velocity line in the winter hemisphere of figure 1 as a boundary region separating mesospheric and lower thermospheric circulation patterns. The wintertime circulation described is an average situation appropriate to a time interval of the order of a month, from which appreciable deviations are indicated on shorter timescales by, for example, the day-to-day wind variations that are observed close to 90 km. Long term variations, such as a solar cycle dependence for wintertime winds, have been identified in mean meridional winds at particular sites and indicate corresponding long term modulations of the circulation.

Meridional circulations in the middle atmosphere cannot be studied observationally without an extensive coverage of global wind data that is far in excess of that available or likely to be available for some time. Evidence of the nature of meridional transport may, however, be indirectly obtained if other observable effects exist that are dependent upon it. Two ways in which wintertime mean meridional circulations may be relevant to middle atmosphere aeronomy are considered to be worthy of further examination: these are the mixing of air of mesospheric–equatorial origin with that of thermospheric–polar origin at about 90 km over mid latitudes, and the effect of such circulations on the internal atmospheric conditions appropriate to the winter anomaly in radio-wave absorption.

14.2. *Diurnal variations*

The principal frequency components of wind variations at meteor ionization heights are solar diurnal and semidiurnal (Teptin 1972): the same components predominate in surface pressure oscillations (Chapman & Lindzen 1970). In §§ 7–13 diurnal and semidiurnal components obtained from short series of rocket launchings are presented and discussed for the intermediate region of approximately 25–60 km.

The results of early series of launchings at White Sands and Ascension Island for diurnal components have previously been reported (Beyers *et al.* 1966; Beyers & Miers 1968): the conclusion reached was that a dominant diurnal tidal oscillation existed at the stratopause at the mid-latitude site whereas at the low-latitude site a well-defined diurnal variation was not evident being masked or distorted by some other short term disturbance. The larger body of data that has now been analysed supports these conclusions with the rider that at low latitudes both diurnal and semidiurnal variations are affected by short term disturbances although only intermittently. Greater clarification of the frequency of such disturbances cannot be given with the very few sets of data available.

Comparisons between phase and amplitude profiles for the same location at different times rarely show similarities and it would seem that at these heights the tidal motions are steady for only short intervals of time, possibly a few days, although no reliable estimate can be given of this variability as data are not available over adequate intervals of time.

It has also been possible to compare results at different stations at the same time including pairs of stations at almost the same latitude in opposite hemispheres (Wallops Island and Mar Chiquita, Kourou and Natal, and Fort Sherman and Ascension Island). Comparisons between phase and amplitude profiles rarely show similarities indicating a detailed horizontal structure to tidal motions including a longitudinal dependence. From the longitudinal separation of the above pairs of stations, variations are significant over distances of *ca.* 10^3 km.

At latitudes greater than about 30°, phase profiles of W–E and S–N wind components usually represent wind vectors that rotate with height (clockwise in the Northern Hemisphere and anti-clockwise in the Southern Hemisphere). Rates of rotation with height are, however, greater than those corresponding to excitation of the leading diurnal or semidiurnal modes, being appropriate to higher-order modes whose latitudinal structure would require significant variations in the source of excitation over *ca.* 10° changes of latitude. Support for the interpretation of observed oscillations in terms of tidal modes is provided by phase profiles at high latitude (Fort Churchill), which differ characteristically from those at lower latitudes for the diurnal component but not for the semidiurnal: the diurnal phases at Fort Churchill are more nearly constant with respect to height than those at lower latitudes in accord with the properties of positive and negative sequences of diurnal models, whereas for the semidiurnal component only one sequence of modes exists.

An area for further investigation is the identification of the relatively localized sources of excitation that seem to be implied by the observations. Significant variations in tropospheric conditions over distances of *ca.* 10^3 km are common enough, but the requirement is for steady or quasi-steady diurnal and semidiurnal frequency components of an adequate magnitude to generate the stratospheric winds observed.

The author gratefully acknowledges the assistance of Miss A. M. Yeung with analysis of data and of Miss U. Campbell and members of the Departmental Photographic Group with preparation of diagrams. Sponsorship has been provided by the Air Force Geophysics Laboratory (A.F.G.L.), United States Air Force under Grant no. AFOSR-77-3224, and by the U.K. Meteorological Office.

References (Groves)

Belmont, A. D., Nastrom, G. D. & Hovland, D. N. 1977 *Periodic wind variations at* 65–118 km *at Saskatoon* (52° *N*). Control Data Corporation Final Report, dated 26 May 1977. Minneapolis, Mn., U.S.A.

Beyers, N. J. & Miers, B. T. 1968 *J. atmos. Sci.* **25**, 155–159.

Beyers, N. J., Miers, B. T. & Reed, R. J. 1966 *J. atmos Sci.* **23**, 325–333.

Chapman, S. & Lindzen, R. S. 1970 *Atmospheric tides: thermal and gravitational.* Dordrecht, Holland: D. Reidel Publishing Co.

C.I.R.A. 1972 *Cospar International Reference Atmosphere.* Berlin: Akademie–Verlag.

Clark, R. R. 1975 *J. atmos. Sci.* **32**, 1689–1693.

Dütsch, H. U. 1974 *Can. J. Chem.* **52**, 1491–1504.

Ebel, A. 1974 *Tellus* **26**, 325–333.

Elford, W. G. 1976 *Nature, Lond.* **261**, 123–124.

E.S.S.A. 1970 *Weekly Synoptic Analyses*, 5-, 2-, *and* 0.4-*Millibar Surfaces for* 1967. E.S.S.A. Technical Report WB 12.

Glass, M., Fellous, J. L., Massebeuf, M., Spizzichino, A., Lysenko, I. A. & Portniaghin, Yu. I. 1975 *J. atmos. terr. Phys.* **37**, 1077–1087.

Gregory, J. B. & Manson, A. H. 1975 *J. atmos. Sci.* **32**, 1667–1675.

Gregory, J. B. & Rees, D. T. 1971 *J. atmos Sci.* **28**, 1079–1082.

Groves, G. V. 1967 *Space Research* **7**, 986–1000.

Groves, G. V. 1969 *J. Br. interplanet. Soc.* **22**, 285–307.

Groves, G. V. 1974 *J. Br. interplanet. Soc.* **27**, 499–511.

Groves, G. V. 1975*a* *J. Br. interplanet. Soc.* **28**, 241–244.

Groves, G. V. 1975*b* *J. Br. Interplanet. Soc.* **28**, 797–809.
Groves, G. V. 1976 *Proc. R. Soc. Lond.* A **351**, 437–469.
Heath, D. F., Mateer, C. L. & Krueger, A. J. 1973 *Pure appl. Geophys.* **106–108**, 1238–1253.
Kellogg, W. W. 1961 *J. Meteor.* **15**, 373–381.
Kreuger, A. J., Heath, D. F. & Mateer, C. L. 1973 *Pure appl. Geophys.* **106–108**, 1253–1263.
Laštovička, J. 1972 *J. atmos. terr. Phys.* **34**, 1233–1239.
Leovy, C. 1964 *J. atmos. Sci.* **21**, 327–341.
Lindzen, R. S. & Hong, S. S. 1974 *J. atmos. Sci.* **31**, 1421–1446.
Lysenko, I. A., Portnyagin, Yu. I., Greisiger, K. M. & Sprenger, K. 1975 *Zeit. f. Meteorologie* **25**, 213–217.
Muller, H. G. & Kingsley, S. P. 1974 *J. atmos. terr. Phys.* **36**, 1933–1943.
N.O.A.A. 1971 *Weekly Synoptic Analyses* 5-, 2-, *and* 0.4-*Millibar Surfaces for* 1968. N.O.A.A. Technical Report NWS 14.
Portnyagin, Yu. I., Kaidalov, O. V., Greisiger, K. M. & Sprenger, K. 1977 *Phys. Solariterr. Potsdam*, **5**, 91–96.
Reed, R. J., Oard, M. J. & Sieminski, M. 1969 *Mon. weath. Rev.* **97**, 456–469.
Roper, R. G. 1975 *Radio Sci.* (*U.S.A.*) **10**, 363–369.
Rose, G., Widdel, H. U., Azcárraga, A. & Sanchez, L. 1972*a* *Phil. Trans. R. Soc. Lond.* A **271**, 529–545.
Rose, G., Widdel, H. U., Azcárraga, A. & Sanchez, L. 1972*b* *Phil. Trans. R. Soc. Lond.* A **271**, 509–528.
Schmidlin, F. J., Yamasaki, Y., Motta, A. & Brynsztein, S. 1975 *N.A.S.A. sp*-3095, Washington, D.C.
Scholefield, A. J. & Alleyne, H. 1975 *J. atmos. terr. Phys.* **37**, 273–286.
Smith, W. S., Theon, J. S., Casey, J. F. & Horvath, J. J. 1970 *N.A.S.A. TR R*-340, Washington, D.C.
Smith, W. S., Theon, J. S., Swartz, P. C., Casey, J. F. & Horvath, J. J. 1969 *N.A.S.A. TR R*-316, Washington, D.C.
Smith, W. S., Theon, J. S., Swartz, P. C., Katchen, L. B. & Horvath, J. J. 1968 *N.A.S.A. TR R*-288, Washington, D.C.
Smith, W. S., Theon, J. S., Wright, Jr., D. U., Ramsdale, D. J. & Horvath, J. J. 1974 *N.A.S.A. TR R*-416, Washington, D.C.
Sprenger, K., Greisiger, K. M. & Schminder, R. 1975 *J. atmos. terr. Phys.* **37**, 1391–1393.
Sprenger, K. & Schminder, R. 1969 *J. atmos. terr. Phys.* **31**, 217–221.
Stubbs, T. J. 1976 *J. atmos. terr. Phys.* **38**, 979–989.
Teptin, G. M. 1972 *Proc. Acad. Sci. U.S.S.R., Atmosph. & oceanic. Phys.* **8**, 139–150.
Vincent, R. A. & Stubbs, T. J. 1977 *Planet. Space Sci.* **25**, 441–455.
Wilkinson, P. J. & Baggaley, W. J. 1975 *Planet. Space Sci.* **23**, 509–522.

Phil. Trans. R. Soc. Lond. A **296**, 41–57 (1980) [41]
Printed in Great Britain

Satellite measurements of middle atmosphere temperature structure

By J. J. Barnett
Department of Atmospheric Physics, Clarendon Laboratory, Oxford, U.K.

Sixteen years have elapsed since the first satellite measurements of atmospheric temperature. These were observations of the lower stratosphere. Techniques have developed rapidly, and observations now extend from the surface to the mesopause. The instruments and techniques are briefly described and a review is given of the wide range of middle atmosphere research that has been based upon these measurements. The Nimbus 6 pressure modulator radiometer has made over 3 years' observations of upper stratospheric and mesospheric temperature, with weighting functions peaking at up to 80 km. The main results from this instrument and their relation to variations at lower levels are discussed. Temperature variations are generally smaller in the upper mesosphere than in the stratosphere. Planetary waves penetrate to this level in winter. There is a strong negative correlation between zonal mean temperature near the mesopause and in the upper stratosphere on both long and short timescales.

1. Spectroscopic remote sounding

Soon after the first artificial satellite was launched it was realized that measurements of infrared radiation emitted by the atmosphere could be used to deduce its temperature structure (Kaplan 1959). At wavelengths sufficiently long (> 2 μm) the radiant energy from the atmosphere is mainly thermal emission by atmospheric constituents, with little or no contributions from airglow, fluorescence or reflected solar radiation. For nadir observations from space, i.e. observations looking vertically down at a single wavelength at which just one gas has significant absorption, the radiation observed originates from a layer. Radiation from below this layer is absorbed by the atmosphere above, so little reaches space, while radiation from above the layer is emitted by so much smaller a quantity of gas because the density decreases so rapidly with height that it is insignificant. Where Lorentz line broadening is dominant, the variation of linewidth with pressure causes the emitting layer to be still thinner. This varying response to emission from different levels is represented by the weighting function K, such that the radiation observed, I, is given by

$$I = \int_0^\infty K(z)\, J(z)\, \mathrm{d}z, \tag{1}$$

where $J(z)$ is the source function and z is a height or pressure-like variable, most conveniently ln (pressure). Figure 1 gives nadir view weighting functions for various instruments viewing carbon dioxide emission at 15 μm. These weighting functions are integrated over a range of frequencies, since no measurement can be monochromatic, and so are broader than single frequency weighting functions. When given in pressure coordinates, weighting functions depend mainly upon the mixing ratio profile of the emitting gas, but also to a varying extent upon the temperature profile because of the temperature dependence of Lorentz and Doppler line broadening. When given in height coordinates the temperature dependence is greater because the hydrostatic equation introduces an additional component of variation. For carbon dioxide the mixing ratio is sufficiently constant up to the thermosphere, and the weighting functions are, to a good extent, fixed except for a little temperature dependence in the mesosphere.

Whereas the weighting function describes the pressure range of the observation, the measured intensity is determined by the source function at these pressures. When possible, measurements are made of spectral bands where there is local thermodynamic equilibrium between the translational energy states of gas molecules and the vibrational and rotational states responsible for the emission. In this case the source function equals the Planck function, which is a simple

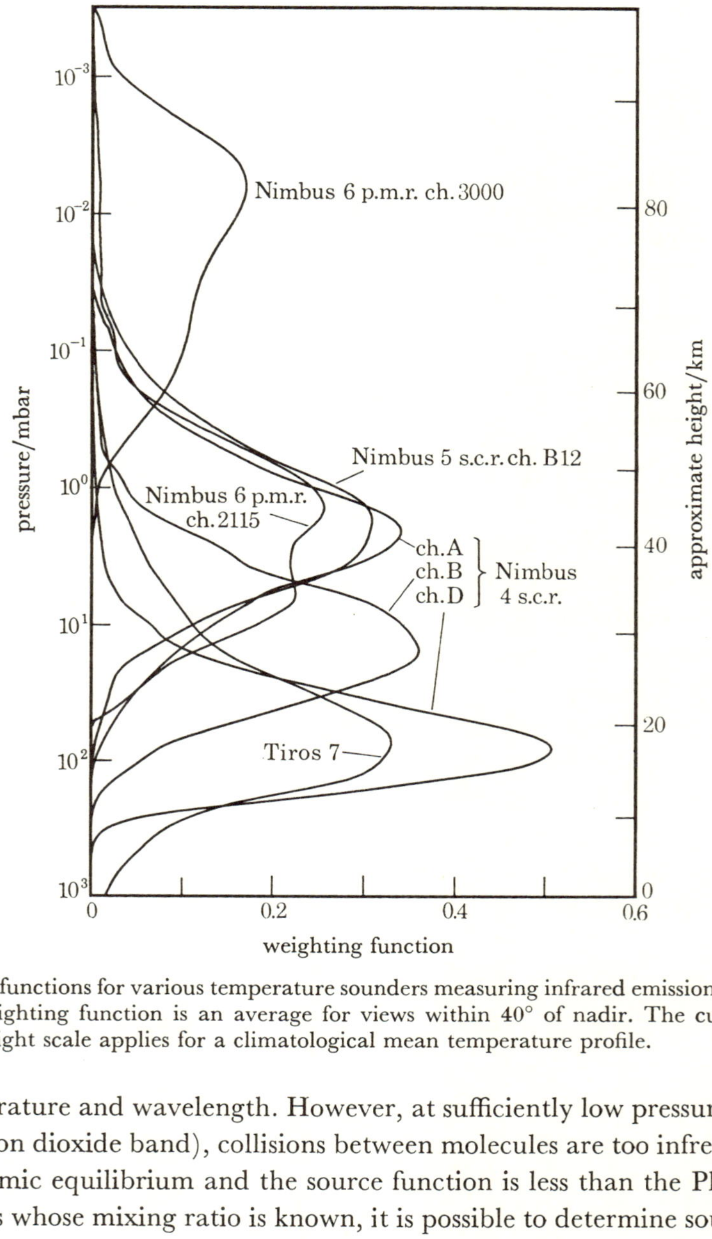

FIGURE 1. Weighting functions for various temperature sounders measuring infrared emission from carbon dioxide. The Tiros 7 weighting function is an average for views within 40° of nadir. The curves are functions of pressure; the height scale applies for a climatological mean temperature profile.

function of temperature and wavelength. However, at sufficiently low pressure (10^{-2} mbar† for the 15 μm v_2 carbon dioxide band), collisions between molecules are too infrequent to maintain local thermodynamic equilibrium and the source function is less than the Planck function.

Thus with a gas whose mixing ratio is known, it is possible to determine source function, and hence temperature, for various layers by measuring at appropriate wavelengths. This is the basis of temperature sounding in most instruments launched so far. Alternatively, given the temperature profile, observations of a gas at various wavelengths can be used to determine the vertical distribution of that gas except in regions where it is isothermal.

† 1 mbar = 10^{-2} Pa.

2. Temperature sounding instruments

The first instrument to use these principles was the medium resolution infrared radiometer (m.r.i.r.) carried by the Tiros 7 satellite launched in 1963. One channel measured emission from carbon dioxide in the 14.8–15 μm region. The satellite was spinning; this caused the instrument to scan the Earth at varying angles to the nadir. By averaging observations within 40° of the nadir over regions about 5° latitude by 5° longitude, and over a few days, almost complete analyses were obtained for an emitting layer in the lower stratosphere (see figure 1). With this instrument many important and now well known phenomena were seen for the first time in such detail, and they were described by Kennedy & Nordberg (1967) in what is surely a classic paper.

The m.r.i.r. was carried by several subsequent Tiros satellites, and an improved version but with a broader filter and a slightly lower weighting function was carried by Nimbus 2 in 1967 (Nordberg *et al.* 1967). The Nimbus series of satellites are more convenient than the early Tiros were, having three-axis stabilization relative to the vertical and the direction of travel. Orbital heights are normally about 1000 km. The orbits are Sun-synchronous, passing over the tropics and mid-latitudes near local noon and midnight, and reach 80° N and 80° S. Orbital periods are about 106 min. Each successive orbit is 26° longitude further west, and as the Earth rotates beneath the orbital path, the satellite passes within 13° longitude of each place twice per day. The data are recorded within the satellite and replayed at high speed when passing ground stations, so that data for much of the globe are routinely obtained. The recorders normally degrade faster than other parts of the satellite, and after about a year of operation, coverage starts to deteriorate. Nevertheless, Nimbus satellites have an excellent record of long life: Nimbus 4 launched in 1970 is still used, and Nimbus 5 launched in 1972 can still give almost complete coverage each day.

The m.r.i.rs had only one carbon dioxide channel. It is fortuitous for stratospheric research that this, the simplest temperature sounder to construct, which measures a broad part of the 15 μm carbon dioxide emission, receives most of its radiation from the lower stratosphere. Developments since then fall into three main categories:

(*a*) improvement of spectral resolution, by several orders of magnitude in some cases, to allow a choice of the height observed, from the surface to the mesopause;

(*b*) finer horizontal resolution, which is particularly important in the troposphere where it is often necessary to view through gaps between clouds;

(*c*) improved accuracy.

The satellite infrared spectrometer (s.i.r.s.) carried by Nimbus 3 (1969) and Nimbus 4 (1970) was the first instrument to have channels sounding at more than one level and so be able to resolve vertical structure. The instrument used eight channels 5 cm^{-1} wide in the 15 μm (668 cm^{-1}) carbon dioxide band. The highest two channels measure emission from layers 15–30 km thick covering the middle and lower stratosphere. More elaborate instruments with approximately the same weighting functions, but using filters, were carried on later satellites, namely the v.t.p.r. on the N.O.A.A. 2 to N.O.A.A. 5 series of satellites (1972–6), the i.t.p.r. on Nimbus 5 (1972) and the h.i.r.s. on Nimbus 6 (1975). These instruments were intended primarily for tropospheric sounding, the stratospheric weighting functions being necessary to eliminate from the tropospheric weighting functions any contribution from the stratosphere. However, the data are used extensively for stratospheric studies, including operational monitoring of sudden winter warming events, by the Upper Air Branch of the U.S. National Oceanic and Atmospheric Administration.

The main problem of sounding above the troposphere is obtaining sufficient spectral resolution. Conventional infrared spectrometers or filter radiometers can give broad weighting functions in the mid and low stratosphere, but nothing higher, since spectral lines becoming narrower and non-overlapping cause transmission to vary much more rapidly with wavelength. Furthermore, the energy available in any sufficiently small spectral interval would be so small as to require long observation times. These problems are overcome by a method used by Oxford and Heriot-Watt Universities in the selective chopper radiometer (s.c.r.) on Nimbus 4 and extensively developed by them since. The radiation is passed through a filter to select radiation only from the 15 μm carbon dioxide band, and is then passed through a cell containing carbon dioxide before falling onto the detector. The carbon dioxide acts as a filter to attenuate radiation near the centre of each line, the overall attenuation depending upon the cell length and carbon dioxide pressure. The Nimbus 4 s.c.r. selected the radiation by chopping the beam between a cell containing carbon dioxide and an identical but empty cell, the difference signal being the intensity near the line centres summed over all of the lines. This gave a weighting function peaking at about 2 mbar (43 km) with a 15 km half-width. Another channel chopped between two cells containing carbon dioxide at different pressures to obtain the radiation in narrow intervals either side of each line, and had a weighting function peaking near 20 mbar. The Nimbus 5 s.c.r. (1972) used the same basic method but, instead of chopping between cells, the radiation was measured in turn through four identical cells, one being empty and the other containing carbon dioxide at different pressures. By subtracting radiances during data analysis, four different weighting functions were obtained, peaking at levels up to 2 mbar. These signals typically differed by only 10%, so care was necessary to obtain an accurate difference signal, and for this reason these designs were unsuitable for higher weighting functions where much smaller differences would occur. The main problems arise from lack of 'balance' between the two paths because of (*a*) inevitable small differences between cells, and (*b*) the imprecision of chopping. These problems were overcome by using one cell containing carbon dioxide whose pressure could be modulated by a piston. This technique is called pressure modulation and was first used in the pressure modulator radiometer (p.m.r.) carried by Nimbus 6 (1975). A refinement was to include a molecular sieve and heater in the cell to allow the mean pressure to be varied if necessary to obtain various mean weighting function heights from a single cell. The p.m.r. has a top weighting function peak near 80 km (see figure 1), for which the spectral resolution is about 1 part in 10^6. A potential problem of sounding at such heights is that the Doppler shift due to just 5% of the satellite velocity is comparable with the Doppler line width. Thus at heights where Doppler broadening is dominant (in this case above about 50 km), moving the direction of view away from nadir along the direction of travel moves the atmospheric emission spectrum relative to the lines in the cell, causing the weighting function to move down. This was used to advantage on the Nimbus 6 p.m.r., where the instrument scans to 15° either side of nadir along the direction of travel, causing the weighting functions to scan vertically. A similar method, without Doppler scanning, is used on the Tiros and N.O.A.A. series of operational satellites, the first being carried by Tiros N (launched in 1978). Three pressure modulator cells have weighting functions peaking in the stratosphere. A similar instrument is carried by the Pioneer Venus orbiter.

In the instruments described so far, selection of the height of observation is carried out spectroscopically. The limb sounder views the limb of the Earth with a very narrow field of view and the height viewed is determined geometrically, most of the radiation originating near the tangent height, with a weighting function that may be as thin as 5 km. However, the height viewed must be known accurately, since an error of 1 km (0.02° angular deviation for a Nimbus orbit where the horizon is 3000 km away) leads to a radiance error of up to 15% for viewing carbon dioxide at 15 μm. Fortunately, the viewing level can be determined sufficiently accurately from radiance measurements alone, but there remains a difficult requirement that the spacecraft remains stable to seconds of arc during a complete observation. The method has two main advantages:

(*a*) thinner weighting functions can be obtained than with nadir sounding;

(*b*) constituents can be sounded more precisely and unambiguously, partly because of the greater optical thickness of the atmosphere for a limb path.

Three limb sounders have been carried on satellites: the Nimbus 6 l.r.i.r., the Nimbus 7 l.i.m.s. (both basically the same design) and the Nimbus 7 s.a.m.s. which uses pressure modulation. They each measure temperature in the stratosphere and mesosphere and the concentrations of various constituents. They are fully described elsewhere in this volume, and so will not be referred to here again.

Data from other instruments have been used for stratospheric studies. The Nimbus 5 n.e.m.s. and Nimbus 6 s.c.a.m.s. sounders measured microwave thermal emission from oxygen, and thereby obtained narrow (*ca.* 10 km wide) weighting functions in the troposphere and lower stratosphere. The Eole constant density balloon experiment (1971–2) provided temperature and wind fields for the Southern Hemisphere at about 200 mbar. A large number of balloons were tracked by satellite, which was used to relay their measurements and to determine their velocities. These data have been used in several stratospheric studies since they provide a reference from which winds in the stratosphere can be determined. A similar experiment known as TWERLE was performed with Nimbus 6.

3. Interpretation of data

One of the shortcomings of satellite temperature sounding is undoubtedly the limited vertical resolution. With a vertical sounding instrument the weighting functions have a half-width of 10 km or more, and although measurements from different channels can be linearly combined to give a narrower weighting function, a width of about 10 km is a practical limit in the middle atmosphere. Thus disturbances with a vertical wavelength of 20 km or less are severely attenuated or are invisible. Fortunately the gross features of the stratospheric and mesospheric temperature structure, i.e. zonal mean variations and planetary waves, have vertical distance scales greater than 10 km. However, some wave motions that determine these features have wavelengths that are too short to resolve, including gravity waves which may be important in the energy, heat and momentum budgets above the middle mesosphere, and equatorial modes which are believed to produce the quasi-biennial oscillation.

Approaches to data interpretation fall into three categories:

(*a*) direct use of radiance data, possibly converted to brightness temperature by using the inverse Planck function;

(*b*) derivation of thickness for layers well described by the measurements;

(*c*) derivation of temperature profiles as continuous functions of height by a process known as retrieval.

There are several methods of temperature retrieval. Insufficient information is measured to determine the temperature at each level which might be desired, so there is an infinity of temperature profiles that will fit a given set of observations, each differing in detail over a distance scale less than the resolution and also in ways allowed by the measurement uncertainty. Retrieval methods attempt to derive the most reasonable profile, although different retrievals from the same data may differ substantially, depending upon how reasonableness is built into the method. A common retrieval method is to determine the most probable profile, given the climatological mean and covariance of deviations from the mean for the appropriate latitude, day of the year, and the probable error in the measurements. The solution is given by a simple linear combination of measured radiances. Mathematically this method is similar to a linear regression between temperature and the observed radiance. A review of retrieval methods was given by Rodgers (1976*b*).

Temperature retrievals are almost essential for some applications such as budget calculations, but they suffer from the problem that more information is present than is actually measured. Thus it is never clear whether medium scale features on the borderline of the vertical resolution are present in the measurement or are artefacts of the retrieval process or originate from other information used by it, such as atmospheric statistics. Hence where temperature retrievals are not essential, radiances or thicknesses are often preferred because they display no more information than is measured. Thickness may be determined by the same type of retrieval methods as temperature. The simplest method is to regress, for a wide range of temperature profiles, thickness against radiance or brightness temperature measured by one channel whose weighting function averages over approximately the same layer. A high correlation usually results, so allowing radiance analyses to be relabelled as thickness for this layer. A more sophisticated approach is to use multilinear regression with several channels, and possibly different coefficients for different seasons and latitude bands.

One of the great advantages of satellite data is their completeness and uniformity of data coverage. This greatly simplifies the mapping task and there is no vital need for objective analysis methods able to assimilate data for arbitrary places, although such methods are frequently used. In Oxford we have used linear interpolation along the orbit for filling small gaps and zonally between adjacent orbits but not over larger gaps, to obtain data on a grid of intervals 10° longitude and 4° latitude. This is done separately for north- and south-bound parts of orbits and the two are combined yielding as a final product a grid which may still have gaps. Only where gaps are unacceptable, such as when zonal Fourier components are to be derived, are they filled by linear interpolation and then only if sufficiently small. Temperature structures observed by satellites in the stratosphere and mesosphere have long distance scales such that they are adequately represented by the lowest four zonal waves, possibly because shorter-scale horizontal variations have vertical scales too small to be observed. Temperature fields can change substantially during a day, but rarely is it necessary to use analyses at more than daily intervals. However, it is frequently a cause of concern that simple analyses as described above are not for any single synoptic time, so that in principle there should be a discontinuity representing the change during a day along the line between the first and last orbits of the day. Rodgers (1977*b*) overcomes this problem by using Kalman filtering to give statistically rigorous time interpolation and smoothing with a long data series, and thereby obtains analyses at any desired synoptic times.

4. Accuracy of measurements

The problems of calibrating an infrared radiometer have been discussed by Houghton (1977). Errors fall into four main categories:

(*a*) Errors that are random from one observation to the next, usually detector noise; although they limit the accuracy they are readily measured and can be taken into account.

(*b*) Errors that vary more slowly but cannot be corrected. For example, it may be necessary to correct calibration parameters for the temperature of various parts of the instrument, but this is likely to fail where rapid temperature changes cause thermal gradients across those parts; since the pattern of temperature variation is likely to repeat each orbit, this problem is liable to produce a pattern of errors which also repeats from orbit to orbit.

(*c*) Constant or slowly varying errors such as those caused by imperfect mirrors, black bodies or filters, or stray radiation, or errors in transmission data used to calculate weighting functions. Pre-launch testing should provide information to eliminate these errors, but there are bound to be residual components.

(*d*) Errors that arise from numerical short cuts when processing the data, such as taking a spectral interval to be monochromatic when integration over a range of wavelengths would be desirable, or neglecting the effect of weighting function variation with temperature, or non-local thermodynamic equilibrium effects. Most of these sources of error can, in principle, be made as small as desired, so there results a compromise between the effort required and the resulting gain of accuracy.

Observation of changes of atmospheric temperature from place to place are of much greater importance than measurement of absolute temperature. For observations from a single instrument, constant errors (category (*c*)) cancel when taking temperature differences from place to place at the same pressure and to some extent when comparing measurements of different pressure levels. However, this does not mean that such errors can be ignored because different instruments, even when using the same basic techniques, may have quite different forms of error and results can only be combined if they have absolute accuracy or are adjusted to be compatible.

Detailed comparisons between s.c.r. and rocket measurements were given by Barnett *et al.* (1975). For the Nimbus 5 s.c.r. they found mean differences between the radiances observed by the s.c.r. and those calculated from coincident rocket temperature profiles equivalent to about 2 K in the upper stratosphere and less than 1 K in the lower stratosphere. Standard deviations from these mean differences were of similar size, and correlation coefficients between the two forms of measurement were typically 0.95. The set of comparisons included a variety of stations and seasons. An indication of the random error was given by the standard deviation of radiance at one location (0° N, 0° E) during July 1973. This is composed of atmospheric plus instrumental variation, so gives an upper limit to either. Values corresponded to approximately 0.1 K for a channel measuring in the lower stratosphere, rising by an order of magnitude towards the upper stratosphere. Such accuracy is impressive and shows that the instruments can provide excellent observations of waves; however, it should be remembered that retrieval methods are able to improve vertical resolution by combining weighting functions, but at the expense of noise, so that final retrievals may be considerably less accurate than this. Miller in an accompanying paper gives an account of the errors in Tiros N s.s.u. measurements.

5. Studies using satellite data

Satellites have extended our measurement of atmospheric temperature in three main ways:

(*a*) Daily radiosonde analyses were previously possible up to 10 mbar for the Northern Hemisphere and weekly analyses up to 0.2 mbar, although the latter depended upon a network of rocket stations which is very sparse. Satellites now give us complete daily analyses throughout the middle atmosphere.

(*b*) Satellites give us complete Southern Hemisphere coverage of the middle atmosphere where previously there were just radiosonde observations at relatively few stations on land.

(*c*) Satellites provide homogeneous data from the same instrument, so allowing waves of as little as 0.3 K amplitude to be detected.

Most studies have dealt with the winter disturbed stratosphere and mesosphere, which is dominated by planetary waves and stratospheric warmings.

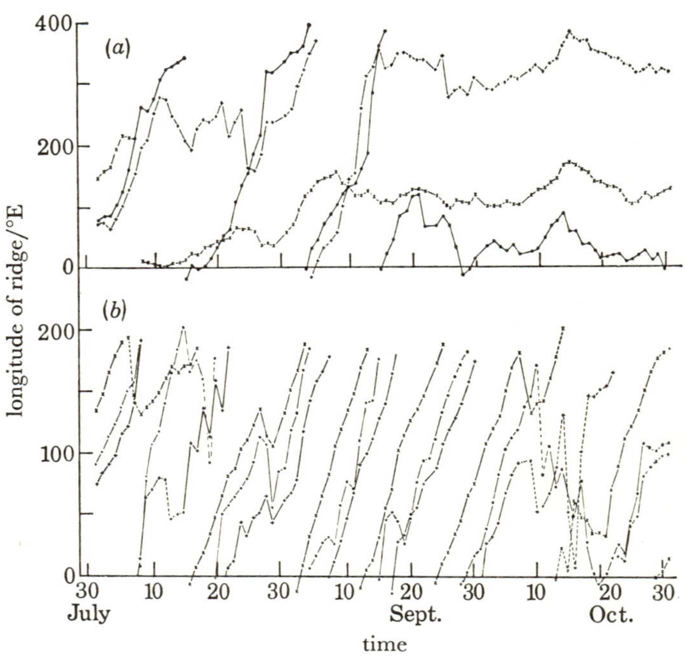

FIGURE 2. Phases (longitude of the ridge) of temperature waves at 52° S during 1971 measured by the Nimbus 4 s.c.r. +, channel A; •, channel B; ×, channel D (see figure 1 for their weighting functions). (*a*) wavenumber one – mainly stationary with a westward tilt; (*b*) wavenumber two – eastward travelling waves with a westward tilt with height. From Harwood (1974).

5.1. *Planetary waves*

The general behaviour of planetary waves has been investigated by several workers. Harwood (1975) studied the 1971 Southern Hemisphere winter and observed a marked wavenumber 2 pattern of up to 12 K amplitude in the stratosphere, which rotated uniformly eastwards about the pole through five cycles in 3 months (figure 2). He also noted an inverse relation between zonal and eddy available potential energy in the lower stratosphere. During this time wavenumber 1 phase was relatively stationary. Leovy & Webster (1976) studied the same data set but also Eole derived temperatures in the upper troposphere, and found excellent coherence between the 200 mbar level and the upper stratosphere. They noted that wavenumber 2 amplitudes have their peak in the late winter in both hemispheres, and that, at least in the Northern Hemisphere, this follows tropospheric behaviour. They asked what determines the meridional extent of planetary waves and pointed out that, although it seems to follow the strength of the zonal mean westerly wind field, the eddies are more alike in the two hemispheres during their respective winters than the wind fields, suggesting that another mechanism, possibly photochemical damping at low latitudes, might be important. Deland (1973, 1977) studied s.i.r.s. and s.c.r. data and found westward travelling planetary waves in the tropics and mid-latitudes, apparently coupled and forced from outside the tropics, and eastward travelling

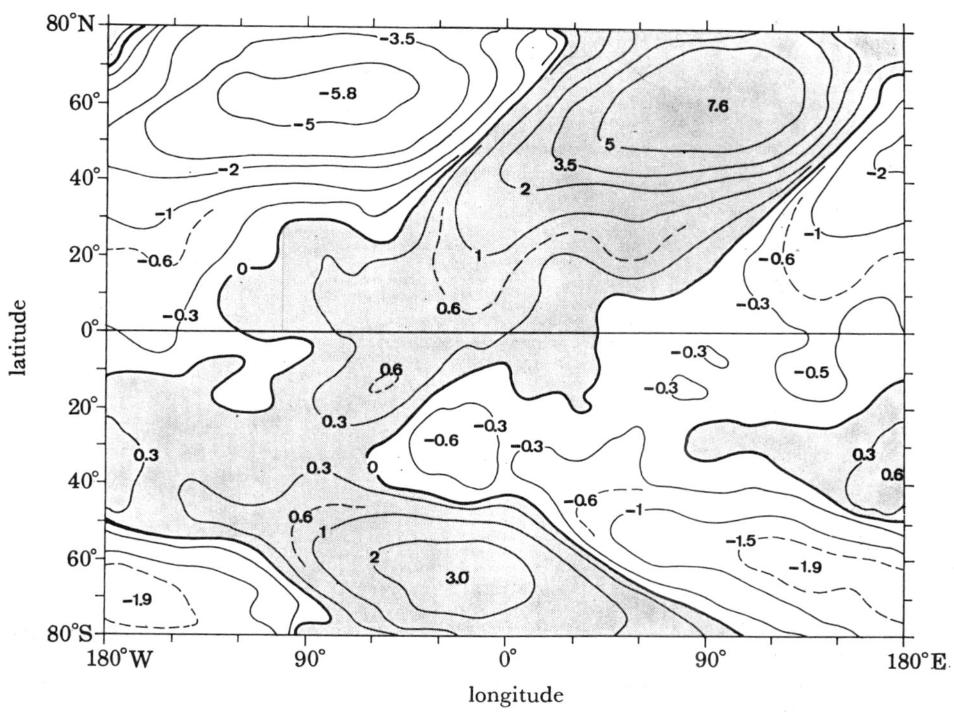

FIGURE 3. Deviation of Nimbus 5 s.c.r. channel B12 brightness temperature (/K) from the zonal mean averaged over 9–23 March 1973. Regions of positive deviation are dotted. Large stationary waves, mainly of wave-number one, were present in both hemispheres, but that of the Northern Hemisphere appeared to penetrate into the Southern as indicated by the westward tilt with latitude. From Barnett (1975).

waves in the Southern Hemisphere winter. He was able to separate them into two classes, one clearly that seen by Harwood, Leovy and Webster (but for different years) which tilted westward with increasing height suggesting forcing from below, and a class of faster moving waves which tilted eastward with height, suggesting downward energy propagation. The temperature and geopotential height structure of Southern Hemisphere travelling waves have also been described by Hartmann (1976) in an extensive study using a data set retrieved by N.O.A.A. combining the Nimbus 5 n.e.m.s., s.c.r. and i.t.p.r. instruments. Similar results were obtained by Stanford & Dunkerton (1978), although their method of analysis differed by using spherical harmonics.

Hirota (1976, 1978) has made extensive use of s.c.r. and p.m.r. data to study travelling waves. He has shown that transient planetary waves penetrate into the tropics near the equinoxes (although amplitudes are less than a degree), a result similar to a finding by Barnett (1975*a*) that stationary planetary waves appear to propagate across the equator at such times to mid-latitudes of the other hemisphere (see figure 3). It is not clear whether this behaviour relates to the semiannual zonal wind oscillation in the tropics, and which is the controlling process. Hirota (1976) used power spectral analysis to obtain the spectrum of phase velocities for wavenumber one during various seasons. This gave convincing support to the simple theory that planetary waves can only propagate westwards relative to the zonal flow (see figure 4).

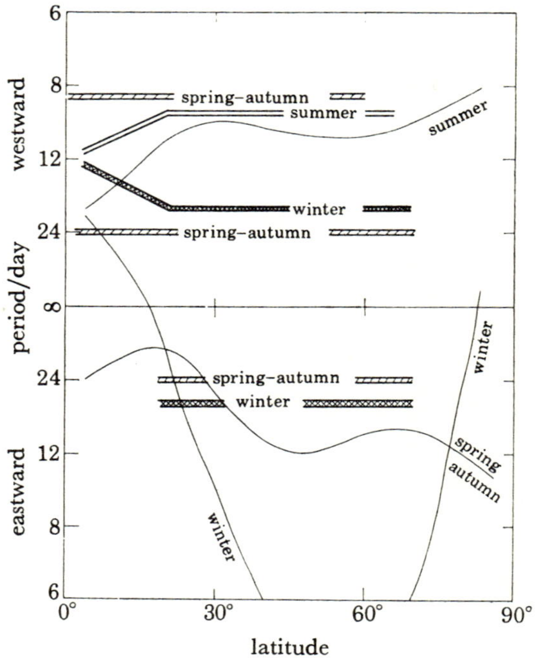

FIGURE 4. Schematic diagram of seasonal characteristics of transient waves for wavenumber one observed by Nimbus 5 s.c.r. channel B12. Thin full lines give the zonal wind velocity at 45 km height. Bold horizontal lines mark regions of high wave amplitude, which are seen always to be at phase velocities westward relative to the zonal flow. (From Hirota 1976.)

Peckham and colleagues at Heriot-Watt University have also made extensive studies of travelling waves using spectral analysis methods with s.c.r. data (see, for example, Chapman *et al.* 1974). Their work is left to Dr Peckham to describe in the accompanying paper.

Rodgers (1976*a*) discovered, by using s.c.r. data, westward travelling wavenumber 1 waves of global extent, period about 5 days, amplitude 0.2–0.6 K and symmetric about the equator with phase independent of latitude. This he identified as a global Rossby mode which had been predicted theoretically and observed in surface pressure data.

Stationary planetary waves have been studied by Labitzke (1977) and many other authors. When averaged over a month or more the temperature field shows little or no eddy activity during summer. The winter field is mainly composed of wavenumber 1, which is in a similar phase for different months and different years. The phase tilts westwards with increasing height, having a vertical wavelength of about 100 km, and towards the equator, with a change of 180° between low and high latitudes in the stratosphere. This is illustrated by figure 2, where wavenumber 1 was fairly stationary for the whole period, and tilted westward with height, while there was little stationary component to wavenumber 2.

5.2. *Stratospheric and mesospheric warmings*

Planetary waves vary in amplitude by an order of magnitude during winter, and sometimes large amplifications are closely followed by rises of the polar cap temperature (an alternative description is that the hot centre or centres of the wave move to the pole). In the Southern Hemisphere these perturbations rarely exceed a few degrees and occur mainly after midwinter, but in the Northern Hemisphere at the stratopause, polar temperature sometimes rises by more than 50 K accompanying planetary waves of 100 K peak to peak. Events of this magnitude are described as major stratospheric warmings. For a few days winter polar temperature may exceed that of midsummer, and the normal winter westerly mean flow is replaced by easterlies. They occur in late December or January, in which case they are midwinter warmings, and the high polar temperature is followed by rapid cooling and very low temperatures, or they occur at the end of winter (March) in which case the circulation and temperature fields remain summerlike. These features can be seen on figure 5 (I). They are very dramatic events and have received much study with satellite and conventional data, notably by Labitzke and colleagues in Berlin and the Upper Air Branch of N.O.A.A.

Fritz & Soules (1972) used s.i.r.s. data to study the winter behaviour and noted cooling in the tropics and summer hemisphere during warmings at the winter pole. Barnett (1974, 1975*b*) saw similar behaviour at higher levels including a notable example for the Southern Hemisphere where the largest zonal mean temperature rise was at mid-latitudes, and the cooling was visible to almost the North Pole. Quiroz (1975) gave a detailed account of the movement and development of temperature anomalies during four winters with the use of s.i.r.s. and v.t.p.r. data. The major warmings during this period seemed to arise from a superposition of a standing wave with a travelling wave. Although links between winter activity and the tropical quasi-biennial oscillation had been suspected, none was evident in these data. Labitzke (1972, 1974, 1977) has made many studies of the winter stratosphere and mesosphere, by using all types of data including s.i.r.s. and s.c.r. She has placed much emphasis on comparison between the hemispheres, on timescales ranging from a few days to the seasonal, and by amassing data from many years has given the climatology of winter behaviour.

5.3. *Seasonal variations and climatology*

This section covers variations on a monthly or longer time-scale, which, although often studied for only a single year, can be reasonably taken to show the climatology. Professor Labitzke's paper in this volume should also be referred to.

Studies of the annual and semiannual cycles have been made by Fritz & Soules (1972) and Barnett (1974), both using 1 year of data, and McGregor & Chapman (1978) using 2 years. Each study used data from a different time and a different instrument, but the major findings were in agreement. The annual cycle has amplitude maxima at the poles near the stratopause of about 20 K in the Northern Hemisphere and 30 K in the Southern Hemisphere. The times of maxima are close to the summer solstice. The Southern Hemisphere amplitude is larger, probably because the eccentricity of the Earth orbit around the Sun causes the Southern Hemisphere to experience colder winters and warmer summers than the Northern, and because the Northern Hemisphere has a warmer winter because of larger stratospheric warmings. Small separate maxima of the annual wave of about 2 K amplitude occur near the Equator in the tropical lower and upper stratosphere. The mean temperature of the whole planet for

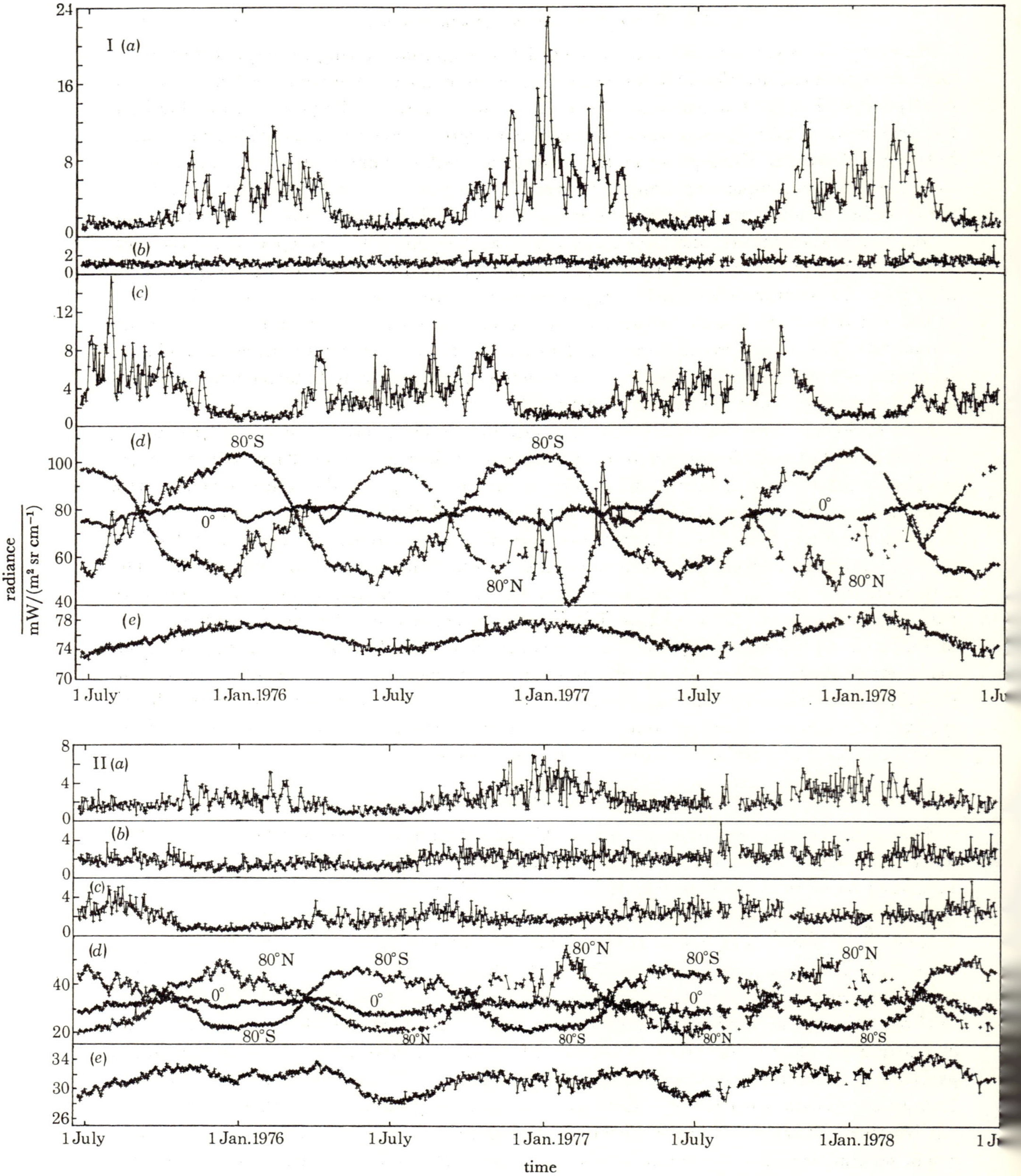

FIGURE 5. Results from the Nimbus 6 p.m.r. for 21 June 1975–22 June 1978. I. channel 2115 (near 45 km – see figure 1 for weighting functions); II. channel 3000 (near 80 km). (*a*), (*b*), (*c*) Standard deviation of radiance around the globe at 64° N, the equator, 64° S respectively; (*d*) zonal mean around globe at 80° N, equator, 80° S; (*e*) global mean.

stratospheric layers undergoes a substantial annual variation, apparently because of the ellipticity of the Earth orbit about the Sun (see figure 5). Insolation is about 7 % greater in December than June; whereas the global mean has this phase, the variation is less than expected from very simple considerations, while a detailed study does not appear to have been published. For the semiannual wave there are amplitude maxima near the pole which apparently are just a result of midwinter warmings giving a warm period in the middle of winter. For this reason the semiannual wave can be expected to vary substantially from year to year. There is also a pronounced 4 K amplitude maximum in the tropical upper stratosphere which dominates the temperature variation there. Maxima occur near the equinoxes, and being out of phase with the polar cycles, the cycle may be forced dynamically from the winter hemisphere through the out-of-phase relationships discussed in § 5.2. The twice yearly passage of the Sun overhead is in correct phase and may also play a part. Fritz (1974) discusses these mechanisms in detail for the lower stratosphere.

Crane (1979*b*) has extended these analyses to the mesopause by using p.m.r. data for 18 months, some of which are shown in figure 5. The annual cycle near 80 km is much as below except for a 180° phase change in mid and high latitudes (maximum temperature in mid-winter), as is well known. The global mean in the upper mesosphere is in phase with lower levels and with the solar forcing. The semiannual wave dominates in the tropics (2 K amplitude) with maxima near the equinoxes, but an important difference is that the semi-annual wave has this phase at all latitudes; consequently the global mean has a strong semi-annual component.

Retrieved zonal mean cross sections have been given by various authors (usually for individual days and not seasonal means) e.g. Barnett (1974), Rodgers (1977*a*), Labitzke & Barnett (1979). They show substantial deviations from the climatologies based on rocket data, and substantial differences between the hemispheres when data six months apart are compared.

5.4. *Budget calculations*

Quite distinct from work described so far are calculations from retrieved temperature fields of such quantities as meridional and vertical wind velocities, eddy fluxes, energy transformations, etc. Several studies have used the measurements in this way and in some cases have attempted to derive substantially complete budgets of energy, heat or momentum. These studies require high data quality, since extensive manipulation of the temperature field is required, and for some products the errors are greatly magnified. Errors are difficult to assess even for an experimenter who has intimate knowledge of the instrument design, operation, calibration, retrieval and final calculations, so error bounds are rarely given and the reader must assess the reliability of a study by comparison with others. A height analysis is normally required at one level to enable the geopotential field to be derived from the temperature field. A lower stratospheric height is often used, but in the Southern Hemisphere the only sufficiently complete analysis may be at the surface and tropospheric retrievals must be used, despite problems caused by cloud, to build up to greater heights.

Adler (1975) calculated meridional circulations in the lower stratosphere of both hemispheres by using s.i.r.s. data. Northern Hemisphere measurements were compared with fields derived by radiosondes and this gave credence to the new Southern Hemisphere results. A 1000 mbar height field was used as a basis for the geopotential field. There were significant hemispheric differences, the most notable being that the centre of ascending motion in the

Southern Hemisphere was at about 70° S compared with at the pole in the Northern Hemisphere.

Hartmann (1976, 1977) used a data set derived by N.O.A.A. from the Nimbus 5 n.e.m.s.. s.i.r.s. and i.t.p.r., and made extensive calculations of various fluxes including potential vorticity, with special emphasis on the Southern Hemisphere and comparison between the hemispheres. In particular, he found similarities between the energy exchanges of the two hemispheres during minor warmings, despite the greater total energy involved in Northern Hemisphere events. Crane (1979*a*) made a detailed study of the energetics of the early 1973 major warming by using s.c.r. data. There was no evidence of vertical phase propagation, the warming occurring simultaneously at all levels. Near the pole the eddy kinetic energy maxima were mainly caused by vertical energy flux convergence, but at mid-latitudes by barotropic conversion. A large part of the work was to develop a satisfactory retrieval method, which illustrates how important this problem is thought to be. Johnson & Rodenhuis (1977) studied the vorticity budget for the same warmings with the N.O.A.A. n.e.m.s., s.c.r., i.t.p.r. data set. In particular they noted extensive counter-gradient flux of potential vorticity before and during the warming, in agreement with Hartmann's (1977) Southern Hemisphere study. They also noted that maximum eddy amplitudes tend to follow the maximum zonal wind westerlies, a common but important observation from the theoretical viewpoint.

6. Satellite studies of the mesosphere

The Nimbus 6 l.r.i.r. and p.m.r. were the first satellite instruments to observe temperature in the mesosphere. The Nimbus 7 s.a.m.s. and l.i.m.s. are now adding to those measurements. Being from a nadir sounder the p.m.r. data are easier to interpret than those of the other instruments, which probably explains why only p.m.r. data have been extensively published. The satellite was launched in June 1975, and 40 months of data were taken until data acquisition was largely discontinued with the launch of Nimbus 7. All of these data have been processed, then reprocessed with optimum calibrations, and calculation of parametrized temperature dependent weighting functions taking non-l.t.e. deviations into account has been recently completed. The instrument was described by Curtis *et al.* (1974).

Figure 5 gives various quantities for the first 3 years of p.m.r. data, 21 June 1975–22 June 1978 for the highest and lowest weighting functions peaking at about 80 km and 43 km respectively (see figure 1). Most of the features of the lower level have already been referred to. The upper level has similarities with these but also differences. The most disturbed times near 80 km are again during winter, but the differences between the winter, summer and tropical eddy amplitudes are less marked than at lower levels. At the mean radiance of the 80 km level, 1 radiance unit change corresponds to about 1.4 K temperature change, compared with 0.84 K change for the lower level. Taking this into account, temperature amplitudes near 80 km are approximately half of those near 43 km during winter. Lower summer and tropical standard deviations are apparent between October 1975 and August 1976 and are caused by the instrument operating in a different mode with one channel not scanning during this period.

Perhaps the most important difference between the levels is the strong negative correlation between the 80° N zonal mean temperatures at the two levels, not only seasonally, which has long been well known, but also in the short term during stratospheric warming events. This is

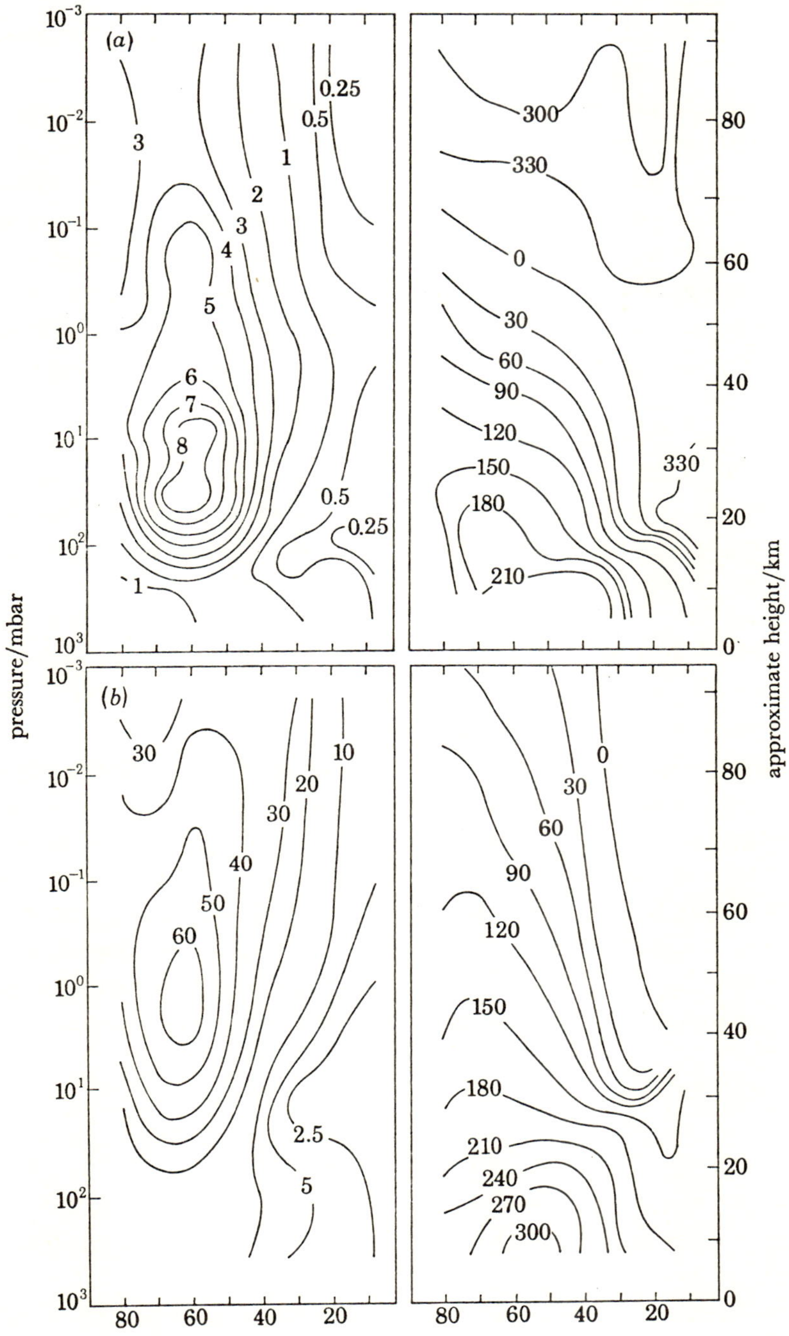

FIGURE 6. (*a*) Amplitude and phase of the wavenumber one component of temperature (K) averaged over December 1975, January and February 1976, derived from observations by the Nimbus 5 s.c.r. and Nimbus 6 p.m.r. (*b*) the geopotential height field (dm) derived from (*a*) by using the climatological 30 mbar height as a reference surface. The phase is the longitude (°E) of the maximum.

most obvious during the 1976–7 winter when large oscillations of the zonal mean occurred at 43 km and were mirrored at 80 km. Houghton (1978) and Hirota & Barnett (1977) discussed these changes and showed that they fit in well with theory, the upper mesospheric cooling during a warming of the stratosphere and lower mesosphere resulting from the updraught forced over the polar cap by eddy heating. As discussed earlier, in the stratosphere tropical minima coincide with polar maxima due to warmings, and are easily seen at 45 km. At the higher level there is no such clear response, particularly during the large 1976–7 midwinter warming, when it might be expected to be most obvious. However, there are strong suggestions of such negative correlations (tropical minima at 80 km coincident with high latitude maxima at the same level) during other warmings, and this is awaiting further analysis.

The structure of planetary waves has been shown by Austen *et al.* (1976), Houghton (1978), Barnett (1977) and Hirota & Barnett (1977). The westward tilt with increasing height previously seen at lower levels continues up to the mesopause, both in the mean and during large single events, with approximately a 150° phase change between the upper stratosphere and upper mesosphere (figure 6). This is a further example of out-of-phase relationships between these two levels, but there is no suggestion that they are due to the same cause. They were previously observed by Quiroz (1969) and Labitzke (1972) from individual rocket observations. Hirota & Barnett (1977) gave latitude–time sections of wave amplitude and phase and also means for a month in midwinter 1975–6. Temperature variations for wavenumber 1 are approximately twice as large near 45 km as near 80 km, both for short term variations and for the monthly mean. Short term variations were not correlated except that particularly large amplitude maxima at 43 km were accompanied by maxima at 80 km.

The phase variations of the geopotential height (figure 6) have direct physical meaning. Westward tilt with height implies eddy heat flux towards the poles and, if the wind is westerly, upward eddy energy flux. Westward tilt towards the Equator implies eddy momentum flux towards the poles and for westerly wind, energy flux towards the Equator. Since the wind is normally westerly during winter, the phase tilts shown indicate forcing of the wave from the high latitude troposphere and absorption at low latitudes. It is interesting to note that the eddy maxima lie in the upper stratosphere or lower mesosphere, showing that the stationary winter low wave number planetary wave is essentially a middle atmsophere phenomenon.

7. Conclusion

A large number of measurements of the middle atmosphere temperature now exist, and although there have been many studies of the data, they have barely started to explore the information which has been gathered.

Until now instruments have been experimental and we have relied upon the goodwill of N.A.S.A. to operate them longer than the nominal 1–2 years design lifetime, and so to provide an unbroken series of measurements. However, the observations are on a firmer basis with the start of the Tiros N series of operational sounders having temperature sounding channels as high as the upper stratosphere. The importance of a long sequence of observations cannot be overemphasized. One reason is the large year-to-year variations of the winter atmosphere. Major warmings occur on average about once per year, and each seems very different from the others. Whether there is a pattern to these variations, such as a link with the 26-month oscillation or the solar cycle, or whether these cycles appear in other ways are open questions and may need decades of data for their answers.

Mesospheric sounding will still rely on experimental instruments and with greater emphasis on constituent measurements in the stratosphere, mesospheric temperature may not be measured continuously for many more years. Given that the mesosphere is just as important a part of the middle atmosphere as the stratosphere with many processes taking place in both, a case can be made for the next generation of operational sounders after the Tiros N series to carry mesospheric sounders.

References (Barnett)

Adler, R. F. 1975 *J. atmos. Sci.* **32**, 893–898.

Austen, M. D., Barnett, J. J., Curtis, P. D., Morgan, C. G., Houghton, J. T., Peskett, G. D., Rodgers, C. D. & Williamson, E. J. 1976 *Nature, Lond.* **260**, 594–596.

Barnett, J. J. 1974 *Q. Jl R. met. Soc.* **100**, 505–530.

Barnett, J. J. 1975*a* *Q. Jl R. met. Soc.* **101**, 835–845.

Barnett, J. J. 1975*b* *Nature, Lond.* **255**, 387–389.

Barnett, J. J. 1977 *Phil. Trans. R. Soc. Lond.* A **279**, 247–259.

Barnett, J. J., Harwood, R. S., Houghton, J. T., Morgan, C. G., Rodgers, C. D. & Williamson, E. J. 1975 *Q. Jl R. met. Soc.* **101**, 423–436.

Chapman, W. A., Cross, M. J., Flower, D. A., Peckham, G. E. & Smith, S. D. 1974 *Proc. R. Soc. Lond.* A **338**, 57–76.

Crane, A. J. 1979*a* *Q. Jl R. met. Soc.* **105**, 185–206.

Crane, A. J. 1979*b* *Q. Jl R. met. Soc.* **105**, 511–522.

Curtis, P. D., Houghton, J. T., Peskett, G. D. & Rodgers, C. D. 1974 *Proc. R. Soc. Lond.* A **337**, 135–150.

Deland, R. J. 1973 *Mon. Weath. Rev.* **101**, 132–140.

Deland, R. J. 1977 *Proc. IAGA/IAMAP Assembly*, Seattle, 1977.

Fritz, S. & Soules, S. D. 1972 *Mon. Weath. Rev.* **100**, 582–589.

Fritz, S. 1974 *J. atmos. Sci.* **31**, 813–822.

Hartmann, D. L. 1976 *J. atmos. Sci.* **33**, 1141–1154.

Hartmann, D. L. 1977 *Space Res.* **17**, 167–174.

Harwood, R. S. 1975 *Q. Jl R. met. Soc.* **101**, 75–91.

Hirota, I. 1976 *Q. Jl R. met. Soc.* **102**, 757–770.

Hirota, I. & Barnett, J. J. 1977 *Q. Jl R. met. Soc.* **103**, 487–498.

Hirota, I. 1978 *J. atmos. Sci.* **35**, 714–722.

Houghton, J. T. 1977 *Appl. Opt.* **16**, 319–321.

Houghton, J. T. 1978 *Q. Jl R. met. Soc.* **104**, 1–29.

Johnson, K. W. & Rodenhuis, D. R. 1977 *Proc. IAGA/IAMAP Assembly*, Seattle, 1977.

Kaplan, L. D. 1959 *J. opt. Soc. Am.* **49**, 1004–7.

Kennedy, J. S. & Nordberg, W. 1967 *J. atmos. Sci.* **24**, 711–719.

Labitzke, K. 1972 *J. atmos. Sci.* **29**, 756–766.

Labitzke, K. 1974 *J. geophys. Res.* **79**, 2171–2175.

Labitzke, K. 1977 *Cospar Space Research* **17**, 159–165.

Labitzke, K. & Barnett, J. J. 1979 *Cospar Space Research* **19**, 97–106.

Leovy, C. & Webster, P. J. 1976 *J. atmos. Sci.* **33**, 1624–1638.

McGregor, J. & Chapman, W. A. 1978 *J. atmos. terr. Phys.* **40**, 677–684.

Nordberg, W., McCulloch, A. W., Foshee, L. L. & Bandeen, W. R. 1967 *Bull. Am. met. Soc.* **47**, 857–872.

Quiroz, R. S. 1969 *Mon. Weath. Rev.* **97**, 541–552.

Quiroz, R. S. 1975 *J. atmos. Sci.* **32**, 211–224.

Rodgers, C. D. 1976*a* *J. atmos. Sci.* **33**, 710–714.

Rodgers, C. D. 1976*b* *Rev. Geophys. Space Phys.* **14**, 609–624.

Rodgers, C. D. 1977*a* In *Dynamical and chemical coupling* (ed. B. Grandal & J. A. Holtet), pp. 3–16. Dordrecht: Reidel.

Rodgers, C. D. 1977*b* In *Inversion methods in atmospheric remote sounding* (ed. A. Deepak), pp. 117–138. New York: Academic Press.

Stanford, J. L. & Dunkerton, T. J. 1978 *Beitr. Phys. Atmos.* **51**, 174–188.

Phil. Trans. R. Soc. Lond. A **296**, 59–63 (1980) [59]
Printed in Great Britain

Spectral analyses of wave motions in the middle atmosphere

By W. A. Chapman and G. E. Peckham
Heriot-Watt University, Edinburgh, U.K.

Data obtained from the selective chopper radiometer on the Nimbus 5 satellite provide a four dimensional (space and time) view of the temperature structure of the middle atmosphere. Spectral analyses of these observations enable the structure of the large scale planetary waves (zonal wavenumbers 1–6) to be deduced.

Analyses of two years of data (1973–4) reveal the important differences that exist between the wave motions in the two Hemispheres. In the Northern Hemisphere the wave motions are dominated by stationary waves, whereas in the Southern Hemisphere, eastward travelling waves are more in evidence. A detailed investigation of data during December–February 1973–4 reveals westward travelling wave 1 components with periods of 5.8, 8.3 and 12.3 days.

Introduction

In this paper we present the results of spectral analyses of stratospheric temperatures derived from the selective chopper radiometer (s.c.r.) flown on the Nimbus 5 meteorological satellite. The radiances measured by this instrument are weighted means over a vertical height range of 10–15 km (see Ellis *et al.* 1973). The three channels used in this study are B12, A1, and A2 centred at approximately 2, 50 and 100 mbar. The black body temperatures appropriate to the measured radiances are used rather than those derived from a retrieved temperature profile. The period analysed, divided into seasons, is taken from 2 years of s.c.r. data, from December 1972 to December 1974.

As the Earth rotates under an orbiting satellite, observations at a particular latitude are equally spaced in both longitude and time. (This is in contrast to the more usual meteorological measurements which are made globally at fixed times.) This relation between time and longitude allows travelling wave spectra to be obtained by a single Fourier transform of the data.

The temperature field T over a period of time corresponding to m orbits can be expanded as a Fourier sum

$$T(\phi, z, \lambda, t) = \sum_{l=-\infty}^{\infty} \sum_{n=0}^{\infty} A_{ln}(\phi, z) \exp\{i(2\pi lt/m\tau - n\lambda)\}, \tag{1}$$

where ϕ is latitude, λ longitude, z height, t time, τ orbital period of satellite, n zonal wavenumber (wave), $l = n\tau\omega/2\pi$, and ω is angular frequency of the travelling wave (positive for eastward, negative for westward). The jth measurement occurs at a time $j\tau$ and a longitude $-j\tau\Omega$ (Ω = rate of rotation of the earth) so that the series of measurements (at a single latitude) are given by

$$T_j = \sum_{ln} A_{ln} \exp\{ij(2\pi l/m + n\Omega\tau)\}. \tag{2}$$

It is apparent that the same series of observations, T_j, would be obtained, for instance from a wave 1 ($n = 1$) travelling eastward with a frequency of 0.5 cycle/day ($l = 0.5\ \Omega\tau$), as would be obtained from a wave 2 ($n = 2$) travelling westward at the same frequency ($l = -0.5\ \Omega\tau$).

If we assume that the waves are of planetary scale with frequencies less than 0.5 c/d then the ambiguity disappears. (This conclusion is not affected by the fact that $\Omega\tau$ is not in general an integer.) Further details are given by Chapman *et al.* (1974). The spectra show peaks clearly concentrated around frequencies of 0, 1, 2, 3...c/d which can be unambiguously assigned to the zonal mean and waves 1, 2, 3.... Peaks at the integer frequencies correspond to stationary waves, while an eastward travelling wave appears displaced to higher frequency and a westward to lower (e.g. in figure 1, wave 2 is displaced to 2.07 c/d so that the peak represents a wave travelling eastward with a frequency of 0.07 c/d, or a period of 14 days). The orbital period of the satellite restricts the highest zonal wavenumber that can be resolved to wave 6; however, in this paper we concentrate on the largest scale waves (waves 1, 2 and 3).

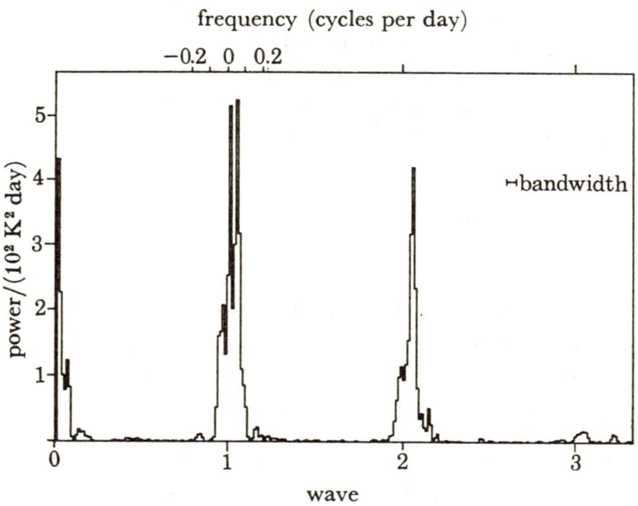

FIGURE 1. Power spectrum of temperature for latitude 60° S, channel B12 (*ca.* 2 mbar) during September–November 1973.

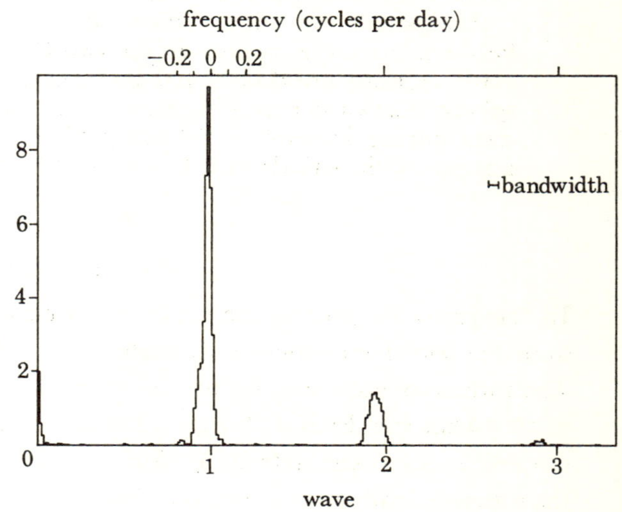

FIGURE 2. Power spectrum of temperature for latitude 60° N, channel B12 (*ca.* 2 mbar) during March–May 1974.

SEASONAL SPECTRA

The spectra described in this paper were obtained by carrying out a discrete Fourier transform of a sequence of measurements from 1024 successive orbital crossings of a particular latitude. The data, covering a period of approximately 76 days, were apodized before transforming to prevent false structure arising. This apodization provides a spectral resolution of approximately 0.03 c/d. The vertical coordinate is spectral power density (K^2 day).

The spectrum for the Southern Hemisphere at latitude 60° during spring 1973 (September–November) for channel B12 (2 mbar) is shown in figure 1. A strong wave 1 stationary peak appears together with displaced peaks representing eastward and westward travelling waves with periods close to 20 days. In contrast, wave 2 consists almost entirely of an eastward travelling component. The spectrum for channel B12 taken at latitude 60° N during spring 1974 (March–May) is shown in figure 2. The wave 1 peak is narrower than that in the Southern Hemisphere, representing a more stable stationary wave. A small westward travelling component can also be distinguished with a period around 5.8 days. Although this peak appears insignificant in this spectrum, its amplitude (*ca.* 0.4 K) is comparable to that of the 5-day wave during May 1973 at 44° N (see Rodgers 1976). The wave 2 component is westward travelling with a period around 3 weeks and it is rather broad, indicating that there is some variation of frequency over the 3-month period of analysis.

The spectra of waves 1 and 2 for each season over the 2-year period are summarized in figure 3. The spectral power (K^2) is divided into stationary waves (periods > 37 days) and westward and eastward travelling waves. This diagram is for channel B12 at latitudes 60° N and 60° S. Wave 1 consists mainly of a stationary component in the Northern Hemisphere in contrast to the Southern Hemisphere where travelling waves predominate. Wave 2 in both Hemispheres shows strong eastward travelling components. Wave amplitudes are generally greater during winter in the Northern Hemisphere and spring in the Southern Hemisphere (but note that data for winter 1973 in the Southern Hemisphere are missing). Similar results were obtained for channel A1 (*ca.* 50 mbar).

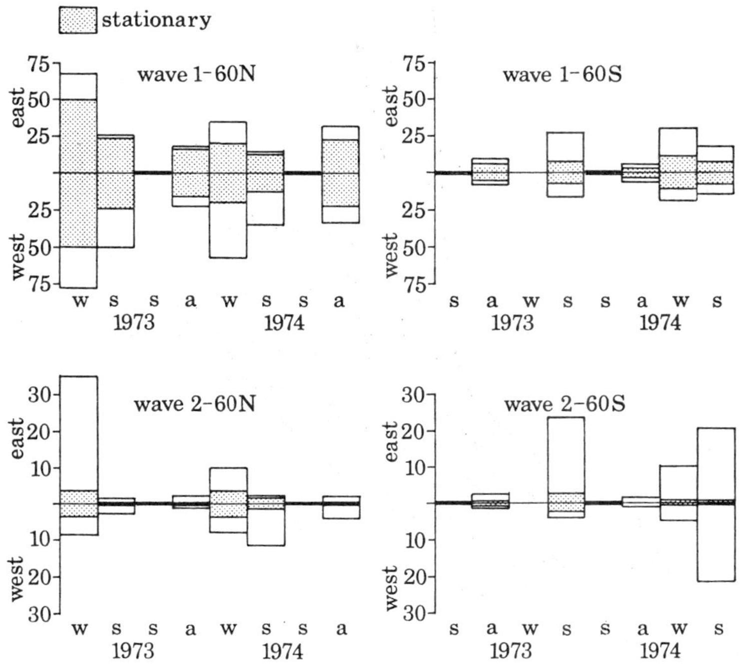

FIGURE 3. Seasonal mean power (1973–4) due to eastward and westward travelling waves and stationary waves (period < 37 days) for latitudes 60° N, 60° S waves 1 and 2, channel B12. (Note that data for winter 1973 at 60° S are missing.)

As an example of the application of this method of spectral analysis we present some preliminary results of a detailed investigation of the zonal wavenumber one structure during December–February 1973–4. Figure 4 shows a projection of the amplitudes of westward travelling waves 1 for channel A2 (100 mbar) plotted against frequency and latitude. The frequency range is from -0.03 to -0.24 c/d, and the latitude range is from 80° S to 80° N. The most obvious feature is the large amplitude stationary component in the winter Northern Hemisphere, which has been truncated in the plot. However, of more interest are the travelling waves which are present in both Hemispheres. The normal modes of oscillation of the Earth's atmosphere have been treated by a number of authors. In the simplest case of an isothermal atmosphere with no zonal winds, these modes are the solutions of the Laplace tidal equations (see, for example, Longuet-Higgins 1968). The s.c.r. will respond only to waves with a vertical wavelength of the order 10 km or greater, and we have already shown that the frequency of observation allows only waves of period greater than 2 days to be unambiguously resolved. This restricts our observations to the planetary scale waves, described by Hough (1898). A number of modes H_k^n are possible, depending on the zonal and meridional structure. The frequencies depend on the indices n and k and on the equivalent depth parameter which also determines the height structure. For large equivalent depths the modes become Rossby waves with stream functions the spherical harmonics P_k^n. This type of wave propagates westward.

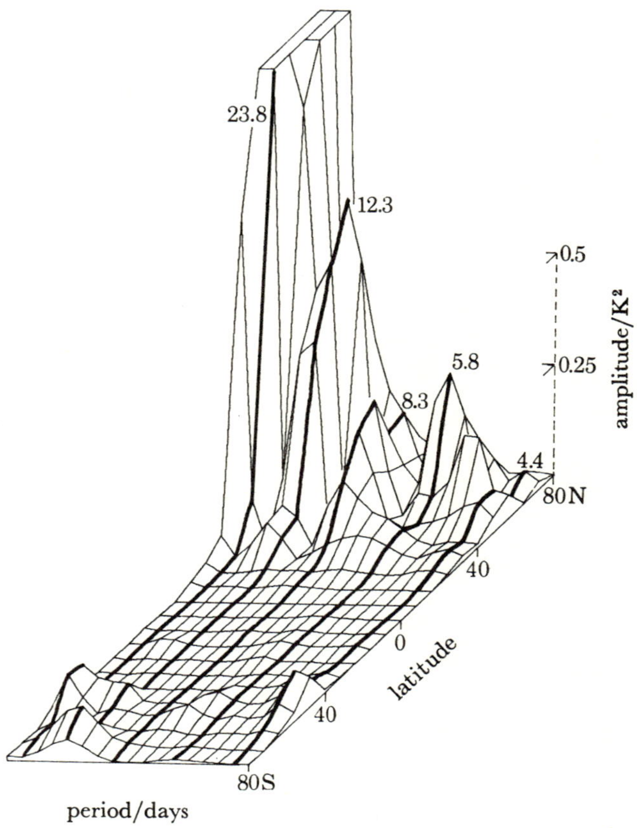

FIGURE 4. Power spectrum of the westward travelling components of wave 1 for channel A2 (*ca.* 100 mbar) during December–February 1973–4.

A number of calculations have been made with more realistic atmospheric conditions. Dickinson & Williamson (1972) derived the normal modes from a two-layer numerical model and found periods of 5, 8.3, 12.3 and 17.3 days for H_2^1 to H_5^1. (These periods are very similar to those obtained from the tidal equations with an equivalent depth of 9.5 km.) These values changed to 5.1, 10.8, 17.5 and 32.3 days when realistic zonal winds and meridional temperature gradients were introduced. The small effect on the 5-day wave has been discussed by Geisler & Dickinson (1976).

The spectra in figure 4 show strong peaks in the Northern Hemisphere at 5.4–6.2, 8.3 and 12.3 day periods. Smaller peaks corresponding to the same periods are visible in the Southern Hemisphere. These periods are very close to those calculated by Dickinson & Williamson. However, before any ambiguous assignment can be made, it will be necessary to examine the phases of the waves as functions of latitude and height. Such waves have also been observed in conventional meteorological data. Recently, Madden (1978) in an analysis extending over many years found evidence for 5- and 16-day waves in Northern Hemisphere sea level pressure and geopotential heights.

Conclusion

We have presented some results of a 2-year spectral analysis of satellite derived temperatures and confirmed the variations that occur in the spectra, both seasonally and hemispherically. In general, stationary waves tend to dominate in the Northern Hemisphere and eastward and westward travelling waves contribute similar powers to the spectra, while in the Southern Hemisphere, especially for wave 2, eastward travelling waves predominate. A preliminary study of the period from December 1973 to February 1974 revealed the presence of global westward travelling wave 1 components with periods of 5.8, 8.3 and 12.3 days (close to periods of the free modes of oscillation obtained from the solution to Laplace's tidal equations). Further analysis is being pursued in order to determine the meridional and height structure of these waves.

We gratefully acknowledge the financial support of N.E.R.C. for this work, which is part of a joint project with the Department of Meteorology at the University of Edinburgh. The data were derived from the selective chopper radiometer on Nimbus 5 satellite. This instrument was jointly developed by groups led by Professor S. D. Smith at Heriot-Watt University and by Professor J. T. Houghton at Oxford University. Members of these groups and N.A.S.A. have made considerable contributions to the data processing.

References (Chapman & Peckham)

Chapman, W. A., Cross, M. J., Flower, D. A., Peckham, G. E. & Smith, S. D. 1974 *Proc. R. Soc. Lond.* A **338**, 57–76.

Dickinson, R. E. & Williamson, D. L. 1972 *J. atmos. Sci.* **29**, 623–640.

Ellis, P. J., Holah, G. D., Houghton, J. T., Jones, T. S., Peckham, G. E., Peskett, G. D., Pick, D. R., Rodgers, C. D., Roscoe, H., Sandwell, R., Smith, S. D. & Williamson, E. J. 1973 *Proc. R. Soc. Lond.* A **334**, 149–170.

Geisler, J. E. & Dickinson, R. E. 1976 *J. atmos. Sci.* **33**, 632–641.

Hough, S. S. 1898 *Phil. Trans. R. Soc. Lond.* A **191**, 139–185.

Longuet-Higgins, M. S. 1968 *Phil. Trans. R. Soc. Lond.* A **262**, 511–607.

Madden, R. A. 1978 *J. atmos. Sci.* **35**, 1605–1618.

Rodgers, C. D. 1976 *J. atmos. Sci.* **33**, 710–711.

Phil. Trans. R. Soc. Lond. A **296**, 65–71 (1980) [65]
Printed in Great Britain

Operational temperature sounding of the stratosphere

By D. E. Miller, J. L. Brownscombe, G. P. Carruthers,
D. R. Pick and K. H. Stewart
Meteorological Office, London Road, Bracknell, Berkshire, RG12 2SZ, U.K.

Observations made by the temperature sounders on the Tiros N series of operational meteorological satellites are being used for generating routine analyses for the stratosphere. One component of the sounder, the stratospheric sounding unit, is described. The methods being used for retrieving stratospheric temperatures and the available products are presented.

1. Introduction

In the early 1970s, the relevant U.S. authorities began planning a third generation of polar-orbiting meteorological satellites. It was envisaged that their payloads should include a vertical sounder to provide measurements of atmospheric temperature for use in numerical models for weather forecasting. Instruments to measure atmospheric radiance in the infrared, with a spatial scale of about 20 km, and in the 60 GHz microwave band of oxygen were expected to provide a substantial improvement in the accuracy of retrieved atmospheric temperatures compared with the existing vertical temperature profile radiometer. Few operational numerical models require data for levels above the 100 mbar surface. However, because of the spread of the individual weighting functions, the radiances observed in the spectral channels used to probe the region between the surface and 100 mbar are influenced, at least in part, by the temperature structure above 100 mbar. Thus it is essential that any sounder should include channels designed to make measurements in this region.

With the development, by Houghton & Smith (1970), of the 'selective chopping' technique – whereby a cell containing carbon dioxide acts as the primary spectral selective element – it became possible to achieve sharp weighting functions within the stratosphere. This technique was successfully demonstrated with the selective chopping radiometer on the Nimbus 4 satellite. The United Kingdom's lead in this field was an important reason for the Meteorological Office to be invited to provide the stratospheric component of the Tiros operational vertical sounders (T.o.v.s.) for the Tiros N satellites. Following a brief description of the instrument and its in-orbit performance, the use of the observations in providing stratospheric analyses is reviewed.

2. The stratospheric sounding unit

The stratospheric sounding unit (s.s.u.) is a 3-channel infrared radiometer designed to measure the radiance emitted by stratospheric carbon dioxide. The instrument exploits the 'selective chopping' principle, by which a signal is derived by viewing the atmosphere in turn through absorption cells containing differing amounts of carbon dioxide. In practice this is accomplished by the pressure modulation technique, first used in the Oxford University's pressure modulator radiometer on Nimbus 6 (Curtis *et al.* 1974). In this case the detector views the atmosphere through a single absorption cell (see figure 1) within which the pressure of the carbon dioxide can be modulated. The cell is connected to a sealed cylinder. Modulation of the pressure is induced by oscillation of a closely fitting piston, suspended within the cylinder on two diaphragm springs. The piston is maintained in oscillation at its resonant frequency (*ca.* 40 Hz) by pulses of current passed through a coil mounted on the piston's shaft and lying in the gap of a magnet. An electronic servo-system, which senses the back-e.m.f. in the coil and controls the length of the current pulse, keeps the amplitude constant. No provision is made for changing the mean pressure within the cell; its long-term stability depends on careful design and choice of materials. However, the frequency of the oscillation provides a measure of the mean pressure.

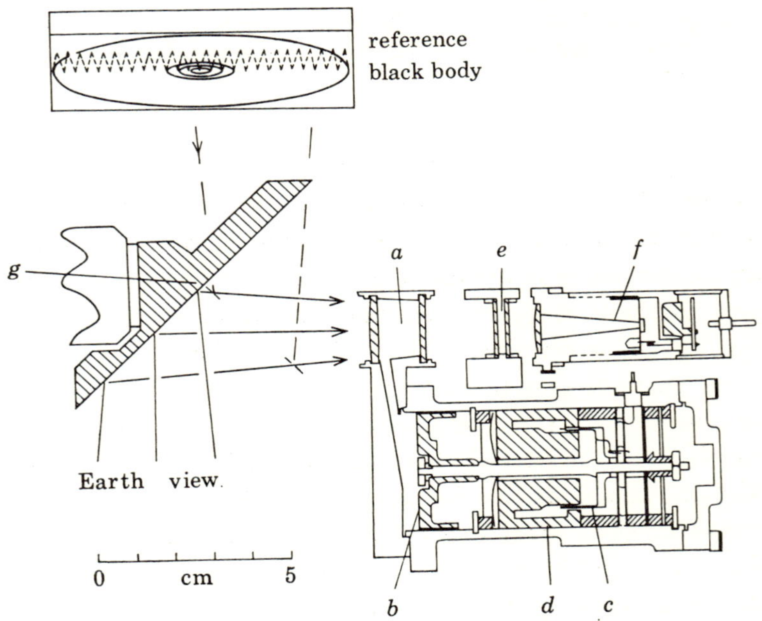

FIGURE 1. The optical system and pressure modulator of a stratospheric sounding unit. *a*, Absorption cell; *b*, piston; *c*, drive coil; *d*, magnet; *e*, interference filter; *f*, detector; *g*, scan mirror. Another two optical channels share the scan mirror.

An interference filter, with characteristics described by Evans *et al.* (1976), confines the spectral response to the 15 μm band. The pyroelectric detector which is a flake of triglycine sulphate attached to a gold-plated light-pipe, was manufactured by Plessey Limited at their Allen Clark Research Laboratory. The signal, at the modulator frequency, is amplified, rectified and integrated for about 3.5 s. The three channels of the s.s.u. differ in the mean pressure of the carbon dioxide within their modulators. Since this pressure, together with the depth of modulation, controls the effective bandwidth of the gas cell filter, the signal to noise ratio is higher for the high pressure cell than for the low pressure cell. The weighting functions are shown by the full lines, in figure 2.

The three optical channels, each of which has a field of view of 10° diameter, view the atmosphere by way of a shared mirror. This mirror can be rotated, in 10° steps, to direct the fields of view through eight positions across the satellite's orbital track, spanning ± 35° from the nadir. The ground resolution is about 200 km. Further details of the scan pattern and its relation to the other components of the T.o.v.s. have been given by Schwalb (1978). The scan mirror also allows the radiometer to view space and a 'black-body' target within the instrument. The temperature of this target is monitored. This two-point calibration is used to interpret the digitized output in terms of radiance.

Laboratory testing of s.s.us includes an exhaustive investigation of their radiometric performance. This calibration would reveal any departure from linearity in the relation between voltage and scene radiance arising from various sources of stray radiation and provides corrections, in the form of small radiance offsets, which can be applied to the in-flight calibration. The weighting functions for the s.s.u. are computed from instrument parameters and spectral line data, using a line-by-line technique and a Voigt line profile.

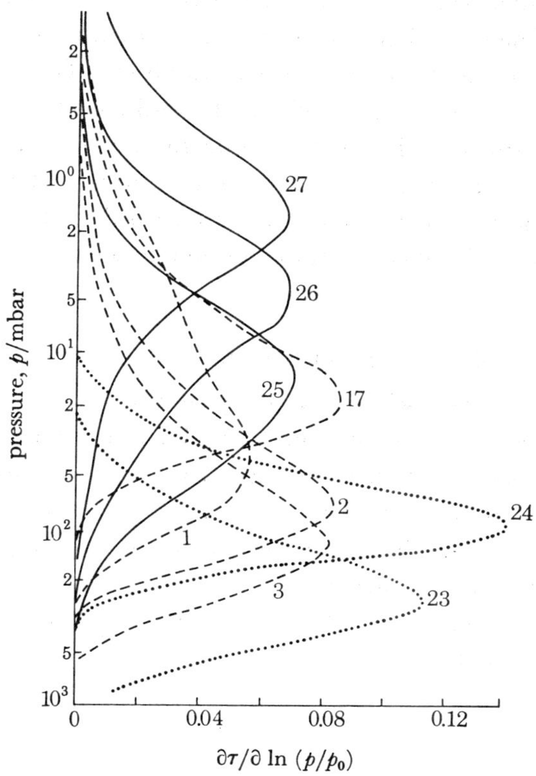

FIGURE 2. Weighting functions for those channels used for stratospheric analysis. The labels indicate the channel designations within the Tiros operational vertical sounder. The associated instruments are: ——, stratospheric sounding unit; ...., microwave sounding unit; --, high-resolution infrared sounder.

To confirm these calculations, the instrument is used to measure the transmission of a 10 m path of carbon dioxide in a White cell over a range of pressures and temperatures. These transmission measurements are used to check the shape of the weighting function and to deduce an effective mean pressure for the absorption cell (which is close to the actual or filling pressure), which brings calculations and measurements into agreement at a transmission value of 0.5.

The detailed design, development and manufacture of the stratospheric sounding units has been done by Marconi Space and Defence Systems Limited, Frimley. The first flight instrument was delivered to the United States in March 1977 and launched on the Tiros N satellite on 13 October 1978.

3. The data flow

Data from the stratospheric sounding unit are recorded on board the spacecraft, replayed to one of two ground stations in the United States and relayed to Suitland, near Washington, D.C. There the data streams from the various instruments are separated and a substantial fraction of the raw telemetry data from the s.s.u. is transmitted to Bracknell via a 2400 baud dedicated telecommunications link. It is intended that these data, which mainly relate to housekeeping parameters but include the radiometer outputs, should be used to monitor the in-flight performance of the instrument. At Suitland the raw data from the various sounding instruments will be processed to derive radiances which will be used to retrieve meteorological parameters, in the form of thicknesses (i.e. mean temperatures) for various layers in the troposphere and stratosphere. These thickness values will be distributed as coded messages to meteorological services, by means of the meteorological Global Telecommunications System, principally for use in numerical forecasting models. The radiances from the s.s.u. channels and for several other T.o.v.s. channels which have significant response at or above the 100 mbar level (i.e. those shown in figure 2) are also to be transmitted to Bracknell on the dedicated link. Data received at Bracknell are up to 6 h old. They are recorded on tape for overnight batch processing on the Meteorological Office's main computer system.

4. The retrieval scheme

A multi-channel linear regression scheme is being used to retrieve atmospheric thicknesses from the radiances. The regression coefficients were derived before launch on the basis of a catalogue of about 1200 rocketsonde temperature profiles (kindly made available by Mr F. Finger of the National Meteorological Centre, Washington). The radiances which one would expect to observe (known as the 'simulated radiances') were computed for each of these temperature profiles from the best knowledge of the weighting functions of the stratospheric sounding unit and the other sounders. In this manner it is possible to process the flight data immediately, without awaiting the accummulation of an adequate collection of co-located rocketsonde and satellite observations.

5. Performance of the stratospheric sounding unit

The weighting functions of the three channels of the s.s.u. are centred at 15, 5 and 1.5 mbar respectively. The r.m.s. noise on the radiances in these channels, as observed in orbit for the instrument on Tiros N, are about 0.3, 0.5 and 40 mW/(m^2 sr cm^{-1}) (a change of 1 mW/(m^2 sr cm^{-1}) at 15 μm is approximately equivalent to 1 K over the range of scene temperatures). The noise levels for the 15 and 5 mbar channels are well within requirements. The pre-launch noise level for the 1.5 mbar channel was about 1 mW/(m^2 sr cm^{-1}); this channel developed a fault during launch (or more precisely before the instrument was switched on in orbit). The cause of the fault, which manifests itself as greatly increased noise and a negative offset, while retaining a substantial fraction of the original sensitivity to radiation, has not yet been isolated. The modulator is apparently functioning perfectly normally. The impact of the fault is to render the 1.5 mbar channel useless and limit the maximum height to which temperatures can be deduced to about 40 km for this first spacecraft. The stability of the in-flight calibration, based on views of space and the internal target at 256 s intervals, has proved better than 0.2 %, both in the short term and as long term drift over the first 2 months in orbit.

Preliminary results of comparisons between simulated radiances, computed from co-located rocketsonde profiles, and observed radiances are presented in figure 3. The differences have been plotted with, as abscissa, the change in simulated radiance corresponding to an increase of 10 % in effective mean pressure within the modulator. Thus the slope of the best-fit line provides a measure of the difference between the actual and the nominal effective mean pressure. In these cases this difference is less than 2 %. The intercepts on the vertical axis give the radiometric offsets; these are less than 0.3 mW/(m^2 sr cm^{-1}). The error bars on each point are dominated by an assumed uncertainty of 1 K in the rocketsonde temperatures.

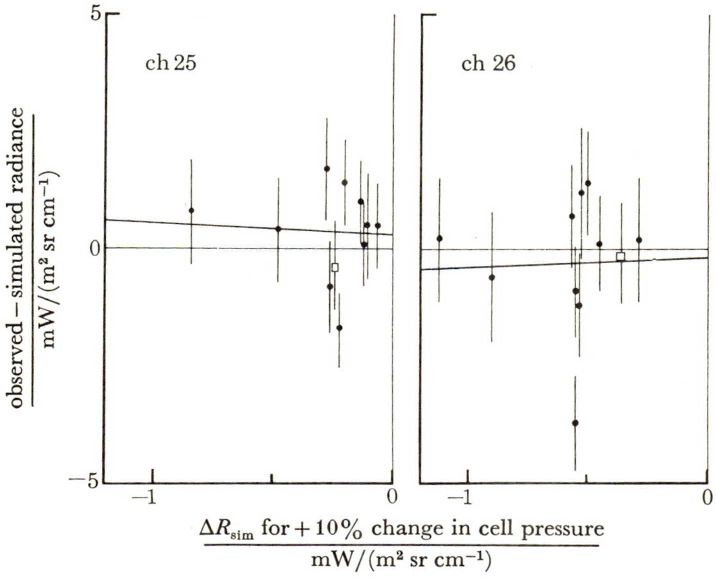

FIGURE 3. Preliminary comparison between radiances observed by stratospheric sounding unit channel numbers 25 (15 mbar) and 26 (5 mbar) and simulated radiances based on rocketsonde temperature profiles. ΔR_{sim}, change in simulated radiance; □, based on West Geirinish rocketsonde; ——, best fit trend line.

Quality control has been exercised to eliminate cases associated with large spatial or temporal changes in radiance. One point corresponds to a flight of the Meteorological Office's Skua rocketsonde from West Geirinish, made within a few minutes of an overpass of Tiros N. These comparisons will be maintained on a regular basis, to accumulate a statistically significant sample.

6. Stratospheric analysis

The aim is to produce global analyses in terms of geopotential height for the 20, 10, 5, 2 and 1 mbar surfaces. These will be based on a combination of analyses of thicknesses retrieved from T.o.v.s. radiances and a 100 mbar global height analysis. The analyses are produced by interpolating within the available observations to obtain grid-point values and subsequently smoothing these, by using an orthogonal polynomial technique, on polar stereographic projections for each of the hemispheres nearer the poles of latitude 30°. Coverage of the tropics is provided on a simple cylindrical projection.

Since the supply of processed radiances from Suitland is not yet fully working, retrievals are at present based on the raw data from the s.s.u. (i.e. on only two out of the ten anticipated radiances). Analyses for the Northern Hemisphere at 10 mbar are being compared with hand-drawn analyses, based on radiosonde data. The mean difference in 10 mbar geopotential height between these analyses is typically 140 m, with a standard deviation of 110 m. There is a considerable latitudinal gradient in these differences; the s.s.u. analysis values are up to 300 m too high south of 45° N and up to 200 m too low around 65° N. Tests made with retrievals from simulated radiances suggest that these differences will be markedly reduced when radiances from the other channels are available.

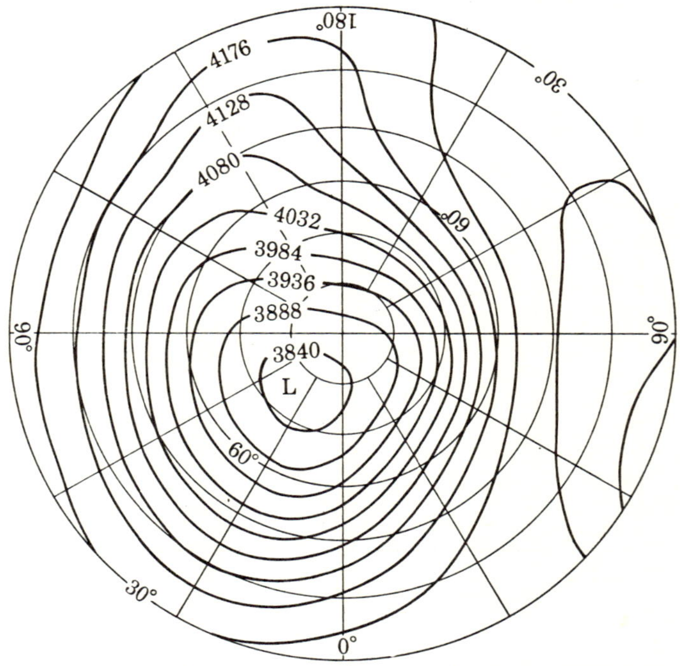

FIGURE 4. A typical analysis, of geopotential height of the 2 mbar surface for 0 U.T. on 4 January 1979, based on observations from the stratospheric sounding unit. Contours are labelled in decameters.

An example of a 2 mbar height analysis is shown in figure 4. The height fields are being used to produce wind fields (by the gradient wind approximation), and to determine the amplitude and phase of the planetary waves and their contributions to total and eddy kinetic energy. These investigations are being expanded to include estimates of transport and eddy fluxes of heat and momentum, by techniques described by O'Neill & Taylor (1979). The analyses and other products are being archived and will be made available on request. Radiance and height analyses are already being transmitted daily to the Free University of Berlin, to assist them in the production of Stratalerts.

7. Conclusion

A scheme of routine analysis for the stratosphere up to the 1 mbar level, based on data from the stratospheric sounding unit and other components of the Tiros operational vertical sounder, has been developed. A further seven spacecraft are planned within the Tiros N series, so continuity of observations and analyses should be possible into the mid 1980s. It is hoped that these analyses will form a valuable contribution to the Middle Atmosphere Programme by forming a basis for studying the dynamical behaviour of the stratosphere and the transport of minor constituents.

We wish to express our gratitude to the National Oceanic and Atmospheric Administration and the National Aeronautics and Space Administration for giving us the opportunity to participate in the Tiros N project and to thank various members of those organizations for their help and co-operation. The development of any satellite instrument involves effort from a large team. We would like to thank those members of the staffs of Marconi Space and Defence Systems Limited and their sub-contractors and colleagues in the Meteorological Office who have contributed to the project. We also wish to thank Professor J. T. Houghton and Professor S. D. Smith for their sound advice in the early stages of the project.

References (Miller *et al.*)

Curtis, P. D., Houghton, J. T., Peskett, G. D. & Rodgers, C. D. 1974 *Proc. R. Soc. Lond.* A **337**, 135–150.
Evans, C. S., Hunneman, R. & Seeley, J. S. 1976 *J. Phys.* D **9**, 309–320.
Houghton, J. T. & Smith, S. D. 1970 *Proc. R. Soc. Lond.* A **320**, 23–33.
O'Neill, A. & Taylor, B. F. 1979 *Q. Jl R. met Soc.* **105**, 71–92.
Schwalb, A. 1978 *N.O.A.A. Technical Memorandum* NESS 95.

Phil. Trans. R. Soc. Lond. A **296**, 73–85 (1980) [73]
Printed in Great Britain

Wave propagation and transport in the middle atmosphere

BY J. R. HOLTON
Department of Atmospheric Sciences, University of Washington, Seattle, Washington 98195 *U.S.A.*

The dynamics of wave propagation and wave transport are reviewed for vertically propagating, forced, planetary scale waves in the middle atmosphere. Such waves can be divided into two major classes: extratropical planetary waves and equatorial waves. The most important waves of the former class are quasi-stationary Rossby modes of zonal wavenumbers 1 and 2 (1 or 2 waves around a latitude circle), which propagate vertically only during the winter season when the mean winds are westerly. These modes transport heat and ozone towards the poles, thus maintaining the mean temperature above its radiative equilibrium value in high latitudes and producing the high latitude ozone maximum. It is shown that these wave transport processes depend on wave transience and wave damping. The precise form of this dependency is illustrated for transport of a strongly stratified tracer by small amplitude planetary waves.

The observed equatorial wave modes are of two types: an eastward propagating Kelvin mode and a westward propagating mixed Rossby–gravity mode. These modes are thermally damped in the stratosphere where they interact with the mean flow to produce eastward and westward accelerations, respectively. It is shown that in the absence of mechanical dissipation this wave–mean flow interaction is caused by the vertical divergence of a wave 'radiation stress'. This wave–mean flow interaction process is responsible for producing the well known equatorial quasi-biennial oscillation.

1. INTRODUCTION

Planetary scale wave motions are essential components of the general circulation in the middle atmosphere. Much of the observed temporal and spatial variability in winds, temperatures, and trace species concentrations in this region is due directly or indirectly to wave motions. Atmospheric waves play major roles in maintaining the zonal mean momentum and temperature budgets as well as the ozone budget. Such waves can be classified according to their horizontal structures, their vertical structures, their sources of excitation, and their modes of interaction with the mean flow.

This classification scheme allows planetary scale waves to be categorized on the basis of the following dualities: (1) extratropical modes against equatorially trapped modes, (2) free modes against forced modes, (3) external modes against internal modes, and (4) modes that interact with the mean flow through wave transience against modes that interact through wave dissipation. Of these various possible wave types the waves of primary importance for middle atmosphere dynamics are forced internal modes which are excited by various processes in the troposphere and propagate vertically into the middle atmosphere. The most significant forced vertically propagating extratropical modes are the quasi-stationary Rossby waves, while the most significant forced vertically propagating equatorial modes are the Kelvin wave and the mixed Rossby–gravity wave. Both the extratropical modes and the equatorial modes are

† Contribution no. 520, Department of Atmospheric Sciences, University of Washington, Seattle.

capable of generating mean flow changes through the process of wave–mean flow interaction. The quasi-stationary Rossby waves interact with the mean flow primarily through the mechanism of wave transience (local changes in wave amplitude) while the equatorial waves interact with the mean flow primarily through the mechanism of wave dissipation (thermal and/or mechanical damping). In both cases the waves transfer momentum vertically from the tropospheric source region into the middle atmosphere. In the following sections these wave–mean flow interaction processes are discussed for both extratropical waves and equatorial waves.

2. Extratropical waves

The zonally asymmetric component of the circulation in the winter hemisphere middle atmosphere is dominated by quasi-stationary Rossby waves of zonal wavenumbers 1 and 2. These waves are merely the upward extensions of tropospheric planetary waves generated by topographic forcing and by land–sea diabatic heating contrasts. Although both types of forcing are strongest during the winter season, the confinement of these waves to the winter hemisphere in the middle atmosphere is not due to seasonal changes in the forcing, but rather to the strong dependence of the vertical wave transmission on the mean zonal wind profile. Likewise, oscillations in wave amplitude and phase may occur not only as a result of oscillations in the forcing, but also in response to changes in the transmission characteristics of the middle atmosphere due to mean zonal wind changes. In addition, travelling free modes may alternately constructively and destructively interfere with the stationary forced modes to produce wave amplitude variability as has been discussed, for example, by Leovy & Webster (1976).

An explanation of the observed confinement of quasi-stationary planetary waves to the winter hemisphere in the middle atmosphere was first provided by Charney & Drazin (1961). Although their analysis was carried out using spherical geometry, the essential physics may be demonstrated using a mid-latitude 'β-plane' in which the spherical geometry is replaced by Cartesian coordinates with x directed eastward and y northward, but the dynamical effects of the variation of the Corolis parameter with latitude are retained by setting $\mathrm{d}f/\mathrm{d}y = \beta =$ constant. The quasi-geostrophic potential vorticity equation can then be written (Holton 1975)

$$\mathrm{d}q/\mathrm{d}t = S, \tag{1}$$

where

$$q \equiv f + \nabla^2\psi + \frac{f^2}{N^2}\rho_0^{-1}\frac{\partial}{\partial z}\left(\rho_0\frac{\partial\psi}{\partial z}\right),$$

with

$$\nabla^2 \equiv \frac{\partial^2}{\partial x^2} + \frac{\partial^2}{\partial y^2}; \quad \frac{\mathrm{d}}{\mathrm{d}t} = \frac{\partial}{\partial t} + u_\psi\frac{\partial}{\partial x} + v_\psi\frac{\partial}{\partial y};$$

and S designates all sources and sinks of potential vorticity. Here, ψ is a geostrophic streamfunction so that $u_\psi = -\psi_y$, $v_\psi = \psi_x$; f is the Coriolis parameter; N is the buoyancy frequency; $\rho_0 \equiv \rho_0(z)$ is the basic state density; $z \equiv -H\ln(p/p_s)$ is a log-pressure vertical coordinate; p is the local pressure; p_s is a standard reference pressure; H is a constant scale height.

Letting $\psi = \bar{\psi} + \psi'$ and $q = \bar{q} + q'$, where the overbars designate zonal means and primes designate deviations, the zonal mean of (1) can be written as

$$\partial\bar{q}/\partial t = -\partial(\overline{q'\psi_x'})/\partial y + \bar{S}. \tag{2}$$

Assuming that wave perturbations are $O(a)$, where a designates a small amplitude, the perturbation form of (1) correct to $O(a)$ is the linear wave equation

$$\frac{\partial q'}{\partial t}+\bar{u}\frac{\partial q'}{\partial x}+\frac{\partial \psi'}{\partial x}\frac{\partial \bar{q}}{\partial y} = S', \tag{3}$$

where

$$q' = \nabla^2\psi'+\frac{1}{\rho_0}\frac{\partial}{\partial z}\left(\frac{\rho_0 f^2}{N^2}\frac{\partial \psi'}{\partial z}\right)$$

and

$$\frac{\partial \bar{q}}{\partial y} = \beta-\frac{\partial^2 \bar{u}}{\partial y^2}-\frac{1}{\rho_0}\frac{\partial}{\partial z}\left(\frac{\rho_0 f^2}{N^2}\frac{\partial \bar{u}}{\partial z}\right).$$

(a) *The Charney–Drazin criterion*

If it is assumed that $\bar{u} \equiv \bar{u}(z)$ the linear wave equation (3) has separable solutions of the form

$$\psi' = \Psi(z)\,\rho_0^{-\frac{1}{2}}\exp\left[\mathrm{i}(kx+ly-kct)\right]. \tag{4}$$

Substitution from (4) into (3) and neglecting perturbation sources and sinks by setting $S' = 0$ yields the vertical structure equation

$$\Psi_{zz}+n^2\Psi = 0, \tag{5}$$

where

$$n^2 \equiv \frac{N^2}{f^2}\left[\frac{\bar{q}_y}{(\bar{u}-c)}-(k^2+l^2)\right]-\frac{1}{4H^2}.$$

For $n^2 > 0$ (5) has solutions in the form of internal (vertically propagating) waves while for $n^2 \leqslant 0$ the solutions are external (vertically trapped) waves. For stationary waves ($c = 0$) vertical propagation is thus possible only for mean zonal winds satisfying

$$0 < \bar{u} < U_c, \tag{6}$$

where $U_c \equiv \bar{q}_y(k^2+l^2+f^2/4N^2H^2)^{-1}$ is the *Rossby critical velocity*. Thus, vertical propagation for stationary waves can occur only if the mean winds are westerly but less than a critical velocity U_c. This criterion, first obtained by Charney & Drazin (1961), is of course based on a highly simplified model of the mean zonal wind. Simmons (1974) and others have shown that it is important to consider the meridional variation of $\bar{u}$ and that for realistic mean wind conditions vertical propagation is possible for somewhat larger mean winds than indicated by the Charney–Drazin criterion. However, qualitatively, equation (6) provides a valid guide for propagation of stationary Rossby waves. In particular, the strong dependence of U_c on wavelength has been confirmed by many models and is undoubtedly the reason for the observed dominance of quasi-stationary zonal wavenumbers 1 and 2 in the winter season.

The behaviour of stationary Rossby waves in the vicinity of a *critical level* where $\bar{u} = 0$ (and hence $n^2 \to \infty$) has been explored in a number of theoretical studies. Dickinson (1970), on the basis of linear theory, found that Rossby waves propagating meridionally towards the $\bar{u} = 0$ line near the equator would be absorbed near the critical line and hence produce easterly accelerations of the mean flow. More recent calculations by Béland (1976) indicate that under most conditions the Rossby wave critical level is a nonlinear phenomenon and that the waves are reflected rather than absorbed. Observations (Van Loon *et al.* 1973) do, however, indicate that in the Northern Hemisphere winter, zonal wavenumber 1 has phase lines that tilt with latitude in the vicinity of the equatorial critical level, thus suggesting absorption at the critical line. The Rossby wave critical level problem has, therefore, not yet been completely resolved.

The question is an important one, however, because Tung & Lindzen (1978) have shown that if the critical line behaviour is nonlinear, then for certain distributions of $\bar{u}(y, z)$ stationary Rossby waves might become resonant and amplify anomalously to produce sudden stratospheric warmings as well as changes in the planetary wave structure in the troposphere.

(*b*) *Wave–mean flow interaction*

The dynamical processes through which quasi-stationary Rossby waves influence the mean circulation in the stratosphere and mesosphere can be elucidated through consideration of potential vorticity conservation. According to the zonal mean quasi-geostrophic potential vorticity equation (2), in the absence of mean sources or sinks ($\bar{S} = 0$), $\bar{q}$ can be changed only if there is a meridional flux of eddy potential vorticity, $\overline{q'v'_\psi}$. This eddy flux can be expressed in terms of wave transience and wave dissipation for the case of linear perturbations by multiplying (3) by q' and averaging zonally to obtain

$$\overline{q'v'_\psi} = [\overline{q'S'} - \partial(\tfrac{1}{2}\overline{q'^2})/\partial t]/\bar{q}_y. \tag{7}$$

Thus, provided that $\bar{S} = 0$,

$$\frac{\partial \bar{q}}{\partial t} = \left[-\frac{\partial}{\partial y}(\overline{q'S'}) + \frac{\partial^2}{\partial t\, \partial y}(\tfrac{1}{2}\overline{q'^2})\right]\left(\frac{\partial \bar{q}}{\partial y}\right)^{-1}. \tag{8}$$

It is then immediately obvious that, in the absence of wave transience, $(\partial \overline{q'^2}/\partial t = 0)$ and wave dissipation $(S' = 0)$ $\bar{q}$ is constant in time so that there is no mean flow forcing by the waves. This result is the famous Charney–Drazin non-interaction theorem (Charney & Drazin 1961), which has recently been greatly generalized by Andrews & McIntyre (1976*a*, 1978). From (7) it is clear that waves which are increasing in amplitude or undergoing damping $(\overline{q'S'} < 0)$ will produce potential vorticity fluxes *down* the gradient of mean potential vorticity. Decaying waves, on the other hand, will produce countergradient fluxes of potential vorticity. Thus, the eddy flux of potential vorticity cannot in general be parametrized as a simple eddy diffusion process.

The potential vorticity flux due to wave transience is of special significance for the general circulation of the middle atmosphere. A number of modelling studies have indicated that wave transience plays an essential role in the development of sudden stratospheric warmings (see, for example, Matsuno 1971; Holton 1976; Holton & Mass 1976). The primacy of the wave transience contribution to mean flow forcing in the sudden warming situation has been specifically demonstrated by Holton & Dunkerton (1978), who explicitly computed the two terms on the right in equation (8) for a β-plane channel model of forced wave–mean flow interaction. Because of the large vertical scale and consequent small temperature perturbations in the observed quasi-stationary Rossby waves of the winter hemisphere, thermal damping processes are not very important on the rapid (*ca.* 7 days) timescale of the sudden warmings.

(*c*) *Wave transport of vertically stratified tracers*

Quasi-stationary Rossby waves serve not only to transport potential vorticity, and hence to drive changes in the mean flow, but also transport trace constituents such as ozone. It turns out that the *net* wave transport again depends on wave transience and dissipation. A particularly simple and instructive analysis of the wave transport problem can be carried out for the special, but important, case of a strongly vertically stratified tracer. This analysis should be approximately valid for the ozone mixing ratio below its peak near 30 km, and is also applicable to the transport of potential temperature.

As in the previous subsection the analysis is based on quasi-geostrophic flow on the mid-latitude β-plane. Thus, the horizontal flow is quasi-non-divergent;

$$\partial u/\partial x = \partial v/\partial y \approx O(\epsilon), \tag{9}$$

where $\epsilon \ll 1$ is the Rossby number. It is assumed that the tracer follows a conservation law

$$\mathrm{d}X/\mathrm{d}t = S, \tag{10}$$

where X is the mixing ratio (alternatively, X could denote the potential temperature) and S designates the sum of all sources and sinks.

For strong vertical stratification it is convenient to define a basic state concentration dependent only on height such that

$$X(x, y, z, t) = X_0(z) + \chi(x, y, z, t). \tag{11}$$

In the following analysis it is assumed that

$$\frac{\partial \chi}{\partial y} \bigg/ \frac{\mathrm{d}X_0}{\mathrm{d}z} \approx O(\epsilon\delta),$$

where $\delta \equiv D/L$ is the aspect ratio, that is, the ratio of the vertical scale D to the meridional scale L. With the aid of (9) and (11) the conservation equation (10) can then be written in flux form correct to $O(\epsilon)$ as follows:

$$\frac{\partial \chi}{\partial t} + \frac{\partial}{\partial x}(\chi u) + \frac{\partial}{\partial y}(\chi v) + w\frac{\mathrm{d}X_0}{\mathrm{d}z} = S. \tag{12}$$

The concentration is now separated into zonal mean and eddy components by letting $\chi = \bar{\chi} + \chi'$, where χ' is assumed to be an $O(a)$ deviation associated with the planetary wave disturbances. Taking the zonal mean of (10) then yields

$$\frac{\partial \bar{\chi}}{\partial t} = -\frac{\partial}{\partial y}(\overline{\chi' v'}) - \bar{w}\frac{\mathrm{d}X_0}{\mathrm{d}z} + \bar{S}, \tag{13}$$

while the perturbation equation correct to $O(a)$ becomes

$$\frac{\partial \chi'}{\partial t} + \bar{u}\frac{\partial \chi'}{\partial x} + v'\frac{\partial \bar{\chi}}{\partial y} + w'\frac{\mathrm{d}X_0}{\mathrm{d}z} = S'. \tag{14}$$

According to (13) the zonal mean tracer concentration for a strongly stratified tracer is determined by a balance among three processes: (1) horizontal eddy flux convergence, (2) vertical advection of the basic state, and (3) zonal mean sources and sinks. (If χ designates potential temperature these terms represent the eddy heat flux convergence, the adiabatic heating, and diabatic heating, respectively.) It is important to note that the transport effect of the waves is not limited to the eddy flux term because the Eulerian mean vertical motion $\bar{w}$ is itself partly driven by the waves. In fact, if $\bar{S} = 0$ and wave transience and dissipation vanish, the advection term $\bar{w}\mathrm{d}X_0/\mathrm{d}z$ exactly balances the horizontal eddy flux convergence so that there is no net transport. Thus, wave transience and dissipation play the same important roles for transport of conservative tracers as they play in the momentum and heat budgets.

In order to examine explicitly the relation of transport to transience and dissipation it is convenient to follow Andrews & McIntyre (1976*a*) by introducing particle displacement fields $\xi'(x, y, z, t)$, $\eta'(x, y, z, t)$, $\zeta'(x, y, z, t)$, such that

$$D_t\xi' = u' + \eta'\,\partial\bar{u}/\partial y; \quad D_t\eta' = v'; \quad D_t\zeta' = w', \tag{15}$$

where $D_t = \partial/\partial t + \bar{u}\,\partial/\partial x$ is the rate of change following the mean zonal flow. The displacement vector (ξ', η', ζ') thus gives the location of a fluid parcel relative to the position which the same parcel would have had in the absence of the wave motion (u', v', w'). Corresponding to this displacement field is a generalized Lagrangian mean $(\bar{\ })^{\mathrm{L}}$ (Andrews & McIntyre 1978) which differs from the Eulerian mean $(\bar{\ })$ in that $(\bar{\ })^{\mathrm{L}}$ is the average along the wavy material line defined by the particle displacement field. Thus, as pointed out by Matsuno & Nakamura (1979), the Lagrangian mean motion is the motion of the centre of mass of a wavy material line which would be parallel to the x-axis in the absence of the wave motion. For small amplitude waves the Lagrangian mean can be related to the Eulerian mean by the Taylor series expansion

$$\bar{A}^{\mathrm{L}} = \bar{A} + \overline{\xi'\frac{\partial A'}{\partial x}} + \overline{\eta'\frac{\partial A'}{\partial y}} + \overline{\zeta'\frac{\partial A'}{\partial z}} + O(a). \tag{16}$$

For quasi-nondivergent motion the term $\overline{\zeta'\partial A'/\partial z}$ is $O(\epsilon)$ and from (9) and (15) it is clear that $\xi_x + \eta_y \approx O(\epsilon)$. Thus, correct to $O(\epsilon)$,

$$\bar{A}^{\mathrm{L}} = \bar{A} + \frac{\partial(\overline{\eta'A'})}{\partial y} = \bar{A} + \bar{A}_{\mathrm{S}}. \tag{17}$$

Here $\bar{A}_{\mathrm{S}} = \partial(\overline{\eta'A'})/\partial y$ is the *Stokes correction* which relates the Eulerian mean to the Lagrangian mean. This Lagrangian mean formalism provides a powerful tool for analysing wave transports.

The eddy flux $\overline{v'\chi'}$ can now be written in terms of wave transience and dissipation by multiplying (14) through by η', averaging zonally and using the fact that $\overline{\chi'D_t\eta'} = \overline{\chi'v'}$ to obtain

$$\overline{v'\chi'} = -\overline{\eta'S'} + \frac{\partial}{\partial t}(\overline{\chi'\eta'}) + \overline{v'\eta'}\frac{\partial\bar{\chi}}{\partial y} + \overline{\eta'w'}\frac{\mathrm{d}X_0}{\mathrm{d}z}. \tag{18}$$

Defining a Lagrangian perturbation χ^{l} by

$$\chi^{\mathrm{l}} \equiv \chi' + \eta'\bar{\chi}_y + \zeta' X_{0_z}, \tag{19}$$

where, from (14) $\qquad D_t\chi^{\mathrm{l}} = D_t\chi' + v'\bar{\chi}_y + w'X_{0_z} = S'$,

correct to $O(a^2)$, and noting that

$$\overline{v'\eta'} = \overline{\eta' D_t\eta'} = \tfrac{1}{2}(\overline{\eta'^2})_t,$$

it is possible to rewrite (18) in the form

$$\overline{v'\chi'} = -\overline{\eta'S'} + \frac{\partial}{\partial t}\left[\overline{\eta'\chi^{\mathrm{l}}} - \tfrac{1}{2}\overline{\eta'^2}\frac{\partial\bar{\chi}}{\partial y} - \overline{\eta'\zeta'}\frac{\mathrm{d}X_0}{\mathrm{d}z}\right] + \overline{\eta'w'}\frac{\mathrm{d}X_0}{\mathrm{d}z} + O(a^2). \tag{20}$$

Substituting from (20) into (13), and using the definition of the Stokes correction (17), then yields the zonal mean budget equation:

$$\frac{\partial\bar{\chi}}{\partial t} = -\bar{w}^{\mathrm{L}}\frac{\mathrm{d}X_0}{\mathrm{d}z} + \bar{S}^{\mathrm{L}} - \frac{\partial^2}{\partial t\,\partial y}\left[\overline{\eta'\chi^{\mathrm{l}}} - \tfrac{1}{2}\overline{\eta'^2}\frac{\partial\bar{\chi}}{\partial y} - \overline{\eta'\zeta'}\frac{\mathrm{d}X_0}{\mathrm{d}z}\right]. \tag{21}$$

In the absence of transience and damping (20) shows that

$$\overline{v'\chi'} = \overline{\eta'w'}\,\mathrm{d}X_0/\mathrm{d}z;$$

thus, the *horizontal* eddy flux is proportional to the *vertical* gradient of the mean concentration not to the horizontal gradient (Clark & Rogers 1979). Furthermore, the eddy flux convergence is in this case simply equal to the advection by the vertical Stokes drift $\overline{w}_S = \partial(\overline{\eta' w'})/\partial y$ (Wallace 1978). However, if transience and damping vanish, the non-interaction theorem of Andrews & McIntyre (1976*a*, 1978) requires that $\overline{w}^L = 0$ (i.e. that fluid parcels not change their mean heights). Thus from (21) there can be no net change in the mean concentration $\overline{\chi}$. In the long term mean, on the other hand, (21) implies a balance of the form

$$\overline{w}^L \mathrm{d}X_0/\mathrm{d}z = \overline{S}^L. \qquad (22)$$

Thus, if $\mathrm{d}X_0/\mathrm{d}z > 0$, the centre of mass for a material line must gradually drift upwards in a region where the Lagrangian mean source, $\overline{S}^L$, is positive. For the case of potential temperature (22) has been used by Dunkerton (1978) to estimate the Lagrangian mean flow in the stratosphere and mesosphere based on computed heating rates for solstice conditions (with the approximation $\overline{S}^L \approx \overline{S}$).

Another interesting approximate form of the wave transport equation (21) can be obtained for transport of a conservative tracer ($\overline{S}^L = \chi^1 = 0$) by planetary waves. Now, η' and ζ' are nearly in phase quadrature for vertically propagating planetary waves. Thus, $\eta'\zeta' \approx 0$ for these disturbances. If the rate of lateral particle dispersion is approximated using the Taylor (1915) eddy diffusion hypothesis,

$$K_{yy} = \tfrac{1}{2}\partial/\partial t(\overline{\eta'^2}),$$

then (21) can be written approximately as

$$\frac{\partial \overline{\chi}}{\partial t} = -\overline{w}^L \frac{\mathrm{d}X_0}{\mathrm{d}z} + \frac{\partial}{\partial y}\left[K_{yy}\frac{\partial \overline{\chi}}{\partial y}\right]. \qquad (23)$$

Thus, the wave transport of a strongly vertically stratified tracer consists of the vertical motion of the centre of mass plus a horizontal 'diffusion' due to the dispersive effects of wave transience. It is important to realize, however, that both the Lagrangian mean vertical advection and horizontal diffusion are wave driven transports which in this case depend on wave transience.

In summary, it should now be clear that calculation of Eulerian eddy fluxes does not in itself give much insight into the net transport of a vertically stratified tracer. The generalized Lagrangian mean formalism of Andrews & McIntyre provides a much better framework for analysis of wave transport processes.

3. Equatorial waves

A particularly fascinating example of wave–mean flow interaction is provided by the interaction of the equatorially trapped Kelvin and mixed Rossby–gravity waves with the mean zonal wind in the tropical stratosphere to produce the quasi-biennial oscillation (Holton & Lindzen 1972). Both Kelvin and mixed Rossby–gravity waves are zonally propagating waves which are excited by forcing in the equatorial troposphere. The nature of this forcing is not known in any detail. However, latent heat release in convective storms appears to be the most likely source for the waves. These waves propagate vertically with comparatively short vertical wavelengths, and interact with the mean flow primarily through the mechanism of wave dissipation. As a result of this interaction process the Kelvin waves produce westerly accelerations of the mean zonal wind while the mixed Rossby–gravity waves produce easterly accelerations.

(a) Kelvin waves

The Kelvin wave is an eastward propagating mode with negligible meridional velocity perturbations, and with zonal velocity and pressure perturbations which are in geostrophic balance and are symmetric with respect to the equator with latitudinal dependence given by

$$\exp\left[-\beta y^2/2\hat{c}\right],$$

where y is the distance from the equator, $\beta \equiv 2\Omega/a$ is the value of $\mathrm{d}f/\mathrm{d}y$ at the equator, and $\hat{c}$ is the phase speed relative to the mean flow. For small dissipation rates the vertical wavelength of the Kelvin wave, $L_z \approx 2\pi\hat{c}/N$, where N is the buoyancy frequency, depends only on the Doppler shifted phase speed $\hat{c}$, not on the zonal wavenumber and frequency separately. (An extensive discussion of the properties of Kelvin waves and other equatorial modes is given in Holton (1975).)

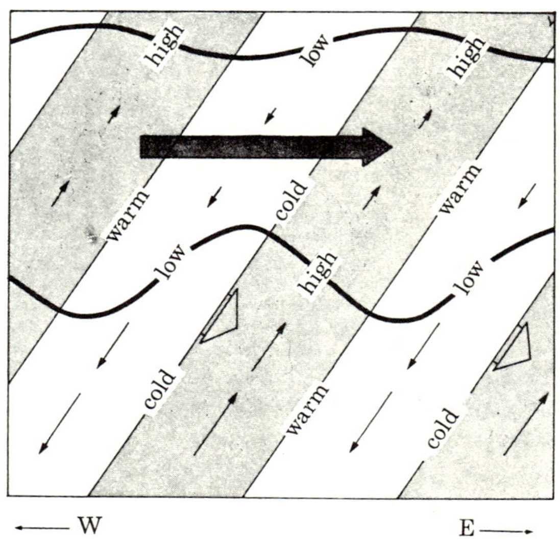

FIGURE 1. Longitude–height section along the equator showing pressure, temperature, and wind perturbations for a thermally damped Kelvin wave. Heavy wavy lines indicate material lines, short blunt arrows show phase propagation. Areas of high pressure are shaded. Length of the small thin arrows is proportional to the wave amplitude, which decreases with height owing to damping. The large shaded arrow indicates the net mean flow acceleration due to the radiation stress divergence.

The observed Kelvin waves in the stratosphere seem to be limited to zonal wavenumbers 1 and 2. The waves have periods in the range of 10–20 days and vertical wavelengths of 6–12 km (Wallace 1973). A longitude–height cross section showing the structure of a Kelvin wave in the presence of weak thermal dissipation is shown in figure 1. The phase relations among the zonal velocity, vertical velocity, pressure, and temperature perturbations are just those of an eastward propagating internal gravity mode. In particular, phase lines propagate downward, consistent with an upward energy propagation.

The role of thermal dissipation in wave–mean flow interaction can be elucidated in a particularly simple manner for the Kelvin wave case. In the absence of mechanical dissipation and horizontal shear of the mean flow the mean zonal flow acceleration due to Kelvin waves is given by

$$\frac{\partial \bar{u}}{\partial t} = -\frac{1}{\rho_0}\frac{\partial}{\partial z}(\rho_0\,\overline{u'w'}). \tag{24}$$

Eliassen & Palm (1961) showed that for waves of steady amplitude in the absence of dissipation and critical levels (levels where the mean zonal wind speed equals the wave phase speed), $\rho_0\overline{u'w'}$ must be independent of height so that the mean flow forcing vanishes. To examine the role of thermal dissipation it is again convenient to introduce the linearized particle displacement $\zeta'(x, z, t)$, which satisfies $D_t\zeta' = w'$, where for steady amplitude waves with zonal phase speed c the operator $D_t = -(c-\bar{u})\,\partial/\partial x$. The perturbation zonal momentum equation

$$D_t u' = -(1/\rho_0)\,\partial p'/\partial x,$$

thus implies that $\rho_0 u' = p'/(c-\bar{u})$. Therefore, the vertical momentum flux may be written

$$\rho_0\overline{u'w'} = -\overline{p'\,\partial\zeta'/\partial x},$$

so that the mean flow forcing is just

$$\frac{\partial\bar{u}}{\partial t} = -\frac{1}{\rho_0}\frac{\partial}{\partial z}\left[-\overline{p'\frac{\partial\zeta'}{\partial x}}\right]. \tag{25}$$

The wavy material lines $\zeta'(x, t)$ are shown at two different levels by the heavy solid lines in figure 1. It should be noted that for adiabatic waves

$$D_t\theta' + w'\bar{\theta}_z = 0,$$

where θ is potential temperature. Thus $\zeta' = -\theta'/\bar{\theta}_z$ so that the material lines are also lines of constant potential temperature. However, in the thermally damped case depicted in figure 1, the potential temperature isolines should actually be shifted slightly eastward relative to the material lines since thermal damping destroys the quadrature relationship between θ' and w'. (No attempt has been made to portray this slight phase shift in the figure.)

Mean flow forcing by the Kelvin waves can be given a particularly simple physical interpretation by using the mean flow equation (25). Now, $-\overline{p'\zeta'_x}$ simply represents the net eastward stress due to the pressure force exerted by the fluid below a wavy material line on the fluid above the material line. In the case shown in figure 1 this so called 'radiation stress' is positive since $p' > 0$ where $\zeta'_x < 0$, and decreases with height because the waves are damped. Thus, the fluid contained between the two wave material lines is subject to a net eastward force due to the action of the radiation stress. The magnitude of this net force is, as indicated in (25), proportional to the rate of decrease of the radiation stress with height, which in turn depends on the rate constant for the damping. Observations indicate that the Kelvin waves in the equatorial stratosphere do have sufficient amplitude to drive the westerly phase of the quasi-biennial oscillation through forcing due to wave dissipation. Kelvin waves damped thermally are also the only plausible mechanism yet suggested for generating the westerly phase of the semiannual zonal wind oscillation in the equatorial upper stratosphere and mesosphere (Dunkerton 1979).

(b) *Mixed Rossby–gravity waves*

The mixed Rossby–gravity mode is a westward propagating wave which has meridional wind perturbations symmetric about the Equator and zonal velocity and pressure perturbations antisymmetric with respect to the Equator. The latitudinal dependence of the pressure and zonal velocity perturbations in the mixed Rossby gravity wave turns out to have the form (Holton 1975)

$$y\exp\left[-\frac{(1-k\omega/\beta)\,\beta^2 y^2}{2\omega^2}\right],$$

where k and ω are the zonal wavenumber and frequency, respectively. These must satisfy the condition $0 < k\omega < \beta$ in order that valid solutions exist. Observed mixed Rossby–gravity waves have periods in the range of 3–5 days, zonal wavelength 10000 km and vertical wavelengths in the range of 4–8 km. These waves have their greatest amplitudes in the lower stratosphere during the westerly phase of the quasi-biennial oscillation. A longitude–height cross section showing the structure of a mixed Rossby–gravity wave subject to weak thermal dissipation is shown in figure 2. The diagram is for a latitude north of the equator. As in the corresponding Kelvin wave figure, wavy material lines are again denoted by heavy solid lines.

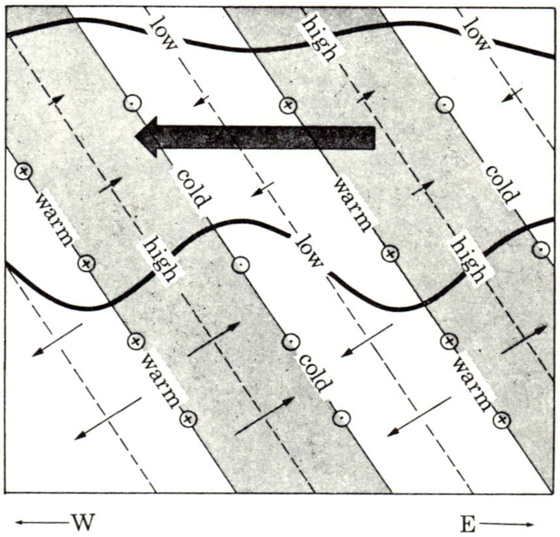

FIGURE 2. Longitude–height section along a latitude circle north of the equator showing pressure, temperature, and wind perturbations for a thermally damped mixed Rossby–gravity wave. Areas of high pressure are shaded. Small arrows indicate zonal and vertical wind perturbations with length proportional to the wave amplitude. Meridional wind perturbations are shown by arrows pointed into the page (northward) and out of the page (southward). The large shaded arrow indicates the net mean flow acceleration due to the radiation stress divergence.

Again the radiation stress concept may be used to elucidate the wave-mean flow interaction process. Andrews & McIntyre (1976*a*) have shown that for small amplitude equatorial waves that have vertical scales small compared to the vertical scale of the mean flow, the mean flow acceleration to lowest order is given by

$$\frac{\partial \bar{u}}{\partial t} = -\frac{\partial}{\partial y} S_{xy} - \frac{\partial}{\partial z} S_{xz}, \tag{26}$$

where

$$S_{xy} = \overline{u'v'} - \bar{u}_z \overline{v'\Phi'_z}/N^2,$$

$$S_{xz} = \overline{u'w'} + (\bar{u}_y - f)\, \overline{v'\Phi'_z}/N^2.$$

Here $\Phi' = p'/\rho_0$, and the Boussinesq approximation has been used for simplicity. However, Eliassen & Palm (1961) showed that for steady inviscid waves

$$S_{xy} = \overline{v'\Phi'}/(c-\bar{u}), \quad S_{xz} = \overline{w'\Phi'}/(c-\bar{u}). \tag{27}$$

Again, introducing the particle displacements $\zeta'(x, y, z, t)$ and $\eta'(x, y, z, t)$, and observing that for steady waves

$$\zeta'_x = -w'/(c-\bar{u}); \quad \eta'_x = -v'/(c-\bar{u}),$$

the relationships of (27) can be rewritten in the form

$$S_{xy} = -\overline{\eta'_x \Phi'}; \quad S_{xz} = -\overline{\zeta'_x \Phi'}.$$

But, in the absence of mechanical dissipation, η'_x and Φ' are in phase quadrature for mixed Rossby–gravity waves. Thus $S_{xy} = 0$ and (26) becomes

$$\partial \bar{u}/\partial t = +\partial(\overline{\Phi' \zeta'_x})/\partial z, \tag{28}$$

which is identical to (25) when the Boussinesq approximation (ρ_0 = constant) is applied. Thus, just as in the case of the Kelvin waves the mean flow forcing for mixed Rossby–gravity waves subject only to weak thermal damping can be expressed in terms of the vertical derivative of the radiation stress $-\overline{p'\zeta'_x}$. In the present case, however, it is clear from figure 2 that $p' > 0$ where $\zeta'_x > 0$ so that the radiation stress exerts a net westward force on the fluid contained between the two wavy material lines shown in the figure. Thus, the mixed Rossby–gravity waves interact with the mean flow to produce the easterly phase of the quasibiennial oscillation.

(c) The influence of mechanical dissipation and horizontal mean wind shear

In the above discussion it was assumed that wave dissipation was due to thermal damping only. Addition of mechanical dissipation causes large changes in the wave–mean flow interaction process, especially in the case of the mixed Rossby–gravity wave. Since both the pressure and vertical displacement fields for a mixed Rossby–gravity wave are antisymmetric with respect to the equator, the mean flow acceleration, $\bar{u}_t$, for a thermally damped wave must, according to (28), be zero at the equator and have two easterly maxima symmetrically distributed on either side of the equator. However, as first shown by Andrews & McIntyre (1976*a*), addition of even a small mechanical dissipation drastically changes the distribution of $\bar{u}_t$. There are two reasons for this sensitivity. In the first place, mechanical dissipation introduces a non-zero viscous contribution to S_{xy}, which for Rayleigh friction dissipation has the form $-\lambda\overline{\eta' u'}$, where λ is the rate coefficient for the damping. Secondly, mechanical dissipation destroys the quadrature relationship between η'_x and Φ' so that S_{xy} can no longer be neglected in (26). In fact, mechanical dissipation produces a strong divergence of the lateral radiation stress, $-\overline{\eta'_x \Phi'}$, at the equator so that when mechanical dissipation is comparable in magnitude to thermal dissipation, $\bar{u}_t$ has a strong maximum in amplitude at the equator. This is just the distribution of $\bar{u}_t$ that is observed in the quasi-biennial oscillation.

Horizontal mean wind shear also has strong effects on the distribution of $\bar{u}_t$. Calculations by Andrews & McIntyre (1976*b*) and by Simmons (1978) have shown that when only thermal dissipation is present there is a tendency for the mean flow acceleration pattern to enhance an existing cross equatorial horizontal shear. However, Boyd (1978) has shown that when mechanical dissipation is included the mean flow acceleration pattern tends to diminish cross equatorial shear. This tendency for the waves to symmetrize the mean flow with respect to the equator has also been demonstrated in a numerical model by Holton (1979).

One striking aspect of the above studies is the finding that the structure of the Kelvin wave is determined almost entirely by the mean zonal wind at the equator. Even a rather strong cross equatorial shear has very little influence on the zonal wind and pressure perturbations. This may be demonstrated by comparison of figures 3 and 4, which show the Kelvin wave structures in two numerical simulations, one with mean zonal flow symmetric about the equator and the other with strong linear shear across the equator. The principal difference

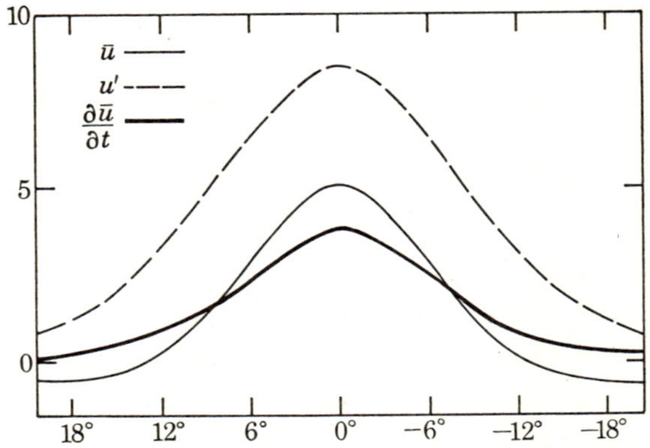

FIGURE 3. Perturbation zonal wind amplitude (m s^{-1}) and mean wind acceleration (10^{-6} m s^{-2}) for a mechanically and thermally dissipated Kelvin wave in the presence of an equatorially symmetric mean flow.

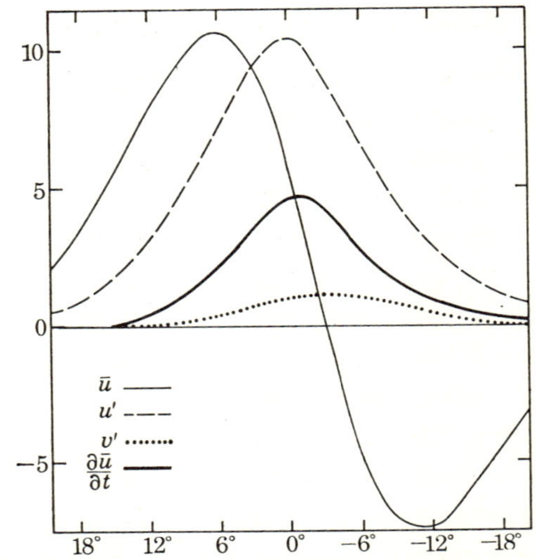

FIGURE 4. Perturbation zonal wind amplitude (m s^{-1}) and mean wind acceleration (10^{-6} m s^{-2}) for a mechanically and thermally dissipated Kelvin wave in the presence of a mean flow with strong cross equatorial shear.

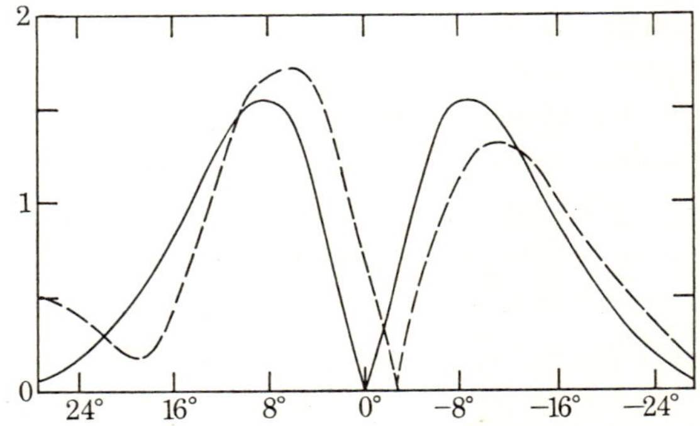

FIGURE 5. Perturbation zonal wind amplitude (m s^{-1}) for a mixed Rossby–gravity wave in the presence of a uniform mean flow (solid line) and a mean flow with linear shear of -0.5 (m s^{-1})/100 km (dashed line).

between these two cases is that a weak asymmetric meridional wind perturbation is present in the latter case, and the maximum mean flow acceleration is shifted slightly into the hemisphere with the larger mean momentum deficit.

The structure of the mixed Rossby–gravity wave is more significantly affected by cross equatorial shear. The amplitudes of the zonal wind perturbation for two simulations, one with no mean wind shear, and the other with a weak horizontal shear $\bar{u}_y = -0.5\ (\mathrm{m\ s^{-1}})/100$ km, are shown in figure 5. In the sheared case the zero line is shifted into the hemisphere where the Doppler shifted frequency is largest while the peak amplitude is depressed in that hemisphere and enhanced in the other hemisphere. With larger horizontal shears, more dramatic distortions occur in the wave structure. However, in all mixed Rossby–gravity wave forcing cases investigated by Holton (1979) the wave driven mean flow acceleration acted to reduce the initial cross equatorial horizontal shear, and the mean flow eventually evolved as an easterly jet centred on the equator. Thus, it appears that as long as there is substantial mechanical damping as well as thermal damping, the equatorial wave–mean flow interaction process will tend to produce mean flow changes which are symmetric about the equator despite the asymmetric horizontal shears introduced by the annual cycle. This tendency may explain the remarkable equatorial symmetry observed for the quasi-biennial oscillation.

I wish to thank Mr Timothy Dunkerton for helpful conversations. This research was supported by the National Aeronautics and Space Administration, N.A.S.A. Grant NSG-2228.

References (Holton)

Andrews, D. G. & McIntyre, M. E. 1976*a* *J. atmos. Sci.* **33**, 2031–2048.
Andrews, D. G. & McIntyre, M. E. 1976*b* *J. atmos. Sci.* **33**, 2049–2053.
Andrews, D. G. & McIntyre, M. E. 1978 *J. Fluid Mech.* **89**, 609-646.
Béland, M. 1976 *J. atmos. Sci.* **33**, 2065–2078.
Boyd, J. 1978 *J. atmos. Sci.* **35**, 2259–2267.
Charney, J. G. & Drazin, P. G. 1961 *J. geophys. Res.* **66**, 83–109.
Clark, J. H. E. & Rogers, T. G. 1978 *J. atmos. Sci.* **35**, 2232–2235.
Dickinson, R. E. 1970 *J. atmos. Sci.* **27**, 627–633.
Dunkerton, T. 1978 *J. atmos. Sci.* **35**, 2325-2333.
Dunkerton, T. 1979 *J. atmos. Sci.* **36**, 32–41.
Eliassen, A. & Palm, E. 1961 *Geophys. Publ.* **22**, no. 3, 1–23.
Holton, J. R. 1975 *The Dynamic Meteorology of the Stratosphere and Mesosphere*, Meteorological Monograph no. 37. Boston: The American Meteorological Society.
Holton, J. R. 1976 *J. atmos. Sci.* **33**, 1639–1649.
Holton, J. R. 1979 *J. atmos. Sci.* **36**, 1030–1040.
Holton, J. R. & Dunkerton, T. 1978 *J. atmos. Sci.* **35**, 740–744.
Holton, J. R. & Lindzen, R. S. 1972 *J. atmos. Sci.* **29**, 1076–1080.
Holton, J. R. & Mass, C. 1976 *J. atmos. Sci.* **33**, 2218–2225.
Leovy, C. B. & Webster, P. J. 1976 *J. atmos. Sci.* **33**, 1624–1638.
Matsuno, T. 1971 *J. atmos. Sci.* **28**, 1479–1494.
Matsuno, T. & Nakamura, K. 1979 *J. atmos. Sci.* **36**, 640–654.
Simmons, A. J. 1974 *Q. Jl R. met. Soc.* **100**, 76–108.
Simmons, A. J. 1978 *Q. Jl R. met. Soc.* **104**, 595–614.
Taylor, G. I. 1915 *Phil. Trans. R. Soc. Lond.* A **215**, 1–26.
Tung, K. K. & Lindzen, R. S. 1979 *Mon. Weath. Rev.* **107**, 714–734.
Van Loon, H., Jenne, R. L. & Labitzke, K. 1973 *J. geophys. Res.* **78**, 4463–4471.
Wallace, J. M. 1973 *Rev. Geophys. Space Phys.* **11**, 191–222.
Wallace, J. M. 1978 *J. atmos. Sci.* **35**, 554–558.

Phil. Trans. R. Soc. Lond. A **296**, 87–102 (1980) [87]
Printed in Great Britain

Interaction between the troposphere and stratosphere

By R. J. Murgatroyd and A. O'Neill†
Meteorological Office, London Road, Bracknell, Berkshire, RG12 2SZ, U.K.

Three aspects of the interaction between troposphere and stratosphere are considered: (i) mass transports, (ii) radiative processes, (iii) dynamical effects.

(i) The transport of constituents across the tropopause is a major feature in determining the detailed composition of the stratosphere since the principal sources and sinks of many of its chemical species are located in the troposphere or at the Earth's surface. The nature of this transfer by the mean circulations and smaller scale phenomena is described.

(ii) Since the radiation balance throughout the atmosphere is affected by the composition at all levels, changes in radiatively important constituents in one atmospheric region may affect conditions in another. The possible importance of this mechanism in troposphere–stratosphere interaction is discussed.

(iii) It is now well established that upward transfer of mechanical energy from the troposphere by planetary-wave motions plays a major role in driving the general circulation of the stratosphere. Recent studies clarifying some details of the processes involved are outlined. The possibility that the stratospheric circulation could significantly influence tropospheric motions is also briefly considered.

Introduction

The high temperatures around 50 km in the middle atmosphere result from the strong absorption of ultraviolet solar radiation by ozone that takes place there and the general circulation in this region is, to a considerable extent, determined by its net radiative sources and sinks. It has, however, been established that the circulation of the lower stratosphere is mainly driven by that of the troposphere below and that in winter at high latitudes the dynamical effects of the latter may extend at least up into the mesosphere. As a result, the circulations of the troposphere and middle atmosphere cannot be studied comprehensively in isolation.

It is known that variability in the solar radiation, which is well marked in its far ultraviolet wavelengths and below, leads to large changes in the thermosphere. It is, however, difficult to establish with certainty either statistically or through studies of possible physical or chemical mechanisms that any effects from these perturbations are significantly transferred down through the middle atmosphere to cause observable changes in the troposphere. This leads to the belief, which is also supported from considerations of relative masses and energy contents, that lower atmosphere changes can often influence upper atmosphere conditions but not vice versa unless some so far unvalidated transfer or reflexion mechanism is found to be important.

In the discussion below of troposphere–stratosphere relations, transport of mass and constituents, radiative fluxes and dynamical interactions will be considered in turn and attention drawn to areas where further progress might be made within the Middle Atmosphere Program. Since the atmosphere must primarily be regarded as one thermodynamical system rather than a number of separate layers it is desirable first to consider the nature and characteristics of the boundary between the troposphere and the middle atmosphere – the tropopause.

† Present address: Department of Atmospheric Sciences, University of Washington, Seattle, Washington 98195, U.S.A.

Tropopause structure

The tropopause is defined as the level at which the general tropospheric lapse rate is replaced above by an inversion (mainly at low latitudes and in summer at high latitudes), or an isothermal profile or a considerably decreased lapse rate, broadly the level at which it becomes less than 2 K/km (see W.M.O. 1966 for the detailed specification). In the last case, which occurs in high latitudes in winter, the tropopause level may be indistinct and some authors (e.g. Court 1942) suggest that the tropopause disappears at times in winter over Antarctica.

The existence of a tropopause is usually thought to be determined, at least as an annual mean on the global scale, as the demarcation level between the stratosphere where the radiation balance is the principal process determining the thermal structure and the troposphere where convective turbulent exchange is of major importance (see, for example, Goody 1949). It has been proposed (Staley 1957) that divergence of vertical enthalpy flux in the presence of the stable stratosphere above and the less stable troposphere below can lead to the formation of a sharp discontinuity in the temperature profile. Figures 1 (*a*), (*b*) illustrate the broad features of global tropopause levels and temperature distributions of the type that have to be accounted for and figures 1 (*c*), (*d*) the corresponding temperature fields in the troposphere (600 mbar)† and in the lower stratosphere (50 mbar) to which they might be related. These data were extracted from monthly mean I.G.Y. pole to pole cross sections at 75° W, July 1957–December 1958 (U.S. Department of Commerce 1961) and, although local in the sense of including some standing wave features, appear to be generally consistent with the available global climatological data. With reservations regarding their coverage and the somewhat artificial tropopause specification for the high latitudes in winter they show:

(i) Mean tropopause heights are greatest in low latitudes where the equatorial tropopause is around the 17 km or 100 mbar level. The annual variation here shows maximum heights and lowest temperatures (about −82 °C) in January and this pattern extends further into the Southern than the Northern Hemisphere.

(ii) At higher latitudes the polar tropopause at about 10 km or 300 mbar is for the most part everywhere highest in August, i.e. both in the Northern Hemisphere summer and the Southern Hemisphere winter. The temperatures are, however, highest in both their summers. Irregularities occur particularly in the Northern Hemisphere.

(iii) Comparison between figures 1 (*d*) and 1 (*b*) indicates a close correlation between 50 mbar and tropopause temperatures. On the other hand the 600 mbar values (figure 1 (*c*)) show little variation with season at low latitudes, whereas the tropopause temperatures show a well marked minimum in January. The correspondence is better at high latitudes where both figures 1 (*b*) and 1 (*c*) show warm summer and cold winter temperatures, but the detailed agreement in the more variable Northern Hemisphere is not as good as that between the 50 mbar and tropopause temperatures.

In so far as these data are representative they indicate that tropospheric convection has a major effect on tropopause height, particularly at low latitudes and in determining the overall latitude distribution. However, at higher latitudes the tropopause height and temperature appear to be strongly influenced by stratospheric temperatures tending to impose a lid or to prolong the lapse rate region upwards. The annual variations at the equatorial tropopause may be

† 1 mbar = 10^2 Pa.

linked with an annual variation of the strength of the Hadley cell (Reed & Vlcek 1969; Newell *et al.* 1969) and it also appears likely that they are a dynamical consequence of the greater activity in the Northern Hemisphere winter.

The mean tropopause is also not continuous from equator to pole but has a major gap at about 40° N in the jet stream region between the equatorial and polar tropopauses. This gap is thought to be due to dynamical effects, with widespread subsidence in this region linked to the descending branch of the Hadley cell (Manabe & Hunt 1968). Isentropic quasi-horizontal

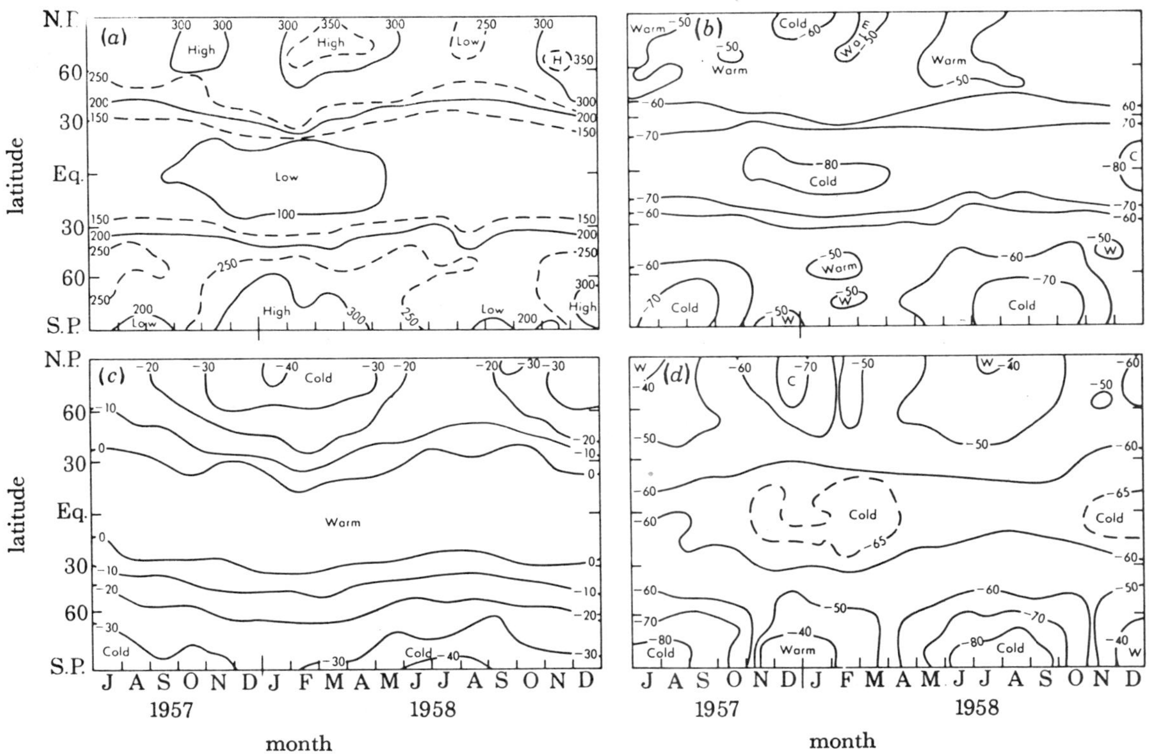

FIGURE 1. Latitude–time sections of (*a*) tropopause pressure (mbar), (*b*) tropopause temperature (°C), (*c*) temperature (°C) at the 600 mbar level, (d) temperature (°C) at the 50 mbar level. Data based on monthly mean I.G.Y. pole to pole cross sections at 75° W.

transfer of atmospheric constituents, radioactive tracers, etc. can readily take place through this gap (see, for example, Danielsen 1959). On the other hand it is not so clear that direct transfer can readily take place through an unbroken tropopause, e.g. in the upward transfer of mass and water vapour, etc. through the equatorial tropopause, although penetrative convection (see, for example, Riehl & Malkus 1958) appears to be important there, changing the problem to one of mesoscale or smaller scale transfer across the tropopause with some diabatic contributions.

There are also difficulties in explaining why the higher latitude tropopause should in general retain so efficiently its sharp change of lapse rate on the small scale in the presence of the large scale dynamical processes which take place in both troposphere and lower stratosphere.

Troposphere–stratosphere mass and constituent transfer

Reiter (1975) has attempted to construct a budget for the annual change of stratospheric mass of a hemisphere. He regarded the possibly important processes as (*i*) seasonal changes in the height of the mean tropopause level, (ii) mean meridional circulations, (iii) large scale eddy transports mainly in jet stream regions and (iv) mesoscale and smaller scale transport across the tropopause. His estimate for (i) was that the seasonal change of tropopause heights accounts

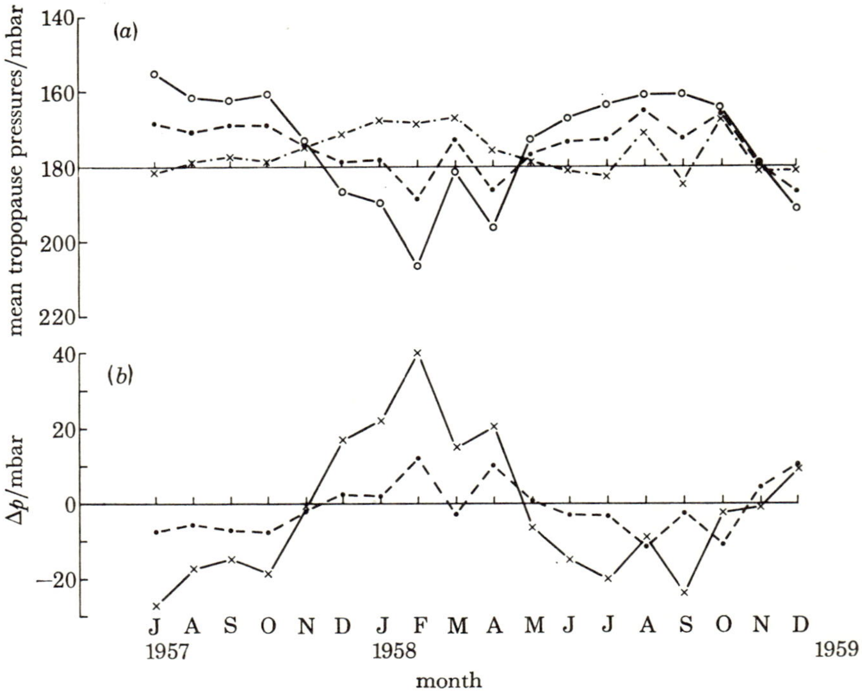

Figure 2. (a) Variation throughout the I.G.Y. of monthly mean tropopause pressure (mbar) at 75° W of the Northern Hemisphere average (○—), the Southern Hemisphere average (×–·–) and the global average (●– –); (*b*) corresponding variation of Northern Hemisphere minus Southern Hemisphere values of the tropopause pressures shown in (*a*) (×—) and the deviations of the monthly global mean values from the global overall mean for the period (●– –).

for a flux of about 10% of the mass of the stratosphere in one hemisphere during the course of 1 year and he concluded that this flux is balanced approximately by the seasonal shift of stratospheric air masses between the Northern and Southern Hemispheres. Figure 2, which is based on the same data as figure 1, indicates that for the whole globe there is about 5% more mass in the stratosphere in the November–May period than in the June–October period, the pattern being the same as that for the Northern Hemisphere but opposite to that for the Southern Hemisphere. These differences between Northern and Southern Hemispheres (figure 2(*b*)) show about 10% more mass in the Northern than in Southern Hemisphere in the November–May period and the opposite in the remainder of the year. A further breakdown into contributions from the higher (> 30°) and lower (< 30°) latitudes indicates that the major cause of the above annual variation is that in the northern high latitudes. Conditions in the other regions show less definite variations and in the Southern Hemisphere the conditions in its summer 1958 were very different in this respect from those in 1957. Reiter also estimated that the Hadley cell is the most effective of the transport mechanisms and that it transfers approximately 38 % of the mass equivalent to one hemispheric stratosphere (which is about 40×10^{19} g) through the tropopause per year. This was based on stream function data computed by Louis (see Reiter *et al.* 1975), which are in general agreement with those of other authors. Louis also calculated that a total mass of $5–8 \times 10^{19}$ g of stratospheric air is shifted between winter and summer hemispheres. Finally, by using estimates of the mass transferences per cyclone and the cyclone frequency in the Northern Hemisphere, Reiter has estimated that systems on this scale transfer about 20 % of stratospheric air through the troposphere each year. His corresponding estimate for the smaller scale systems is only about 1 %.

The above picture of tropospheric–stratospheric mass exchange is consistent with that obtained from studies of the vertical transfer of individual constituents and their estimated stratospheric residence times. The monthly variations of total ozone over the two hemispheres and over the globe appear to be very similar although a few months later in phase than those of the tropopause heights shown in figure 2. According to London *et al.* (1978) the total ozone deduced from ground based data in the Northern Hemisphere is a maximum in March or April and a minimum in September or October. In the Southern Hemisphere the variation is in antiphase with that of the Northern Hemisphere, with the maximum in October or November. The global cycle broadly follows that of the Northern Hemisphere which has 2–3 % more total ozone than in the Southern Hemisphere. This trend is similar to that obtained from satellite results given recently by Keating (1978). Angell & Korshover (1974) have reported that the tropopause pressure and temperature exhibit a quasi-biennial oscillation. The amplitude of the pressure oscillation during the period 1957–71 was about 4 mbar at the equator and there was a pronounced out of phase relationship between equatorial and mid-latitudes. In addition these authors found that during this period there was a long term increase in tropopause pressure with values of about 5–7 mbar per decade in the tropics and 2–4 mbar per decade in higher latitudes in the Northern Hemisphere. The tropopause pressure increases were accompanied by temperature increases which may have led to increases in stratospheric humidity over the period 1964–70, observed by Mastenbrook (1971) at Washington, D.C., if it is controlled by the 'cold trap' mechanism suggested by Brewer (1949). Newell & Wu (1978) have presented evidence that latitudinal distribution of total ozone and lower stratosphere temperature vary with the intensity of the Hadley cell. They also pointed out that the long term trend in observed Northern Hemisphere total ozone amounts, which showed an increase in the 1960s, and Angell & Korshover's results above are compatible with a decrease in the Hadley cell circulation during this period. It also appeared from Newson's three dimensional model (C.O.M.E.S.A. 1975) that on the whole the flux of water vapour in winter is from stratosphere to troposphere mainly in the descending part of the Hadley cell. A mean meridional circulation with descent in middle latitudes in the winter stratosphere has also been inferred from observations of radioactive substances (List & Telegadas 1969). Descent through tropopause gaps of radioactive substances, ozone, potential vorticity and dry stratospheric air has been verified in many observational studies.

It is difficult to obtain from observations a measure of the overall rate of removal of a substance from the stratosphere and it is usually estimated in terms of its apparent residence time in the region tropopause to, say, the 25 km level. For example according to Krey *et al.* (1974), this residence time for Sr^{90} is about 14 months, which Reiter argued is consistent with a combined effect based on his estimates of vertical mass transfer by the Hadley cell and large scale eddy mechanisms together with 15 % by movement between hemispheres annually. Other authors believe the above residence time is an underestimate by a factor of two or more when applied in calculations of atmospheric composition (involving the transfer of species such as N_2O, CH_4, $CFCl_3$, CF_2Cl_2 from troposphere to stratosphere and HNO_3, HCl and O_3 from stratosphere to troposphere) in current studies of possible effects of pollutants on O_3 amounts in the stratosphere. The question is of great importance both in the calculations of the possible duration of any effects and also their magnitudes and presents one of the major uncertainties in this work, particularly as there have been many one and two dimensional studies which have attempted to represent the eddy transports in terms of empirically based flux-gradient relations. Accordingly, although there is now some confidence that the mechanisms and patterns of transfer of these substances have been identified and can be simulated reasonably well in three dimensional and, at least for the mean circulation contributions, in two-dimensional models of the atmosphere, further work is needed to obtain better quantitative estimates of the magnitudes of the different vertical transfer mechanisms.

Radiative interactions

The importance of radiation, particularly the absorption of solar ultraviolet radiation by ozone, in establishing the overall temperature distribution of the stratosphere is apparent from a comparison of its calculated temperature field in radiative equilibrium (see, for example, Manabe & Strickler 1964), with the mean observed temperature distribution (see, for example, Murgatroyd 1969). The heating rates due to this absorption by ozone are calculated to be of the order of 10 K/day near the stratopause and to fall to less than 0.1 K/day below about 20 km. The importance and role in establishing the thermal structure and driving the circulation of the different radiative processes in the lower stratosphere are more difficult to assess. Significant absorption of solar radiation also takes place here in visible and infrared wavelengths by aerosols, ozone (Chappuis bands), NO_2, molecular oxygen, water vapour and carbon dioxide with small contributions from other minor constituents (see Houghton 1963). In a similar manner to the patterns produced by the absorption by ozone at higher levels, the resultant heating rates are larger in the summer hemisphere and increase towards the pole. Hering *et al.* (1967) have estimated that absorption of back-scattered radiation in the O_3 Chappuis bands may increase heating by O_3 by about 10% around the 30 km level. The patterns of cooling and heating due to divergence and convergence of the balancing atmospheric radiation, mainly in the 15 μm CO_2, 9.6 μm O_3, and the water vapour rotational and 6.3 μm bands, show cooling increasing with height above about 30 km, but heating or small cooling above the equatorial tropopause at lower levels. Elsewhere the cooling pattern in the lower and middle stratosphere generally tends to follow the temperature distribution. In the troposphere there is a general radiative cooling with maximum rates of up to 2 K/day occurring in low latitudes. The boundary here between the net heating above, due primarily to the 9.6 μm O_3 band, and net cooling below, mainly due to the water vapour bands, is at about 15 km, i.e. somewhat below the equatorial tropopause and does not completely account for the sharp inversion there. It appears also from the calculations of Manabe & Strickler (1964) that the variation of the concentrations of the radiatively important constituents from equator to pole is not sufficient to explain the latitudinal variation of the mean tropopause height (17–10 km approximately).

The overall distribution of the net heating and cooling rates indicates that their relation to the source–sink mechanisms driving the general circulation of the troposphere and stratosphere and also the establishment of the detailed pattern of the thermal structure is indirect. Since there is net radiative cooling, which is a maximum at low latitudes in the troposphere, the contributions to the overall heating rate due to sensible heat and latent heat release must be a major factor in this region. Since the equatorial tropopause region is very cold and it is also a region of net heating, available potential energy must be destroyed in the lower stratosphere and its circulation driven by that of the troposphere. Although the heating region in the equatorial lower stratosphere is likely to favour upward motion and thereby assist the vertical extension of the rising branch of the Hadley cell, this heating primarily arises because the low-latitude lower stratosphere is a dynamically produced cold region exchanging radiative fluxes with warmer regions above and below. Accordingly in this sense at least the radiative heating region in the lower stratosphere should preferably be regarded as a by-product of the dynamics rather than vice versa.

With this background, it seems that perturbations in the infrared flux from troposphere to stratosphere are most likely to be produced by radiation in the 9.6 μm O_3 band which, as well as being the main agent in causing heating in the low-latitude stratosphere, is situated in an important atmospheric window. Consequently, in contrast to the 15 μm CO_2 and water bands which are strongly absorbent and can only 'see' the upper troposphere above about the 500 m-bar level from the lower stratosphere, the 9.6 μm O_3 band may 'see' either the warm surface of the Earth through the small concentrations of tropospheric ozone or the cold tops of the clouds. According to Hitschfeld & Houghton (1961) the effects of the latter situation could be strong enough to cause all or most of the lower stratospheric heating to disappear. Ramanathan (1977) has suggested that, at least in winter, the circulation of the stratosphere can interact strongly with that of the troposphere through the agency of radiation emitted primarily by the 9.6 μm O_3 band. This downward flux is considerably enhanced during sudden warmings when both lower stratospheric temperature and ozone amounts are considerably increased. A model calculation indicated that the increase of the radiative flux into the troposphere at high latitudes during a major warming was as much as 11 % of the flux toward the poles of total energy by the general circulation and reduced the tropospheric cooling in the arctic by roughly 9 %, possibly equivalent to a temperature increase of about 2 K. Since stratospheric sudden warmings at high latitudes are accompanied by cooling at low latitudes (Fritz & Soules 1972) this mechanism could lead to a net reduction of the tropospheric available potential energy, estimated as much as 9 % in the above example. Such an interaction could lead to a significant effect on the tropospheric circulation. The situation during a sudden warming is an extreme example of this kind of radiative–dynamical feedback which is generally operative to some extent and is initiated by the dynamical forcing of the stratosphere by the tropospheric circulation, so that it appears likely that this mechanism plays a role in the overall equator to pole heat transfer by the circulation of the troposphere and lower stratosphere. Further studies are needed to provide a more detailed quantitative assessment of its mean contribution and variability.

Changes in aerosol content, ozone and water vapour in the stratosphere may affect surface heating through changing the penetration of solar radiation to the surface, altering the balance of the downward and upward fluxes of atmospheric radiation and by changing the heat balance, initiating dynamical interactions. Attempts have been made in the last few years to estimate the possible magnitude of these effects in modelling studies of differing sophistication, mostly using one- or two-dimensional convective–radiative equilibrium types, but at least one in which short integrations were made with a three-dimensional model. However, none of the present models can so far be regarded as satisfactory in their treatment of the radiative phenomena or their simulation of the atmospheric dynamics, clouds and the hydrological cycle, transfer by the oceans, etc. (see, for example, Manabe & Wetherald 1967, C.I.A.P. 1975 and C.O.M.E.S.A. 1975 for details).

For aerosols, which are of particular importance in scattering and absorbing solar radiation, there is evidence that volcanic eruptions can on occasion greatly increase the stratospheric burden. For example, it has been estimated that as a result of the 1963 Mt. Agung eruption the global mass of stratospheric aerosol increased from about 1×10^8 to 3×10^9 kg (Wilson 1970). This may have resulted in an aerosol heating rate near the equatorial tropopause of about 0.1 K/day, leading to a rise of up to 6 K in the temperature there (Newell 1970), but seemed to have had little effect on surface temperatures (Lamb 1970). In general, there appears to be a likelihood of a small decrease in the latter 1 or 2 years after a volcanic eruption (see Mass & Schneider 1977). Most of the model calculations also indicate that large increases of stratospheric aerosol amounts will cause temperature increases of perhaps around 10 K in the lower stratosphere and, probably, decreases of the order of a fraction of one degree Kelvin in mean surface temperatures. The latter would probably result also in some decrease of other forms of atmospheric energy and a decrease of rainfall. However, since it is known that in the last century the mean surface temperature has shown long term variations, whose cause has not been fully explained, of the same general magnitudes as those calculated for large increases in stratospheric aerosol amounts, it will be difficult to validate these model results completely. Considerably larger effects may occur at individual locations and further work is required to refine and expand the current results.

Several model calculations of the effect on surface temperatures of changing stratospheric ozone amounts (either by the effect of pollutants or by natural phenomena) have also been reported (see, for example, Manabe & Strickler 1964; Manabe & Wetherald 1967; C.I.A.P. 1975; C.O.M.E.S.A. 1975). It appears that a 25% reduction might lead to about a 10 K decrease in the low latitude upper stratosphere, and a fraction of a degree Kelvin decrease in mean global temperature at the surface. It seems likely that this decrease in temperature would be enhanced somewhat if the vertical ozone profile were changed to remove more ozone from the lower stratosphere, i.e. from a typical polar ozone profile towards a typical tropical profile.

The model calculations have indicated also that increasing the water vapour in the stratosphere leads to increased infrared cooling and lower temperatures there and to very small increases in the mean surface temperatures. Estimates vary from rises in the latter of less than 0.1 K for a 10% water vapour increase to 2 K for a fivefold increase, with the stratospheric temperature changes being about an order greater than those at the surface. If high cirrus clouds were formed, their effect would also be to cause temperature increases at the surface, perhaps about 1 K for a 25% increase.

It seems unlikely that the large changes of stratospheric composition assumed in the model calculations could arise as the result of variability in solar radiation or other extra-terrestrial effects. Willis (1978) has shown that extra-terrestrial processes will only produce extremely small changes in water vapour in the stratosphere. Keating (1978) estimated a 3% variation of ozone over the last solar cycle, which could be responsible for a 0.05 K change in mean surface temperature when using the C.I.A.P. (1975) results but a negligible amount if the C.O.M.E.S.A. (1975) results are correct.

Dynamical interactions

A striking feature of the stratospheric circulation is its apparent simplicity as compared with that of the troposphere. In summer, the stratospheric flow is almost completely zonal with easterly winds around a warm anticyclone approximately concentric with the pole. The winter mean circulation of the extratropical Northern Hemisphere stratosphere above about the 100 m-bar level consists mainly of planetary waves of zonal wavenumbers 1 and 2 superimposed on a zonal westerly flow around a cold polar vortex, while in the Southern Hemisphere stratosphere the situation is similar except that the wave perturbations are far less pronounced. These planetary waves are quasi-stationary in phase but fluctuate in amplitude with time-scales of the order of 2 weeks. Observational and theoretical studies have shown that their associated horizontal momentum and heat fluxes play an essential role in governing the mean meridional circulation in the winter stratosphere and in inducing sometimes dramatic changes of the zonal flow, e.g. during stratospheric sudden warmings. Additionally, planetary wave motions have been found to be important in transporting various chemical species, e.g. ozone within the stratosphere. These in turn can modify the field of net radiative heating and thus the dynamical motions themselves. The predominance of planetary waves in the winter stratosphere, as compared to motions of a smaller scale, their origin and relative absence from the stratosphere in summer pose fundamental questions for dynamical theory.

There is now a considerable body of evidence (see Holton (1975) for a review) that such wave motions in the stratosphere are not excited by internal hydrodynamic instabilities of a baroclinic or barotropic nature but are produced and maintained by vertical propagation of planetary-wave disturbances generated in the troposphere. Orographic forcing and differential heating are believed to be the principal source mechanisms with the resulting wave motions being of a quasi-stationary nature. In addition, there may be some long wave generation as a result of nonlinear interactions between waves on a smaller spatial scale produced by baroclinic instability in the troposphere.

The predominance of wave motions of a planetary scale in the winter stratosphere and their relative absence from the summer stratosphere were first accounted for theoretically by Charney & Drazin (1961), their work being extended by Dickinson (1968, 1969), Matsuno (1970), Simmons (1974) and others. These studies have shown the importance of the zonal wind distribution in governing the vertical propagation of quasi-stationary waves from the troposphere. The following features have emerged: Quasi-stationary waves on a planetary scale may propagate significantly into the stratosphere during periods when the zonal wind is westerly (hence inhibited propagation in summer) and they do so such that the maximum perturbation amplitude aligns itself approximately along the axis of strongest mean westerly wind component (the so-called waveguide effect). The degree of vertical penetration of the waves depends sensitively on their zonal wavelength in that shorter waves propagate less readily, which accounts for the absence of the smaller synoptic-scale features from middle and upper stratospheric charts. A region marking the transition between westerly and easterly winds (which forms a 'critical line') acts to absorb quasi-stationary waves and to produce zonal flow accelerations in that area. Such a region exists in the equatorial stratosphere separating the westerlies of the winter hemisphere from the easterlies of the summer hemisphere. Dickinson (1968) and Matsuno (1970) have demonstrated that the resulting energy sink for the wave motions causes the upward energy flux from the troposphere to be directed mostly equatorward. Relations of the form derived by Eliassen & Palm (1960) show that, for quasi-geostrophic waves in a steady state when diabatic effects are unimportant, this implies eddy fluxes of momentum and heat directed towards the poles when the zonal winds are westerly and increasing with height, as in the winter stratosphere.

While these fluxes are of vital importance in determining the mean state of the stratospheric circulation in winter, linear perturbation analyses carried out by Charney & Drazin (1961), Boyd (1976), and Andrews & McIntyre (1976, 1978) have established that for the planetary wave disturbances to change the mean flow, wave transience, damping or interaction with a critical line are required. It has been found that transient planetary-wave activity in the troposphere is related to one of the most spectacular manifestations of troposphere–stratosphere interaction, namely the stratospheric sudden warming phenomenon for which both numerical simulations and observational studies of the real atmosphere have shown that the breakdown of the stratospheric polar-night jet, which characterizes 'major events', occurs in conjunction with enhanced upward fluxes of eddy energy from the troposphere. However, it is perhaps significant that no correspondence has yet been found between the magnitude of the energy fluxes and the intensity of the warming.

In view of the above, it is reasonable to suppose that interhemispheric differences in stratospheric circulation characteristics are related to different conditions between hemispheres in the troposphere. This is supported by detailed studies involving general circulation models of the troposphere and stratosphere (see, for example, Manabe & Mahlman 1976). In particular, numerical experiments by Kasahara *et al.* (1973) and Manabe & Terpstra (1974) have suggested that the large-scale mountain ranges in the Northern Hemisphere, such as the Tibetan Plateau and the Rocky Mountains, are chiefly responsible for maintaining the quasi-stationary waves in the troposphere. The fact that major midwinter warmings on the scale of those observed in the Northern Hemisphere have not been found in the Southern Hemisphere is therefore also likely to be ultimately related to interhemispheric differences in tropospheric forcing.

Some data relevant to the relation between the seasonal evolution of tropospheric planetary waves and those of the stratosphere have recently been obtained in the Meteorological Office from an analysis of the results of an annual integration of a three-dimensional numerical model simulating the seasonal behaviour of the troposphere and stratosphere. This model, which was based on that of Corby *et al.* (1972) and extended in the vertical by Newson to contain thirteen levels, eight of which were in the stratosphere, has been briefly described in C.O.M.E.S.A. (1975) and a fuller description of the model and its simulated climatology is in preparation.

In the examination of the seasonal variation and latitudinal distribution of stationary (monthly mean) planetary-wave amplitudes in the model atmosphere, the 5 mbar level was chosen to be representative of the middle stratosphere with the 300 and 100 mbar levels taken to illustrate variations in the upper troposphere and lower stratosphere. Figure 3 (*a*) shows that stationary wave 1, found to be the dominant mode in the stratosphere, exhibited a pronounced seasonal cycle in amplitude at the 5 mbar level with maximum amplitude in winter and minimum in summer. This is an outstanding feature of the seasonal variability of the real stratosphere and is consistent with the theoretical predictions of Charney & Drazin (1961) regarding the propagation characteristics expected for planetary waves in westerly and easterly wind regimes. However, figure 3 (*c*) shows that, in the model simulation, there was a further important seasonal variation likely to influence stratospheric wave amplitudes and thus the seasonal evolution of its circulation, namely the seasonal cycle of stationary wave 1 amplitude in the upper troposphere. This cycle was approximately in phase with that of the stratosphere. In broad agreement with observations (Labitzke & Van Loon 1972), the seasonal variation of wave 1 amplitudes in the Southern Hemisphere stratosphere was less pronounced, with amplitudes considerably smaller than those prevailing during the Northern Hemisphere winter. Figure 3 (*c*) suggests that these differences may be related to the fact that large stationary wave 1 amplitudes were sustained over a considerably longer period in the upper troposphere in the Northern Hemisphere than in the Southern Hemisphere.

A study of the corresponding fields in the model for stationary wave 2 shows that this wave also exhibited a seasonal cycle in amplitude at 5 mbar in approximate phase with that of the upper troposphere (figure 4). In the Northern Hemisphere, peak amplitudes were considerably

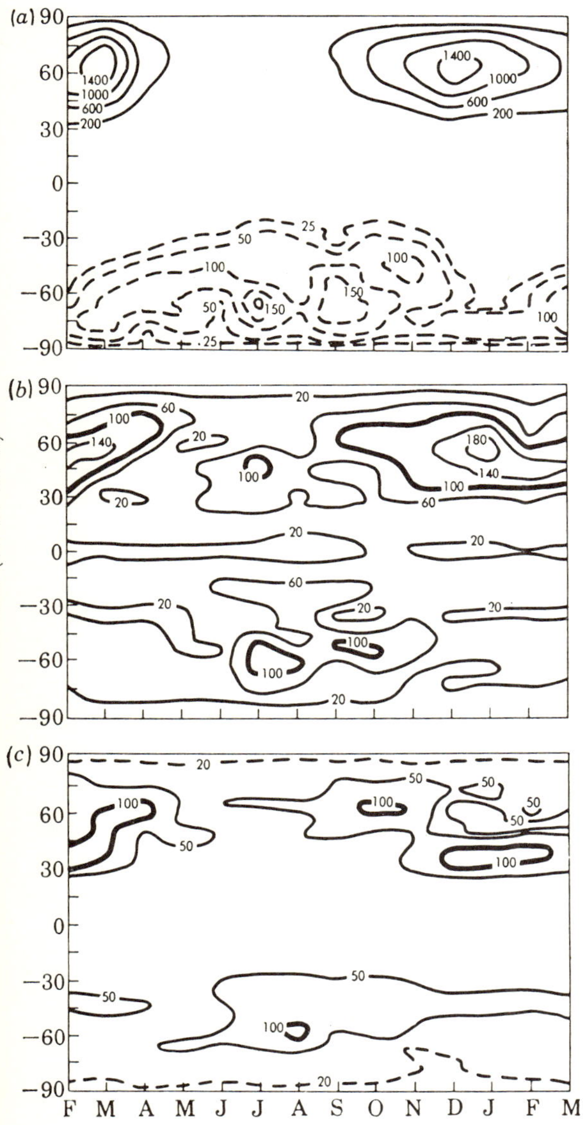

FIGURE 3. Stationary wave 1 amplitude of geopotential-height (m) at (*a*) 5 mbar, (*b*) 100 mbar, (*c*) 300 mbar.

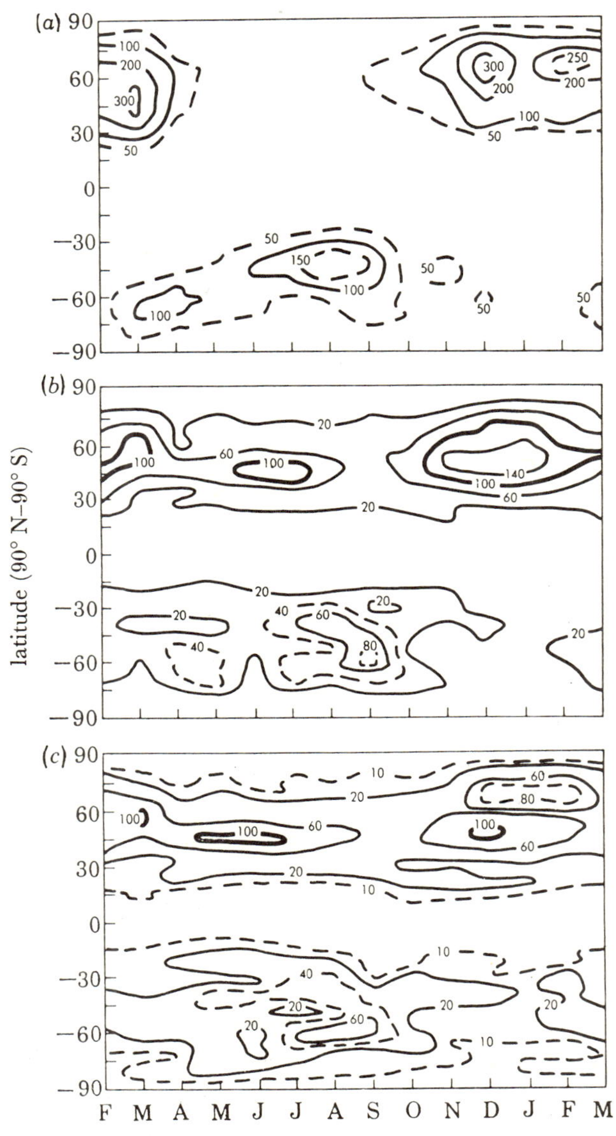

FIGURE 4. Stationary wave 2 amplitude of geopotential-height (m) at (*a*) 5 mbar, (*b*) 100 mbar, (*c*) 300 mbar.

smaller than those of wave 1 but the two waves were of roughly comparable amplitude in the Southern Hemisphere. Stationary waves of higher wavenumber than 2 were of progressively smaller amplitude in the middle stratosphere. The upward energy fluxes associated with stationary wave 1 and 2 in the model at 100 mbar are shown in figure 5, further illustrating the pronounced differences between hemispheres. Since linear wave calculations (see, for example, Matsuno 1970) indicate the preferred vertical propagation of wave 1 in the polar westerly jet, these results provide further evidence that interhemispheric differences in the stratospheric circulation, particularly in its wave 1 aspects, are directly related to those of the troposphere.

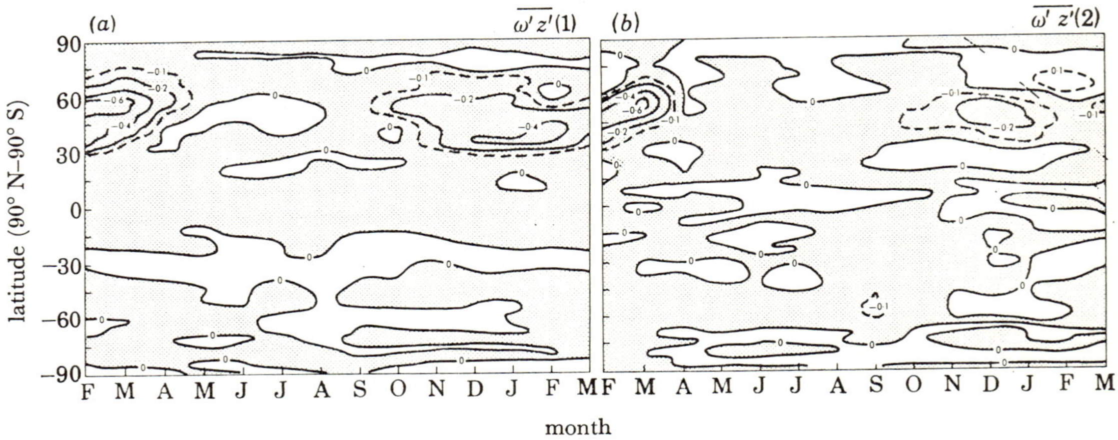

FIGURE 5. Vertical eddy flux of mechanical energy (W m^{-2}) at 100 mbar as determined by the 'pressure-interaction term' (*a*) associated with zonal wave 1, $\overline{\omega' z'}$ (1), (*b*) associated with zonal wave 2, $\overline{\omega' z'}$ (2). Stippling indicates upward flux, ω is the vertical velocity in pressure coordinate terms, z is geopotential-height, the overbar denotes a zonal average on a constant pressure surface and a prime the local deviation from that average.

Transient wave components (here taken to include perturbations with amplitude or phase differing from the monthly mean value) were of much smaller amplitude than their stationary counterparts in the Northern Hemisphere but were of comparable amplitude in the Southern Hemisphere and also exhibited a pronounced seasonal cycle in the middle stratosphere in approximate phase with that of the upper troposphere (figures 6 and 7).

In addition to the interaction between tropospheric and stratospheric circulation systems which apparently exists on a seasonal timescale, Hirota & Sato (1969) have made observations that suggest a relation between quasi-periodic oscillations in the strength of planetary waves in the winter stratosphere and the day-to-day changes in the strength of the mean zonal wind in the troposphere, the so-called index cycle. Quiroz (1969) showed oscillations of this index at 700 mbar that were in phase with the oscillations in the upward energy flux through 100 mbar, preceding thermal oscillations in the stratosphere. In a recent study of a major stratospheric warming during the 1976–77 winter in the Northern Hemisphere, O'Neill & Taylor (1979) investigated in some detail the nature of dynamical interactions between troposphere and stratosphere. They demonstrated the continuity of the planetary wave amplitudes from upper troposphere to middle stratosphere and concluded that the circulation reversal, which occurred at all levels up to at least 10 mbar, could be accounted for mainly by planetary wave 1 developments in the troposphere. In addition, they argued that Matsuno's (1971) mechanism for sudden warmings did not provide an adequate dynamical description of the event and that the appearance of eddy momentum fluxes towards the equator in the troposphere and stratosphere, with the consequent export of westerly momentum from high latitudes, was central to its dynamical evolution. They noted a number of other occasions during which areas of eddy momentum fluxes towards the equator in the troposphere appeared to spread upwards into the stratosphere during periods of enhanced planetary wave activity in the troposphere. Because quasi-geostrophic theory provides an adequate description of motions on this scale, it may readily be shown that this process requires that the planetary waves change their meridional structure to one in which the trough and ridge lines lie in a SE–NW direction. O'Neill & Taylor proposed that such a reorientation and amplification of planetary waves in the

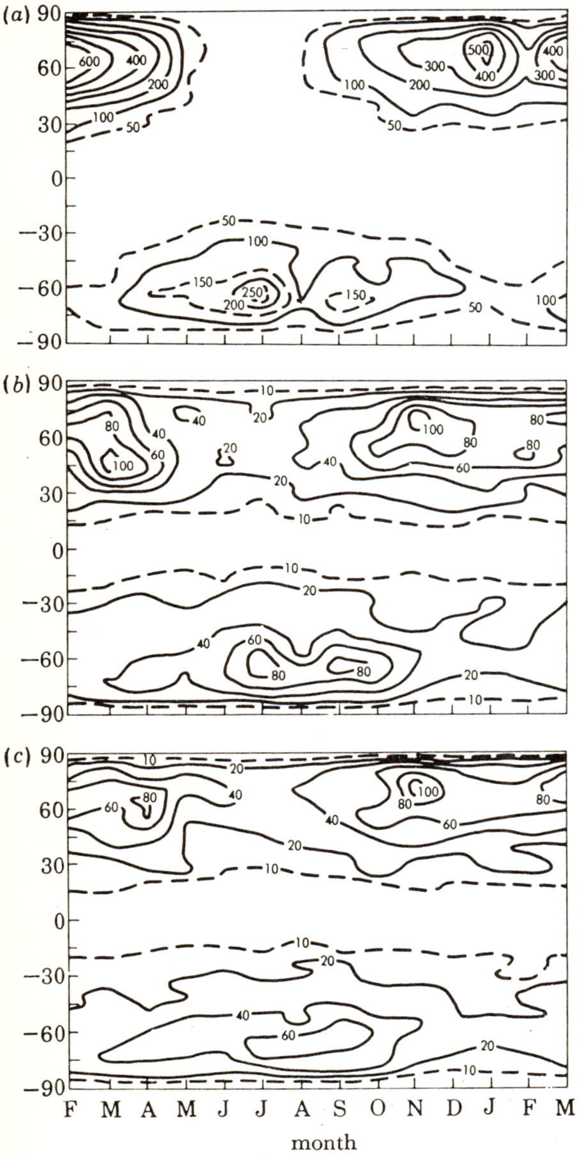

FIGURE 6. Transient wave 1 amplitude of geopotential-height (m) at (*a*) 5 mbar, (*b*) 100 mbar, (*c*) 300 mbar.

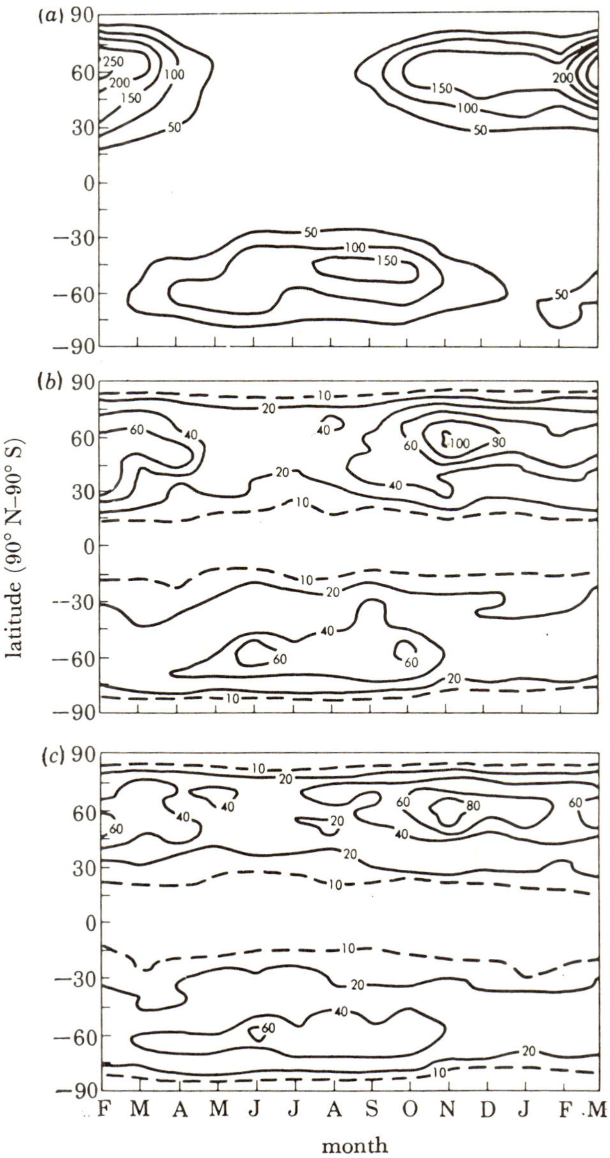

FIGURE. 7 Transient wave 2 amplitude of geopotential-height (m) at (*a*) 5 mbar, (*b*) 100 mbar, (*c*) 300 mbar.

troposphere, which they found to be associated with the appearance and subsequent evolution of a blocking-type circulation pattern, could have important consequences for the stratospheric circulation. Furthermore, both in this study and in the analysis of the above model results, it was found that major planetary wave developments in the middle stratosphere could be associated with enhanced planetary wave activity at high latitudes in the upper troposphere, low-latitude perturbations being less important in this respect. This is not unexpected in view of the location at high latitudes of the strong westerly wind waveguide formed by the polar night jet in the stratosphere. The particular tropospheric synoptic patterns associated with such developments were found by O'Neill & Taylor (1979) and to some extent in the above model simulations to include at some stage intensification of a ridge in the mid-Atlantic sector which, in conjunction with an extensive low-pressure area over east Asia, gave rise to a pronounced wave 1 component in the circulation. However, the complexity of the interactions is such that no unique relationship has yet been established between the intensity and duration of a given planetary-wave disturbance and the degree to which the zonal flow in the stratosphere is disrupted.

In the above discussion, the connection between planetary-wave motions in the troposphere and stratosphere at extratropical latitudes has been emphasized because such motions account for most of the eddy kinetic energy in the stratosphere and play an important part in determining its overall circulation. In addition, upward-propagating Kelvin and mixed Rossby gravity waves from the troposphere interacting with the wind systems of the stratosphere are believed to account for the quasi-biennial oscillation of zonal wind in the equatorial stratosphere. The former waves are also thought to act as a source of westerly momentum for the semi-annual cycle of zonal winds in the equatorial stratosphere, with planetary waves propagating upward and equatorward in the winter hemisphere providing the required oscillating source of easterly momentum. Mechanistic models which simulate the proposed interactions leading to the quasi-biennial wind oscillation have been formulated by Lindzen & Holton (1968), Holton & Lindzen (1972) and Plumb (1977).

Although a great deal of progress has been made in identifying dynamical processes by which changes in the tropospheric circulation can affect the stratosphere, there is comparatively little known of the way effects of stratospheric perturbations may be communicated downwards mechanically and perhaps cause significant alterations to the tropospheric flow. One possibly important mechanism by which this might occur is through the effect of stratospheric changes on the vertical propagation of planetary waves generated in the troposphere (Hines 1973; Bates 1977). These waves transfer their energy upwards and might be reflected at higher altitudes to interfere constructively or destructively with the initial systems that led to their generation. Hines argued that in the absence of substantial dissipation the energy density of such a reflected disturbance would be appropriate to the tropospheric level of its origin and would be likely to have some phase coherence with the generating system. It might be expected that this mechanism would be more efficient in producing a significant change in the troposphere than would be the case for an incoherent mechanism. Bates (1977), in an analytical study using a simplified model, found that the structure of its planetary-scale waves and hence their horizontal heat fluxes in the troposphere were very sensitive to his choice of stratospheric wind profile and static stability. Accordingly, he proposed that upper-atmospheric changes that might alter the planetary-wave structure could also lead to changes in the tropospheric circulation and hence an overall climatic effect. However, it was necessary for the model to contain many simplifying assumptions regarding zonal wind distribution and wave structure in order to make the problem mathematically tractable and thus more comprehensive studies are necessary before his conclusions can be generally accepted.

Discussion

Although the nature of the tropopause and its distribution in space and time are reasonably well known, there are considerable uncertainties about the precise nature of the processes by which they are determined and about the transport mechanisms in this region of the atmosphere. Until these are better understood, it will be difficult to construct the detailed budgets of the mass and constituent transports between troposphere and stratosphere necessary to gain a quantitative knowledge of the sources and sinks of the chemical species that determine the composition of the stratosphere in its 'natural' or 'perturbed' (by pollutants) state. Further studies will have to include radiative, chemical and dynamical processes on all scales.

Questions of the likely consequences in the troposphere of perturbing stratospheric composition include both effects on the penetration of solar radiation at different wavelengths and the disturbance of heat balances and hence the general circulation. These can only be answered by gaining a reasonably complete knowledge of all the physical processes that determine the natural atmosphere's composition, temperature structure and general circulation and then constructing adequate models for diagnostic and predictive purposes. It must be realized that this goal is at present far from being attained and that a great deal of work will be needed before it can be said to be reasonably approached. Work along these lines will contribute to an improved appreciation of possible solar–terrestrial relations, which can only be properly studied when the appropriate mechanisms have been quantified.

Both observational and theoretical studies have led to considerable progress in recent years in understanding the dynamical mechanisms involved in a number of stratospheric phenomena. Although it is now established that planetary-wave developments in the troposphere are important in governing the stratospheric circulation, the situation as regards possible dynamical effects of stratospheric changes on the troposphere has yet to be clarified. In the case of the quasi-biennial oscillation, it is now known that this can be observed in a number of tropospheric parameters at middle and high latitudes (Ebdon 1975; A. F. Jenkinson 1978, private communication). This raises the question of the extent to which the quasi-biennial oscillation may be intrinsic to the stratosphere and communicated to the troposphere, for example by the influence of the changing zonal wind distribution in the equatorial stratosphere on the overall planetary-wave structure.

It is hoped that in view of the theoretical and practical importance of the subjects discussed in this paper, their further investigation will form an important part of the Middle Atmosphere Program.

References (Murgatroyd & O'Neill)

Andrews, D. G. & McIntyre, M. E. 1976 *J. atmos. Sci.* **33**, 2031–2048.
Andrews, D. G. & McIntyre, M. E. 1978 *J. Fluid Mech.* **89**, 609–646.
Angell, J. K. & Korshover, J. 1974 *Mon. Weath. Rev.* **102**, 29–34.
Bates, J. R. 1977 *Q. Jl R. met. Soc.* **103**, 397–430.
Boyd, J. 1976 *J. atmos. Sci.* **33**, 2285–2291.
Brewer, A. W. 1949 *Q. Jl R. met. Soc.* **75**, 351–363.
Charney, J. G. & Drazin, P. G. 1961 *J. geophys. Res.* **66**, 83–109.
C.I.A.P. Monograph 4 1975 *Final Report* DOT-TST-75-54. U.S. Department of Transportation, Washington, D.C. 20590.

C.O.M.E.S.A. 1975 *Report of the Committee on Meteorological Effects of Stratospheric Aircraft*, 1972–1975. Bracknell: Meteorological Office.
Corby, G. A., Gilchrist, A. & Newson, R. L. 1972 *Q. Jl R. met. Soc.* **98**, 809–832.
Court, A. 1942 *Bull. Am. met. Soc.* **23**, 220–238.
Danielsen, E. F. 1959 *Arch. Met. Geophys. Bioklim.* A **11**, 293–332.
Dickinson, R. E. 1968 *J. atmos. Sci.* **25**, 984–1002.
Dickinson, R. E. 1969 *J. geophys. Res.* **74**, 929–938.
Dopplick, T. G. 1972 *J. atmos. Sci.* **29**, 1278–1294.
Ebdon, R. 1975 *Met. Mag.* **104**, 282–297. London: Meteorological Office.
Eliassen, A. & Palm, E. 1960 *Geophys. Publikasjoner* **22**, no. 3, 1–23.
Fritz, S. & Soules, S. D. 1972 *Mon. Weath. Rev.* **100**, 582–589.
Goody, R. M. 1949 *Proc. R. Soc. Lond.* A **197**, 487–505.
Hering, W. S., Touart, C. N. & Borden, T. R. 1967 *J. atmos. Sci.* **24**, 402–413.
Hines, C. O. 1973 *J. atmos. Sci.* **31**, 589–591.
Hirota, I. & Sato, Y. 1969 *J. met. Soc. Japan* **47**, 390–402.
Hitschfeld, W. & Houghton, J. T. 1961 *Q. Jl R. met. Soc.* **87**, 562–577.
Holton, J. R. 1975 *The dynamic meteorology of the stratosphere and mesosphere.* Meteorological Monographs, vol. 15, no. 37. Boston, Mass.: American Meteorological Society.
Holton, J. R. & Lindzen, R. S. 1972 *J. atmos. Sci.* **29**, 1076–1080.
Houghton, J. T. 1963 *Q. Jl R. met Soc.* **89**, 319–331.
Kasahara, A., Sasamori, T. & Washington, W. M. 1973 *J. atmos. Sci.* **30**, 1229–1251.
Keating, G. M. 1978 *Nature, Lond.* **274**, 873–874.
Krey, P. W., Schonberg, M. & Tonkel, L. 1974 Rep. HASL-281, pp. I-130 – I-142. New York: U.S. Atomic Energy Commission.
Labitzke, K. & Van Loon, H. 1972 *Met. Monogr.* **13**, no. 35, 113–138.
Lamb, H. H. 1970 *Phil. Trans. R. Soc. Lond.* A **266**, 425–433.
Leovy, C. 1964 *J. atmos. Sci.* **21**, 238–248.
Lindzen, R. S. & Holton, J. R. 1968 *J. atmos. Sci.* **25**, 1095–1107.
List, R. J. & Telegadas, K. 1969 *J. atmos. Sci.* **26**, 1128–1136.
London, J., Bojkov, R. D., Oetmans, S. & Kelley, J. I. 1978 *N.C.A.R. Technical Note*/113+STR. Boulder, Colorado.
Manabe, S. & Hunt, B. G. 1968 *Mon. Weath. Rev.* **96**, 447–502.
Manabe, S. & Mahlman, J. D. 1976 *J. atmos. Sci.* **33**, 2185–2217.
Manabe, S. & Strickler, R. F. 1964 *J. atmos. Sci.* **21**, 361–385.
Manabe, S. & Terpstra, T. B. 1974 *J. atmos. Sci.* **31**, 3–42.
Manabe, S. & Wetherald, R. T. 1967 *J. atmos. Sci.* **24**, 241–259.
Mass, C. & Schneider, S. H. 1977 *J. atmos. Sci.* **34**, 1995–2004.
Mastenbrook, H. J. 1971 *J. atmos. Sci.* **28**, 1495–1501.
Matsuno, T. 1970 *J. atmos. Sci.* **27**, 871–883.
Matsuno, T. 1971 *J. atmos. Sci.* **28**, 1479–1494.
Murgatroyd, R. J. 1969 *Global circulation of the atmosphere*, (ed. G. A. Corby), pp. 154–195. London: Royal Meteorological Society.
Newell, R. E. 1970 *Nature, Lond.* **227**, 697–699.
Newell, R. E., Kidson, J. W. & Vincent, D. G. 1969 *Nature, Lond.* **222**, 76.
Newell, R. E. & Wu, M.-F. 1978 *Q. Jl R. met. Soc.* **104**, 999–1002.
O'Neill, A. & Taylor, B. F. 1979 *Q. Jl R. met. Soc.* **105**, 71–92.
Plumb, R. A. 1977 *J. atmos. Sci.* **34**, 1847–1858.
Quiroz, R. S. 1969 *Mon. Weath. Rev.* **97**, 541–552.
Ramanathan, V. 1977 *J. atmos. Sci.* **34**, 439–447.
Reed, R. J. & Vlcek, C. L. 1969 *J. atmos. Sci.* **26**, 163–167.
Riehl, H. & Malkus, J. S. 1958 *Geophysica* **6**, 503–538.
Reiter, E. R. 1975 *Rev. Geophys. Space Phys.* **13**, 459–474.
Reiter, E. R., Bauer, E., Grobecker, A. & Coroniti, S. 1975 *C.I.A.P. Monograph* 1. U.S. Department of Transportation, Washington, D.C.
Simmons, A. J. 1974 *Q. Jl R. met. Soc.* **100**, 76–108.
Staley, D. O. 1957 *Arch. Met. Geophys. Bioklim.* A **10**, 1–19.
U.S. Department of Commerce 1961 *Monthly mean cross-sections pole to pole along meridian* 75° W *for the I.G.Y. period.*
Willis, D. M. 1976 *J. atmos. terr. Phys.* **38**, 685–698.
Willis, D. M. 1978 *J. atmos. terr. Phys.* **40**, 513–528.
Wilson, C. L. 1970 *Report of the study of critical environment problems* (S.C.E.P.).
W.M.O. 1966 *International Meteorological Vocabulary.* W.M.O./O.M.M./B.M.O. No. 182, TP.91.

Phil. Trans. R. Soc. Lond. A **296**, 103–127 (1980) [103]
Printed in Great Britain

Dynamical models of the middle atmosphere for tracer studies

By R. S. Harwood
Department of Meteorology, The University, James Clerk Maxwell Building, King's Buildings, Mayfield Road, Edinburgh EH9 3JZ, U.K.

The distribution of ozone and other tracers depends on chemical factors and on the atmospheric circulations, which themselves depend, through heat sources and sinks, on the ozone amount. Examples of the sensitivity of ozone to variations in the dynamics are given from a two dimensional model. The total ozone depends greatly on the heating rate in the equatorial lower stratosphere; changes of tens of milli atmosphere-centimetres† arise from alteration of only a few tens of kelvins per day in heating. The effects of variations in momentum flux and mixing-surface slope are also discussed.

The current ability to model dynamics for tracer studies is reviewed. Only three dimensional models provide a consistent formulation; they simulate many observed atmospheric phenomena broadly but not in detail. However, the amount of computing needed precludes all but the simplest treatment of chemistry. Two dimensional models offer a valuable compromise. Although these are rather empirical and not adequately based on sound physical principles, good agreement can be found in practice between observed and modelled ozone, winds and temperatures.

The calculation of model-dependent meridional circulations in two dimensional models is shown to be important, for which a parametrization of the momentum flux is an essential prerequisite. Some observations are given that suggest that a parametrization based on potential vorticity fluxes may be possible. This would be consistent with the usual formulation of tracer transport.

1. Introduction

The spatial abundance of ozone has been known since the very earliest measurements to depend not only on photochemistry, but also on redistribution by the winds (Dobson *et al.* 1928).This became apparent as the latitudinal distribution of total ozone was determined (figure 1), because, contrary to expectations based on photochemistry alone, the minimum column densities are in low latitudes while the maxima occur in high latitudes.

The need for quantiative models of these motion systems has assumed some urgency if potential threats to the ozone layer from manmade pollution are to be properly evaluated. Calculations of the photochemical equilibrium state give equatorial column densities slightly over twice those observed, from which it can be loosely inferred that changes of, say, 10% in the velocities can produce changes in the ozone column densities of between 5 and 10%. This is comparable to the predicted decreases in ozone due to continued release of chlorofluoromethanes at the 1973 rate calculated in simple models (N.A.S. 1976). Accordingly, it is important to understand the extent to which the motion systems themselves depend on the ozone amount (see, for example, Pyle 1978). We are faced with a complicated feedback loop, for the motions are driven to a large extent by the stratospheric heat sources and sinks, the source (the absorption of solar radiation at the stratopause) depending strongly on the ozone amount. The sink (the emission of long wave radiation) is also weakly ozone-dependent. Moreover, stratospheric motions are partially forced by planetary waves propagating from the troposphere and

† 1 milli atmosphere-centimetre (hereafter abbreviated 1m atm-cm) = 10^5 m at s.t.p.

the details of this motion, too, depend on the radiative damping which is intricately connected with the ozone chemistry (Lindzen & Goody 1965; Blake & Lindzen 1973).

This paper reviews the progress that has been made in modelling the dynamical side of these processes, dealing exclusively with models which have been developed primarily for studies of ozone and other tracers. It makes no attempt to cover the extensive literature on analytic (linearized) studies of planetary propagation (see, for example, Holton 1975) and wave–mean flow interaction (see, for example, McIntyre & Andrews 1976).

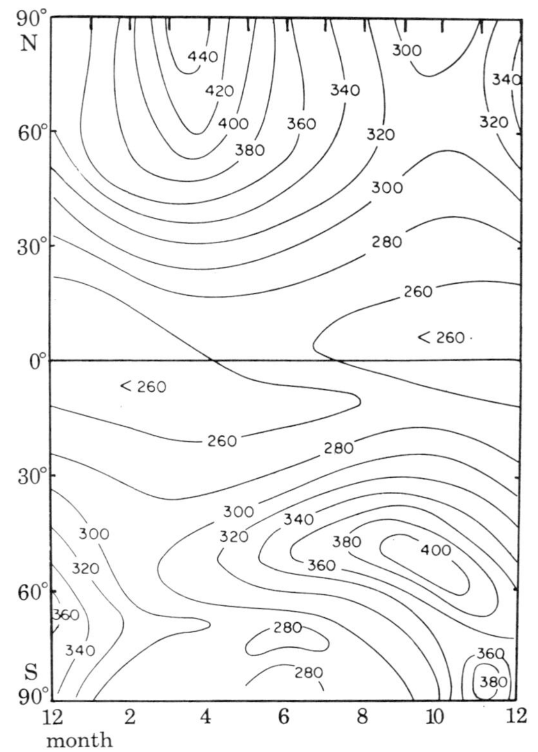

FIGURE 1. The observed zonally averaged distribution of total ozone (m atm-cm) (Dütsch 1971). The low latitude minima and the hemispheric differences immediately reveal the importance of the dynamics of transport processes.

2. THREE DIMENSIONAL MODELS

(a) *Highly truncated spectral models*

The atmosphere is a three dimensional system so it is natural to assume that anything less than a three dimensional model is unlikely to be satisfactory. In practice, however, finite resources mean that some feedback mechanisms need to be curtailed and the price of a three dimensional treatment will be a gross simplification of the physical processes at work. It may not be possible to achieve a detailed treatment of the chemistry or the run may not extend long enough for a quasi-steady state to be reached.

Because the stratospheric circulation is dominated by the largest scale waves, spectral methods with severely truncated spectra appear to offer an economical representation. Instead of being represented at each horizontal level by values at regularly spaced intervals, as in the more conventional grid point methods, the fields are represented by the leading coefficients in their spherical harmonic decomposition. A very large fraction of the variance of stratospheric motion is accounted for by only a few coefficients, so that large savings in computation appear possible. The situation is not straightforward, however, because the waves are forced in the troposphere. The tropospheric forcing mechanism is in part an interaction between the zonal mean wind and the relevant Fourier component of the orography and heating pattern and in part is a non-linear interaction between higher wavenumbers. Both mechanisms involve the higher wave-numbers; the latter obviously, and the former because the energy of the zonal mean wind is maintained by transfer from the baroclinic eddies, which are predominantly of wavenumbers around 6. Consequently, the severe truncation may not be possible for the tropospheric forcing.

A further saving in computer time can be made by using the quasi-geostrophic equations rather than the primitive equations. The main predictive equation here is the vorticity equation in which the vorticities and winds are obtained from the height fields by using what is essentially a geostrophic approximation. This prevents gravity waves from developing and permits considerably longer time-steps than can be taken in primitive equation models. Lorenz (1960) has shown how to develop energetically consistent versions of the quasi-geostrophic equations and his formulation provides the starting point of most spectral models.

One of the first attempts at a spectral model using spherical harmonics was that of Clark (1970), whose main aim was to investigate the ozone budget and energetics of the wintertime circulation (see also Peng 1965). To minimize computing time he employed a rather severe truncation, retaining only the zonal mean field and zonal wavenumbers 1, 2, 3 and 6. Moreover he retained only those harmonics of wind, temperature and ozone which are symmetric across the equator, and with only the lowest order modes in N–S direction. This gave about 20 degrees of freedom for each variable in the horizontal and he employed a six-layer model in the vertical with winds held at 750, 350, 150, 62.5, 15 and 2.5 mbar.† His ozone photochemistry was a linearized form of the Chapman scheme with rate coefficients somewhat arbitrarily adjusted to give moderate agreement with the observed distributions. The model was run for a total of 250 days with a constant (winter) solar declination starting from radiative-photochemical equilibrium.

Notwithstanding its obvious limitations, the overall behaviour of this model was sufficiently encouraging for it to be extended by Cunnold *et al.* (1975) to produce probably the most comprehensive three dimensional joint dynamical, chemical model of the stratosphere to date, their aim being the study of the possible effects on the ozone layer of supersonic airliners.

They removed the requirement of symmetry across the equator and also included *all* zonal wavenumbers less than or equal to 6. Hence the treatment of the troposphere is much improved, although the representation is still very crude compared with that of a typical tropospheric general circulation model. The global mean temperature is specified in the model and only departures from it are calculated. This greatly reduces the computational effort required to deal with convective overturning, latent heat release and surface energy balance in the troposphere. Symmetry between hemispheres was assumed for the distribution of land form and sea-surface, so that the model is unable to reproduce the observed interhemispheric differences in ozone distribution.

Their treatment of ozone photochemistry is also a great improvement upon that of Clark, mainly by not linearizing the photochemistry and by the inclusion of the catalytic destruction of

† 1 mbar = 10^2 Pa.

ozone by the oxides of nitrogen. However, ozone is the only species calculated explicitly in the model, the distribution of NO_x being specified as tables to the model, at first independently of latitude (Cunnold *et al.* 1975), then later (Alyea *et al.* 1975) with the use of distributions obtained from the steady-state two dimensional model of Hesstvedt (1974*a*) (see § 3 (*a*) below).

Two dimensional distributions of NO_x appropriate to both the natural and polluted stratosphere were employed so as to assess the likely effects of supersonic passenger aircraft on the ozone layer. The ozone behaviour in the 'natural' run showed excellent realism though the maximum values were about 10 % higher than those in figure 1, while the minimum values were too low by about 5 %.

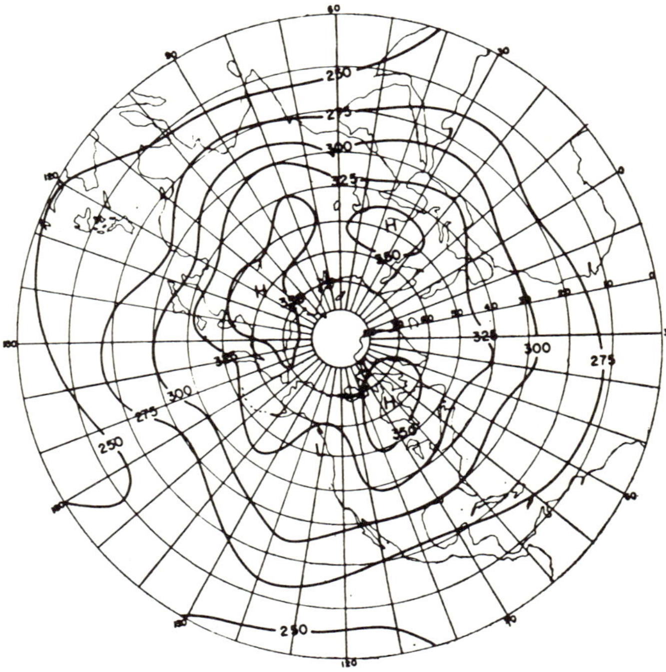

FIGURE 2. Horizontal distribution of total ozone (m atm-cm) for winter calculated in a quasi-geostrophic spectral model (Cunnold *et al.* 1975).

Figure 2 shows the distribution of total ozone for winter, taken from Cunnold *et al.* (1975). Although the polar values in this particular run are considerably lower than observed, owing largely to the latitudinally constant NO_x profile, the displacement of the largest values towards Siberia is in broad agreement with the observations compiled by Wu (1973). The major waves thus appear to be well represented although some difficulties with excessive wavenumber 6 amplitudes in the stratosphere are reported.

The wind and temperature field predicted by the model have a fair measure of agreement with observations. In particular, the strength of the jets in the upper stratosphere are quite similar to the observations of Murgatroyd (1969) and Newell *et al.* (1969). However, the modelled temperature gradient does not reverse above the stratosphere so the wind maxima are not attained within the modelled region. This is probably because vertical velocities are inhibited by the upper boundary conditions so that the adiabatic cooling and warming above the summer and winter polar stratopauses respectively are correspondingly underestimated.

The 'polluted' stratosphere case corresponds to an injection from aircraft of 1.8×10^6 metric tons of NO_x per year. On a global average the ozone reduction amounted to about 12 %, but the high latitude changes were nearer 20 % with corresponding smaller low latitude changes.

It is ironical that under the pressure of the C.I.A.P. schedule coupled with the need to economize on computer time, a full three dimensional distribution of NO_x could not be calculated, because the principal supposed advantage of three dimensional models for pollution studies is thereby lost, namely that the transports of all constituents, natural or pollution, are calculated in a consistent and realistic way. Some indications of the potential importance of this are given by comparing the latudinal distribution of total ozone in the run, where NO_x is independent of latitude, to that given by the two dimensional model. The pole–equator gradient in spring is increased by 60 Dobson units by this modification. While the effects on the longitudinal structure cannot be expected to be so great, they are probably appreciable. Nonetheless this is probably the most dynamically complete model with realistic nitrogen oxide catalytic destruction of ozone yet to have been produced.

Studies of the dynamics of two quasi-geostrophic spectral models with high truncation have emphasized the importance of the feedback loops outlined in § 1 above. Trenberth (1973) identified the cause of minor warmings in his model with horizontal transports of heat by the transient eddies, the mechanism being in broad accord with the theory put forward by Matsuno (1971). The transience was related to transience in the vertical energy flux (correlations between pressure and vertical velocity) which, in turn, Trenberth attributed to the tropospheric index cycle associated with 'beating' between the stationary and travelling disturbances. A number of interesting relations between the stratospheric warmings and tropospheric events, such as blocking and index cycles, are thus highlighted. On the other hand, the strength of both the zonal mean circulation and the eddies exhibit considerable sensitivity to the approximation adopted for the long wave cooling (Ramanthan & Grose 1978; Chen & Ramanthan 1978).

All the models mentioned above show departures from the details of the observed atmosphere, but many features are successfully reproduced. They will continue to provide powerful tools as advances in computer power allow the restrictions, such as the level of truncation or the number of species carried explicitly, to be relaxed.

(*b*) *Primitive equation models*

All the models mentioned so far have been based on the quasi-geostrophic equations because of the reduction in computation afforded by the longer time steps permitted by this approximation. As the approximation requires small Rossby numbers, the flow in the equatorial region may be misrepresented. In particular, the Kelvin and mixed gravity–Rossby waves cannot be treated properly. This in turn means, on current thinking, that the quasi-biennial oscillation and the semi-annual wave at 50 km will not be reproduced (Holton 1975; Hirota 1978). The interhemispheric exchange of stratospheric material may be poorly estimated in consequence. Moreover, it is possible that even the large scale extratropical disturbances may be misrepresented by the quasi-geostrophic equation (Dickinson 1968; Holton 1976), including the tropopause-folding process that is responsible for cross-tropopause transports. Accordingly, great interest attaches to comparing models with and without the quasi-geotrophic simplification. Such a direct comparison is not yet possible as the models differ in many more respects than simply in which dynamical approximations were adopted.

A major series of studies of the stratosphere has been undertaken by Manabe, Hunt, Mahlman and others using primitive equation models based on the G.F.D.L. general circulation model. The earlier studies (e.g. Hunt & Manabe 1968; Hunt 1969) concerned hemispheric versions of the model, often with 'perpetual winter' conditions, but these restrictions have been removed more recently (Manabe & Mahlman 1976; Mahlman & Moxim 1978).

Manabe & Mahlman (1976) give the result of a 3.5 year run of the model with seasonal variation of solar radiation, (prescribed) sea surface temperatures, and cloud amounts. The model has 11 vertical levels, the top level being at 10 mbar; the vertical velocity (strictly speaking, ω in σ-coordinates) is set to zero at $p = 0$; the horizontal resolution is approximately 265 km; the tropospheric hydrological cycle is computed in detail and a very full radiative computation is performed. This uses the modelled water vapour amounts but all other radiatively active constituents, including ozone, are prescribed as tables of latitude, height and date.

The overall behaviour of the model is summarized in table 1, taken from Mahlman & Moxim (1978). This shows a high level of achievement but it also indicates that, even with the very complete treatment of the physical and dynamical processes attained in this model, it is possible to show considerable departure from the observed atmospheric state. In particular, the polar night jet in the model and the associated equator–pole temperature difference are two or three times stronger in the model than in reality. This may be an effect of the lid-like boundary condition or of the limited vertical resolution at the upper level. Moreover, the polar night jet is not sufficiently separated from the subtropical jet in the Northern Hemisphere winter. This is linked to the strength of the mid-latitude descending motion.

Table 1. Summary of the behaviour of the model of Manabe & Mahlman (1976)

category	comments
zonal mean temperature	Very good below 38 mbar. Too cold in polar night middle stratosphere by 20 K. Northern Hemisphere. (N.H.) midwinter stratospheric warming not simulated
tropopause	Agrees well with observed poleward-downward slopes and appears at the proper altitudes.
zonal mean wind	N.H. subtropical jet slightly weaker than observed. Polar night stratospheric jet too strong by a factor of 2. Tropical stratospheric easterlies too strong by 15 m s^{-1}; quasi-biennial oscillation not simulated. Seasonal variation of zonal wind qualitatively satisfactory.
meridional circulation	Qualitatively correct: 3-cell troposphere, 2-cell N.H. stratosphere, 3-cell S.H. stratosphere. Midstratosphere flow from summer to winter hemisphere.
eddy kinetic energy	Somewhat too small in midlatitude troposphere. Good in N.H. stratosphere. Very large seasonal and interhemispheric variations in stratosphere.
stationary disturbances	Structure very good, amplitude somewhat large, particularly in N.H. middle stratosphere winter.
transient disturbances	Cyclonic disturbances somewhat weak in midlatitude troposphere. Transient long waves somewhat weak in midlatitude stratosphere. Tropical transient waves qualitatively correct but vortex disturbances too large in scale.
precipitation	General qualitative agreement with observation. Average continental precipitation somewhat too high. Too little precipitation in southern United States and southwestern China.

One of the major aims of this simulation is to investigate interhemispheric differences. Accordingly, the topography and sea surface temperatures specified in the Southern Hemisphere are appropriate to that hemisphere. Thus the level of planetary wave activity and the consequent forcing of the mean circulation are different in each hemisphere. This is clearly brought out in figure 3. The modelled eddy kinetic energy at 10 mbar in both the standing and the transient eddies is larger in the northern winter than in the southern; the interhemispheric difference in the standing wave energy being a factor of over ten while the difference in

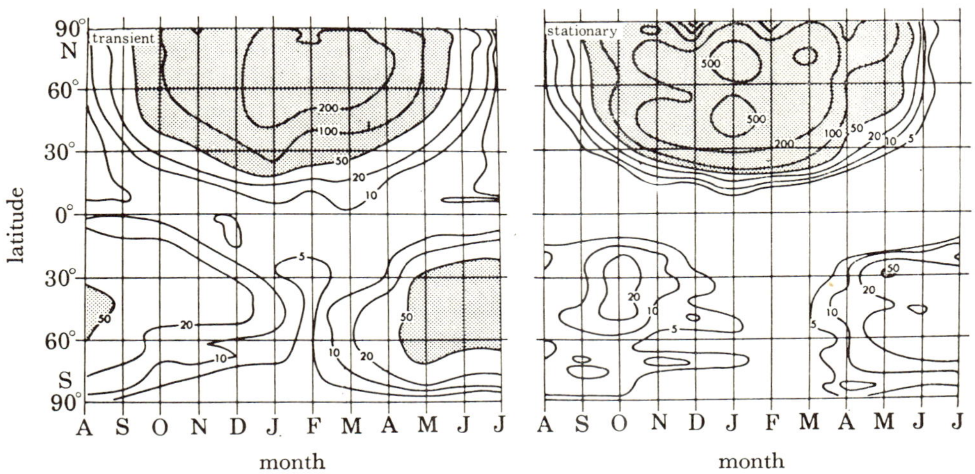

FIGURE 3. Latitude–time distribution of two components of eddy kinetic energy (J kg^{-1}) at 10 mbar in a three dimensional primitive equation model (Manabe & Mahlman 1976). The dominance of standing eddies in the Northern Hemisphere and transient in the Southern Hemisphere is an observed feature.

transient wave energy is nearer to four. In the Northern Hemisphere, standing wave energy is more than two times larger than transient, whereas in the Southern Hemisphere the transient wave energy is larger than the standing. The difference in eddy activity between the hemispheres is greater than is shown by the observations, but the relative importance of the transient disturbances in the Southern Hemisphere has been remarked upon by many authors (e.g. Harwood 1975; Leovy & Webster 1976; Hartmann 1976).

Mahlman & Moxim (1978) have used the winds from this model to simulate the behaviour of radioactive fallout. The initial injection was centred on 65 mbar at 36° N, 180° E and the subsequent spread of tracer was followed in detail over the next four simulated years (the winds from 1 year of the dynamical simulation being repeated four times). The results show that certain basic features such as the interhemispheric exchange are handled well, so that during the first year cross-equatorial transports reduce the ratio of material in the Northern Hemisphere to that in the Southern to about four, further reducing it to about two over the next 3 years, in general accord with the behaviour of $^{14}CO_2$ from 1963 to 1966 as found by Telegadas (1971). Stratospheric residence times of relevance to one-dimensional transport models based on the modelled distribution increase with the height of the centre of mass of tracer by 0.35 year per kilometre, being about 1.6 years at 20 km.

By October of the first year, a large cancellation is established between the zonal mean rates of change of tracer mixing ratio produced by eddies and by mean motions (see figure 4) as is found by most models. This cancellation is bound to arise in steady conditions for any property whose Lagrangian time derivative is near to zero (Andrews & McIntyre 1978). Under these circumstances, averages along the deformed streamlines of the Lagrangian motion show zero rates of change. Transience disturbs the average position of the streamlines in the meridional plane so that the cancellation in the zonal means is no longer exact. Thus the cancellation may be somewhat exaggerated in this particular model since it does not produce significant sudden warmings.

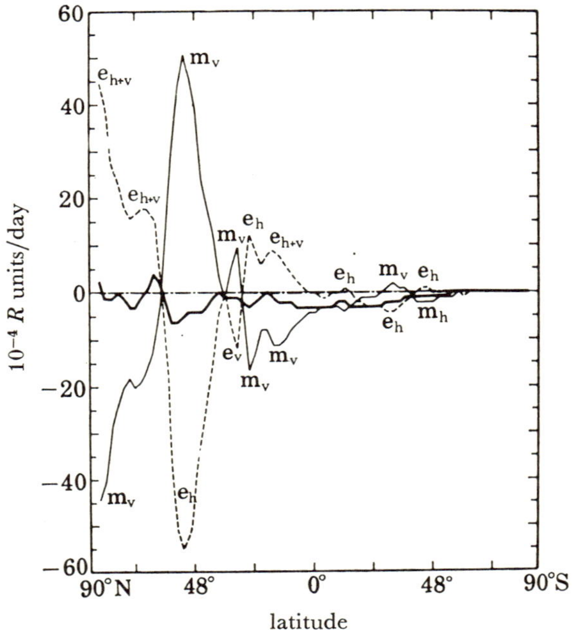

FIGURE 4. Mechanisms leading to the change at 65 mbar during October of the zonal mean mixing ratio of a tracer, *R*, in the same model as in figure 3 (10^{-4} *R* units/day) (Mahlman & Moxim 1978). Note the large cancellation between the mean and eddy contributions. The labels on the extrema denote mean (m) and eddy (e) contributions, with the subscript showing if the horizontal (h) or vertical (v) transport dominates. ——, Net tendency; ----, eddies; ——, meridional circulation; −·−·−, 'filling' + 'diffusion'.

When averaged over the globe, transports across the model tropopause are approximately an order of magnitude more by the eddy transport than by the zonal mean motion. However, at a given latitude the two effects are of similar magnitude. This arises because the eddy fluxes are usually downward whereas the mean flux depends on the local sense of the zonal mean vertical velocity. The calculated downward eddy flux through the tropopause is strongest near 50–60° N in March–May, predominantly associated with developing extratropical cyclones, while the strongest downward zonal mean fluxes occur near 35–45° N during December–February.

A group working in the U.K. Meteorological Office have also made a substantial study by using a primitive equation stratospheric model (Newson 1974; C.O.M.E.S.A. 1975).

The model has 13 σ-levels and was derived, mainly by increasing the vertical resolution, from that of Corby *et al.* (1972). Eight levels are normally in the stratosphere, the highest being near 44 km; the horizontal resolution is close to 300 km; tropospheric physical processes are treated in considerable detail; a smoothed version of the Earth's topography is assumed; and sea surface temperatures vary as seasonal climatological values, but land temperatures adjust according to radiative and convective heat fluxes.

This work is of particular interest since realistic sudden warmings occur in the winter stratosphere. Figure 5 shows the changes in zonal mean wind between day 70 and day 90 of an earlier run of this model (Newson 1974) in which only one hemisphere was treated and 'perpetual January' conditions were adopted. On day 70, the stratospheric polar night jet is too strong by a factor of perhaps two, and the distinction between the polar night jet and the subtropical jet is insufficiently marked. This appears not to be a result of maintaining the January conditions for too long, as the seasonally varying model behaves similarly. After day 78 a dramatic change occurred; by day 90 the temperature at the pole had risen by 60 K at 40 km and the zonal mean wind reversed (figure 5*b*). The polar night vortex began to re-form on about day 100, but even by day 130 it was not as intense as it had been before day 78.

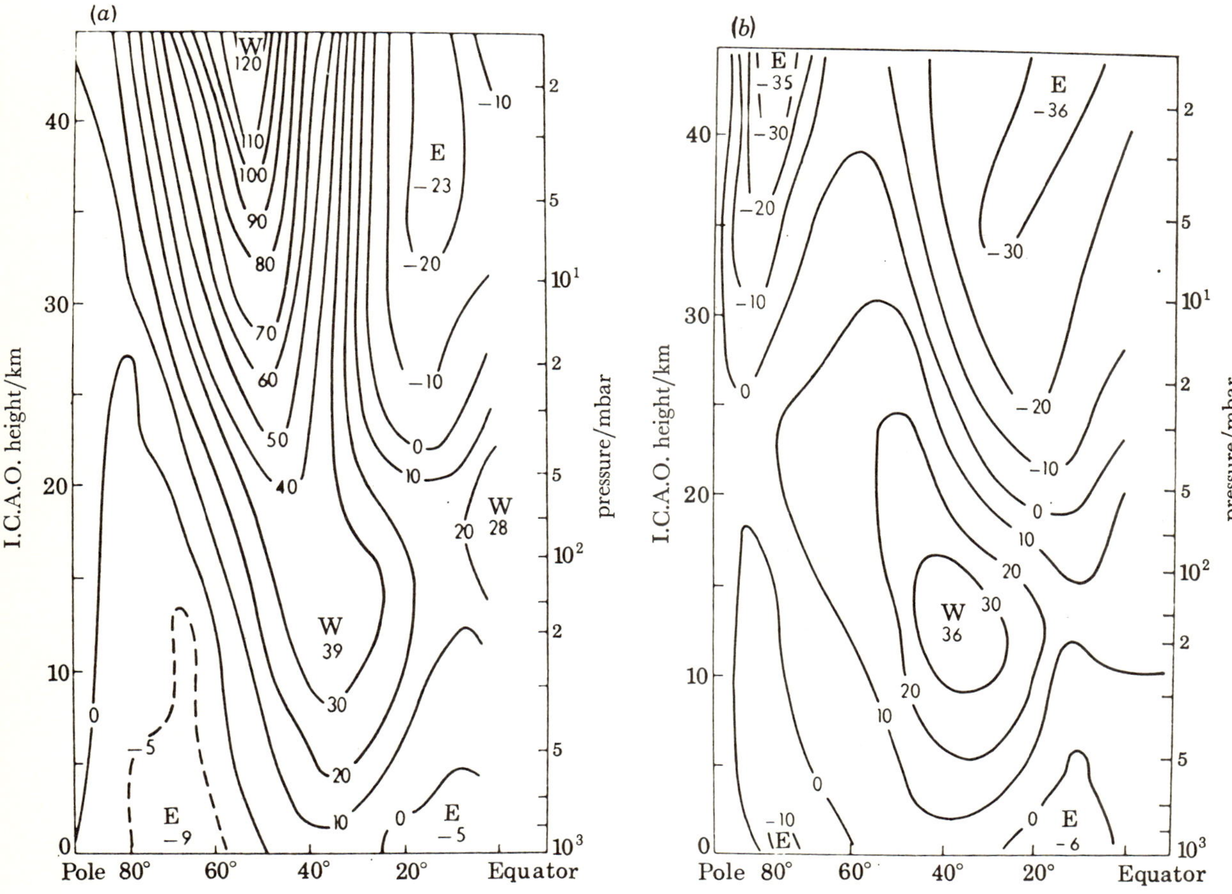

FIGURE 5. A sudden warming simulated in a three dimensional primitive equation model. Zonal mean winds (m s^{-1}) on (*a*) day 70, before the onset, (*b*) day 90, 12 days after the onset of warming (Newson 1974).

It is probable that the major factor in the success of this model in producing a major warming compared with the G.F.D.L. model is the superior resolution of the upper stratosphere. In the G.F.D.L. model, vertical velocities in the upper level must be distorted by the presence of the 'lid' boundary conditions. Furthermore, radiative damping of the planetary waves and critical level interactions will be better handled by the C.O.M.E.S.A. model.

So far the model has been used to study the spread of tracers that were inert or had very simple ozone-like sources and sinks. Runs with a more realistic treatment of ozone are awaited with great interest.

3. Two dimensional models

(a) *Averaging the photochemical and continuity equations*

The large amounts of computing time required to perform the dynamical calculations in three dimensional models has usually meant that only a very simplified treatment can be made of the chemical processes. For instance, Cunnold *et al.* (1975) state that their model requires 40 s central processor time to simulate 1 model day on an I.B.M. 360/90, while Newson's hemispheric calculation required 9 min per day on an IBM 360/195. Both were thereby forced to adopt very simple photochemical schemes in which only one tracer was explicitly calculated. To improve the modelling of the chemistry, therefore, many workers have sought to reduce the amount of computation by constructing two dimensional models, calculating only the zonal mean of the relevant variables as functions of latitude and height.

Two difficulties arise immediately, namely how to calculate the relevant chemical production rates and how to calculate the meridional transports. Consider first the chemical processes. Let s be the local value of some variable and let $\bar{s}$ denote the zonal mean of s. Let s' be the departure of s from the zonal mean, $s' = s - \bar{s}$. If we now consider two reacting substances having mixing ratios a and b, then the zonally averaged rate of the reaction is proportional to $\overline{(ab)}$.

This may be split up thus, $$\overline{ab} = \bar{a}\bar{b} + \overline{a'b'},$$

into a contribution by the mean state $\bar{a}\bar{b}$, and a contribution by the 'anomalies' or 'eddies' $\overline{a'b'}$.

The eddy contribution involves quantities that are not explicit model variables and is usually simply ignored. It is not difficult, however, to conceive of circumstances where this might matter. For instance, if at a certain height and latitude all the 'a' substance is found in air parcels that are rising and all the 'b' substance is in sinking air, then the two substances never mix and the true reaction rate $\overline{(ab)}$ is zero. However, the usual formulation gives instead a finite reaction rate $\bar{a}\bar{b}$ (Tuck 1979).

Photochemical processes likewise present difficulties because the transmission of radiation through the atmosphere is not a linear function of absorber amount. Again it is usually assumed that the zonally averaged rate of photodissociation is that which would obtain for the zonal mean absorber amounts.

The difficulties with transport are analogous. The zonally averaged continuity equations for the explicitly modelled gases require knowledge of meridional transports. If the gas S has mixing ratio s, then the northward transport of S involves expressions of the form $\overline{vs}$ where v is the northward velocity. This may be split up into two terms involving transport by the zonal mean winds $(\bar{v}\bar{s})$ and the eddies $(\overline{v's'})$

$$\overline{vs} = \bar{v}\bar{s} + \overline{v's'}.$$

The greater problem is posed by the eddy transport term $\overline{v's'}$ since s' and v' are not model variables.

The technique almost universally adopted is a parametrization in terms of the zonal mean fields, proposed by Reed & German (1965). They introduced eddy diffusion coefficients K_{yy}, K_{yz} such that

$$\overline{v's'} = -K_{yy}\,\partial\bar{s}/\partial y - K_{yz}\,\partial\bar{s}/\partial z,$$

y being distance northwards and z being height.

A mixing-length hypothesis provides some justification for this formulation and further suggests that all substances having similar mixing processes will have the same diffusion coefficients. In principle, this means that measurements of the fluxes and distributions of two suitable quantities would determine the value of K_{yy} and K_{yz}. However, for the tracers that are most easily measured such as ozone and potential temperature the problem is ill-conditioned because of the high correlation between them (Mahlman & Maxim (1978) gives an example of the build-up of such correlations). Accordingly, less direct methods must be used. Once more the favoured method was proposed by Reed & German, again based on the mixing-length idea, though more tentatively: K_{yy} is assumed to be proportional to $\overline{(v')^2}$, which is readily observed. Adoption of a suitable constant of proportionality completely determines K_{yy} and the measurement of transports and distribution of only one tracer now suffices to find K_{yz}.

Eddy transport in the vertical cannot be neglected and here two new coefficients are introduced by

$$\overline{w's'} = -K_{zy}\,\partial\bar{s}/\partial y - K_{zz}\,\partial\bar{s}/\partial z,$$

where w is the vertical velocity.

Mixing-length theory gives $K_{zy} = K_{yz}$, so one measurement of a vertical tracer flux distribution in principle determines K_{zz}. However, vertical fluxes are in general very poorly known and such direct estimates are not available. Again, mixing-length theory provides an estimate, this time involving the means and variances of the slope of the mixing motions. Even here the relevant quantities are not available and most estimates are little more than intelligent guesses of about the correct order of magnitude.

In spite of all the uncertainties involved – whether mixing-length ideas can adequately represent the processes; whether the same *Ks* are suitable for all substances; what the values of K really should be – it can be said in favour of this method that it works surprisingly well, as we shall see below, and as yet no alternative method is available. Fabian & Libby (1974) tested several combinations of *Ks* and mean meridional circulations by computing the spread of a radioactive tracer in a two dimensional model and comparing the results against observations. They obtained the most consistent results by combining a mean meridional circulation obtained from Gudiksen *et al.* (1968) with a set of *Ks* deduced by Luther (1974), who followed closely Reed & German's method based on heat flux data.

It will be noticed that the K approach is most suited to treatment of the natural stratosphere. The addition of pollution can be expected to induce changes in the eddies either directly by altering the radiative damping or indirectly by altering the zonal mean state through which they must propagate. In terms of mixing-length theory this implies changes in $\overline{v'^2}$ and hence K_{yy}, and in the mean slope of the mixing paths (which turns out to be K_{yz}/K_{yy}). Likewise, the inter-annual variation in atmospheric state should imply the use of different *Ks* for each year but the sophistication involved in altering the *Ks* from year to year probably exceeds the physical basis on which they are founded.

It should be noted in passing that all the problems associated with averaging, which have been mentioned with two dimensional models, occur even more chronically in one dimensional models. The usefulness of all these models lies not in their ability to predict, say, the effects of introducing pollution, but rather in their ability to reveal some of the possible major mechanisms and feedbacks at work.

Transport by the mean motion $\overline{vs}$ is relatively straightforward, but even here care is needed not to attempt modelling experiments beyond the scope of the model, for while it is reasonable to adopt observed values of mean circulation for studies of the unpolluted stratosphere, the same is not true if any pollution affects the radiatively active constituents (such as ozone). Besides the direct effect through alteration of the zonal mean radiative heating, an indirect effect can be anticipated through modification of the eddy heat and momentum fluxes consequent upon alteration of the planetary waves through changes to their radiative damping. As yet we have little knowledge of the likely importance of such effects.

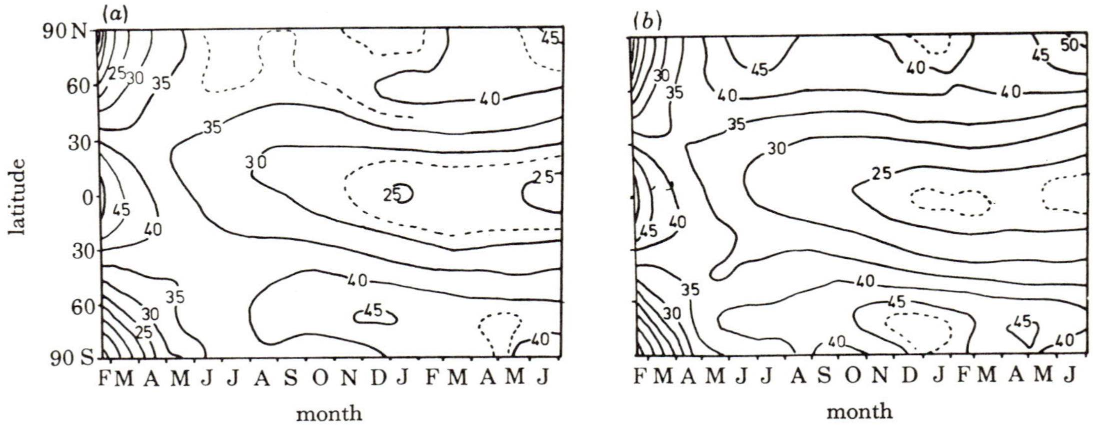

FIGURE 6. The sensitivity of ozone in a two dimensional model to the diffusion coefficients. Total ozone (10^{-2} atm-cm) calculated (*a*) by using Luther's (1974) *K*-values, (*b*) with K_{yz} doubled in mid- and high latitudes (equivalent to doubling the slope of the mixing surfaces) (Harwood 1977).

Most models take the values of the transport parameters ($\bar{v}$, $\bar{w}$ and the eddy coefficients) as external variables supplied to the program as suitable tables from estimates based on observations or three dimensional models. However, recognizing that these transport parameters are but poorly known, several investigators adjust them to produce a good agreement between modelled and observed ozone amounts. Selecting two examples from many, the model of Brasseur & Bertin (C.O.V.O.S. 1976) utilized rather crude spatial distributions of the *K*s with magnitudes adjusted to give realistic O_3 distributions, while the U.K. Meteorological Office's 2-D Comesa model had an adjustment made to the mean meridional circulation above 25 km after two simulated years so as to improve the total ozone distribution (C.O.M.E.S.A. 1975).

Some indication of the magnitude of the response to changes in the diffusion coefficients is given by figure 6 from Harwood (1977). The calculations were performed in a two dimensional general circulation model with a linear approximation to the photochemical source and sink of ozone. The only difference between the two runs lies in the values of K_{yz} adopted in mid and high latitudes in the mid-stratosphere. Doubling this coefficient increased the high latitude ozone amounts by about 10%, associated with a lowering of the height of maximum ozone concentration at the poles. The equatorial values decreased by about a similar fraction, even though K_{yz} was unchanged there.

While it may be that adjustments to obtain a better agreement with observations are necessary because of inadequate knowledge of the transport parameters, there are obvious dangers in the practice. In particular, one may be attempting to use the transports to correct for what is really a deficiency in the chemical scheme. It is salutary in this respect to consider the growth during the last two decades in the number of reactions thought necessary to explain the ozone cycle.

A considerable saving in computation can often be made for some purposes by seeking steady-state solutions with solar declination held constant. It is perhaps surprising that this leads to reasonable distributions for ozone, since the stratospheric residence times of many months appear too long to allow an approach to steadiness within a season. Nonetheless, Rao & Christie (1973) and Hesstvedt (1974*a*) found moderately good distributions of ozone in steady-state models. The reason for their success probably lies in the large cancellatoins between mean and eddy transports already referred to (§ 2.2 above and figure 4): only a small change in the ozone distribution is needed to the steady-state distribution to produce the observed secular rates of change. In this context it is noteworthy that Hesstvedt (1974*b*) reports that a time dependent version of his model gave markedly better results than the steady-state version. Comparison is hindered, however, by the use of different *Ks* in the two versions. Although it is often claimed (see, for example, Widhopf & Taylor 1975) that adoption of mean velocities and eddy coefficients deduced from different data sets will lead to a gross disturbance of the near cancellation between mean and eddy flux convergence, in practice the near cancellation is found. Harwood & Pyle (1977) contend that this is to be expected, because in *K*-theory the eddies are a diffusive system which, since K_{yy} is typically $10^6\ \mathrm{m^2\ s^{-1}}$, has a diffusive timescale of only 10 days for distances of 1000 km. Consequently, transient effects are very rapidly damped out, leaving a quasi-stationary state in which the near cancellation is found. The resulting tracer distribution may not be very realistic, of course, but the cancellation is unavoidable.

(*b*) *Model dependent mean circulations*

The importance of the mean circulation to the ozone amounts has led to several models in which it is calculated in terms of the model state. To exemplify some of the methods, we will outline the technique of Harwood & Pyle (1975).

The relevant dynamical equations are:

(1) the zonal mean momentum equation

$$\frac{\partial(\bar{\tau}F)}{\partial t}+\frac{\partial(\overline{v\tau}F)}{\partial y}+\frac{\partial(\overline{w\tau}F)}{\partial\eta} = H, \tag{1}$$

(2) the zonal mean thermodynamic equation

$$\frac{\partial(\bar{\theta}F)}{\partial t}+\frac{\partial(\overline{v\theta}F)}{\partial y}+\frac{\partial(\overline{w\theta}F)}{\partial\eta} = Q+\bar{q}F, \tag{2}$$

and

(3) the continuity equation

$$\frac{\partial(F\bar{v})}{\partial y}+\frac{\partial(F\bar{w})}{\partial\eta} = 0, \tag{3}$$

where $F = (p/p_0 \cos\theta)$, p is pressure, $p_0 = 1000$ mbar, y = distance north, $\eta = \ln(p_0/p)$, $w = \mathrm{D}\eta/\mathrm{D}t$, θ = potential temperature, τ = angular momentum = $(u+a\Omega\cos\lambda)\,a\cos\lambda$, with u = westerly wind speed, v = northward wind speed, a = radius of the Earth, $\Omega = 2\pi$ radians/sidereal day, λ = latitude. q is the rate of change of potential temperature due to diabatic processes.

$$H = -\frac{\partial(\overline{v'\tau'F})}{\partial y} - \frac{\partial(\overline{w'\tau'F})}{\partial \eta}; \quad Q = -\frac{\partial(\overline{v'\theta'F})}{\partial y} - \frac{\partial(\overline{w'\theta'F})}{\partial \eta}.$$

As with the tracer equations, the major difficulty is presented by the need to calculate the eddy terms H and Q in terms of the zonal mean state. Q is amenable to K-theory treatment; indeed, as we have seen, the *Ks* are usually calculated on this basis from observed potential temperature fluxes. H is less tractable and will de discussed later (§ 3 (*b*)).

The equations so far do not constitute a complete set, the y-momentum equation having been so far omitted. The use of the complete y-equation leads to a kind of primitive equation model which contains gravity waves and thus requires short time steps to satisfy the C.F.D.L. criterion. Moreover, untrapped gravity waves are expected to have velocity amplitude proportional to $\rho^{-\frac{1}{2}}$, which increases by two orders of magnitude between the ground and mesopause, so that small gravity wave disturbances excited in the lower layers may dominate the upper layers. Primitive equation methods are not ruled out, however, and Holton (1976) has used a slightly approximate form of equation (1)–(3) with the complete y-momentum equation to investigate the sudden warming phenomena.

The gravity waves may be filtered out by employing the thermal wind equation

$$\frac{\partial \bar{\tau}}{\partial \eta} = -\frac{R}{2\Omega \tan \lambda}\left(\frac{p}{p_0}\right)^{K}\frac{\partial \bar{\theta}}{\partial y}. \tag{4}$$

White (1979) has shown that there is a slight energetic inconsistency introduced by this form of equation (4) which can be corrected by including a nonlinear term in (4) or making suitable approximations in equation (1). However, the spurious energy source is trivial compared with natural energy exchanges in the model.

Introducing a stream function ψ such that equation (3) becomes

$$v = -\frac{1}{F}\frac{\partial \psi}{\partial \eta}, \quad w = +\frac{1}{F}\frac{\partial \psi}{\partial y},$$

and eliminating the time derivatives from (1) and (2) by using (4), gives

$$\pounds(\bar{\tau}, \bar{\theta}; \psi) = R_1(H) + R_2(Q) + R_2(qF), \tag{5}$$

where $\pounds$ is a linear second order partial differential operator for ψ which is elliptic if the zonal mean state is inertially stable. The coefficients are functions of the mean state $\bar{\tau}$, $\bar{\theta}$. R_1 and R_2 are differential operators, involving principally horizontal gradients of heating effects, R_2, and vertical gradients of eddy momentum flux convergence, R_1. The right-hand sides thus arise from processes which tend to disturb thermal wind balance. Equation (5) determines that circulation which is necessary to maintain the balance, and is solved for ψ at each time step and the value substituted in (1) and (2), which are then used to step forward in time.

Rao's (1973) method has many similarities with the scheme just described, but also several major differences: he holds the solar declination constant and finds the steady solution for $\bar{\tau}$, $\bar{\theta}$ and $\overline{[O_3]}$; he uses a constant value of the coriolis parameter between latitudes 0 and 25° in each Hemisphere in his version of equation (4) so as to ensure that the inertial stability criterion is satisfied. Consequently, it is probable that the cross-equatorial winds and interhemispheric exchange of material is incorrectly estimated. A further difference lies in the parametrization of momentum fluxes discussed in § 4.

The success of schemes outlined above can be judged from figures 7–9 taken from Harwood & Pyle (1977) and Pyle (1976). In the case portrayed, the diabatic processes q were calculated explicitly in terms of model temperatures and absorber amount using a cooling-to-space approximation for long wave radiation around 50 km. Tropospheric heat sources due to radiation and latent heat release were specified from climatology. Q, the eddy heat flux convergence, was calculated by K-theory. The vertical momentum fluxes were calculated from $\overline{w'\tau'} = -k_{\eta\eta}(\partial\bar{\tau}/\partial\eta)$, but the horizontal fluxes were specified from observations. The ozone

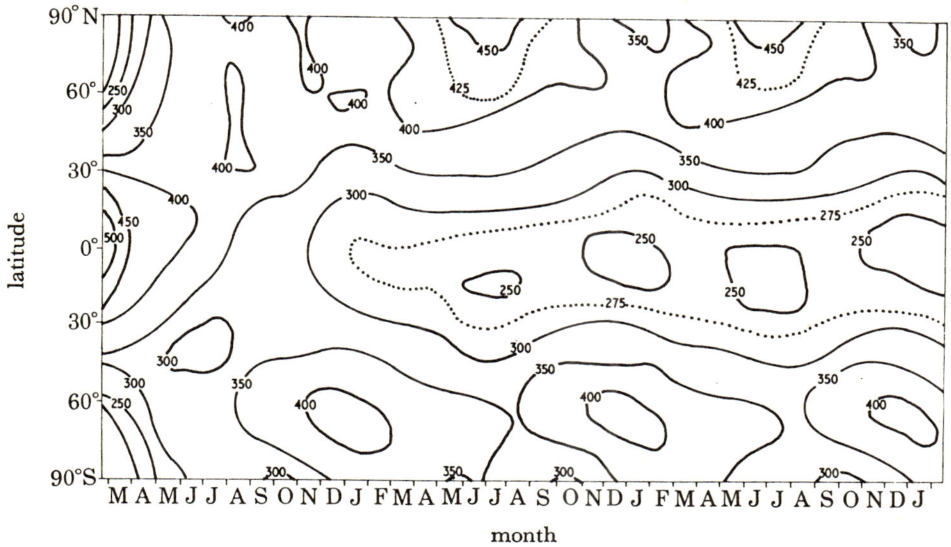

FIGURE 7. The distribution of total ozone (m atm-cm) calculated in a two dimensional time dependent model (Harwood & Pyle 1977). The starting values correspond to photochemical equilibrium. The interhemispheric variations (cf. figure 1) arise from asymmetries in the adopted momentum fluxes.

chemistry was linearized so that the creation or destruction rate is simply proportional to the departure from the equilibrium value. The constants of proportionality (reciprocals of the relaxation times) and photochemical equilibrium concentrations were specified to the program in the form of tables as functions of latitude, pressure and season, having been calculated offline from a reaction scheme that includes the catalytic cycles of NO_x and HO_x. There was a sink of ozone in the model troposphere: the ozone concentration at the bottom level is maintained at zero, thus resembling rapid destruction at the ground. Each model day required 2.5 s of computing time on a CDC 7600.

The latitude–time section of total ozone for the run is given in figure 7. On the starting day (19 February) the ozone takes its spring photochemical steady-state concentration and it is nearly a year before a periodically reproducible distribution is established. At the end of the run there are small year-to-year differences.

Once the model has settled down, many of the observed features of the total ozone distribution are reproduced. The magnitude of the total ozone amount is quite good. The equatorial minimum is modelled well with closed minima of less than 250 Dobson units. The latitudinal variation is about 2–3 months out of phase with the observed behaviour with a minimum in the model in the Northern Hemisphere in December, and an observed minimum at the beginning of October. This is perhaps not unreasonable since the momentum fluxes and K-coefficients are obtained from 1 and 5 years' data respectively and should not necessarily reproduce an ozone distribution averaged over a much longer period of time (since fluctuations from year to year are not insignificant (Dütsch 1971)). Experience with a later version of the model in which the NO_x was calculated as a model variable produced the high latitude ozone maximum in spring, implying that the main cause of the late maxima and minima is the linearization of the photochemical scheme. In particular, the destruction rate is known to be underestimated by this approximation.

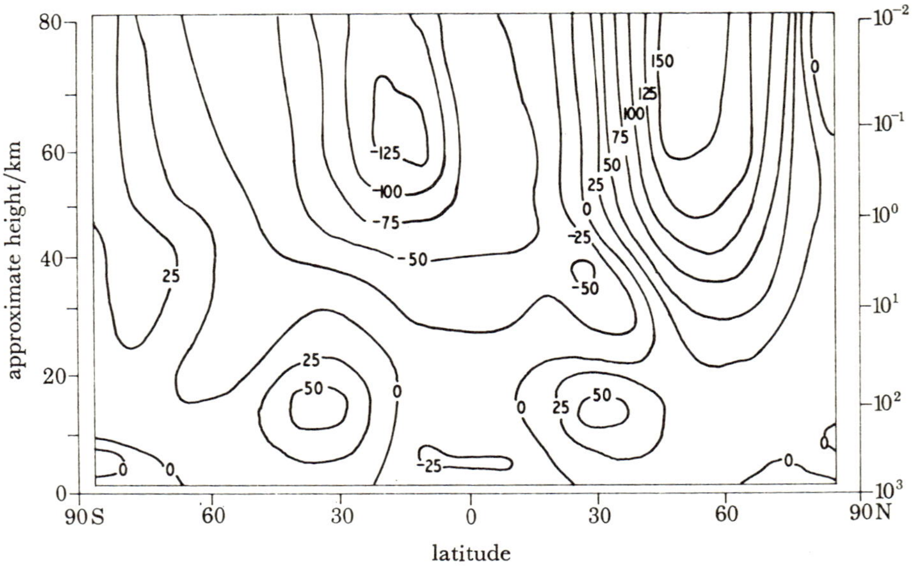

FIGURE 8. Zonal mean winds for 22 December of the second year of the simulation shown in figure 7 (Pyle 1976).

Figure 7 points at once to the importance of the dynamical feedback since the principal asymmetry between the Hemispheres comes from the horizontal eddy momentum fluxes specified in the stratosphere: the diffusion coefficients and the parameters from which the radiation is calculated are seasonally symmetric between the Hemispheres. The amount of ozone in mid and high latitudes of the Northern Hemisphere is 5–12 % higher than that in the Southern Hemisphere. It follows that any prediction of pollution-induced changes of this order of magnitude from a model without dynamics cannot be considered reliable unless it can also be shown that the adjustment of the eddy momentum flux to the new final state is considerably smaller than the currently observed difference between the Hemispheres.

Figure 8 shows the mean zonal wind in the model on day 671, which corresponds to 22nd December. The west–east wind component reproduces the gross features of the observed winds in both the troposphere and stratosphere, but with important differences of detail. In the troposphere the subtropical jets are separated by a region of tropical easterlies, and there are regions of surface easterlies at the poles. The latitudes of the polar night jet in the stratosphere and the easterly jet are in good agreement with the observations. The extension of the stratospheric easterly régime into the winter hemisphere is also a feature of observations. The easterly jet at 30° N and 40 km is not shown on most cross sections. However, Kats (1968) has reported similar features, though of much lesser intensity. The principal jets in figure 8 are larger than the observed circulation by a factor of between 1.5 and 2. It follows that the horizontal temperature gradients are also exaggerated in the model, although the equatorial temperatures are in good accord with observations (Pyle 1976). The excess temperatures at the summer stratopause are probably due to an overestimate of the ozone and hence absorption of solar radiation in that region, but the excessively cold winter pole is not understood.

The mean meridional circulation of the model and the modelled meridional distribution of ozone have been shown (Pyle 1976) to agree sufficiently well with the values of Vincent (1968) and Dütsch (1971), respectively, for confidence to be placed in the values of modelled ozone transport shown in figure 9, partitioned into mean and eddy fluxes. It is of importance to note that the main *vertical* fluxes are carried by the mean motion and not by the eddies: the processes usually represented by a vertical 'eddy' diffusion in one dimensional global models are what two dimensional modellers regard as the mean motion with planetary dimensions, if the fluxes are averaged horizontally over a hemisphere.

This is emphasized in figure 10 from Harwood (1977), which shows the vertical flux of ozone, integrated horizontally over the whole of the Northern Hemisphere. The eddy flux contribution is tiny. Note that the hemispheric average mid-stratospheric flux is downward in the winter but upward in the summer. It follows that the hemispherically averaged ozone flux is sometimes up- and sometimes down-gradient. Similar behaviour can be expected for other constituents also. Indeed, it is easy to see that, given the right latitudinal distribution of material, the hemispheric-average flux at certain levels could always be directed against the hemispheric-average gradient. This further underlines the need for caution in evaluating the results of one dimensional models.

Figure 9 demonstrates that the amount of ozone in high latitudes is greatly dependent on the strength of the sinking motions in mid-latitudes. Again, the wisdom of calculating model-dependent mean motions in pollution studies is indicated.

A striking example of the sensitivity of the ozone amounts to dynamical influences is provided by the response to a small change in the radiation field of the lower stratosphere. The 'control run' is that described above, which produced the ozone distribution already seen in figure 7. In this model the lower stratosphere was forced to be in radiative equilibrium by setting the radiative heating and cooling to zero between about 12 and 24 km at all latitudes. Figure 11 shows the result of changing this assumption. The principal change made from the 'control' run was in the region 12–24 km where the radiative equilibrium condition was replaced by heating rates from computations by Rodgers (1967). This introduces heating rates of a few tenths of a degree per day in low latitudes and cooling of comparable magnitude in high latitudes in winter. (There was also a small change to the bottom boundary condition which has been shown to be insignificant.) The influence of this change in radiation on the mean circulation was marked, there being a far stronger ascent in the stratospheric Hadley cell. This produced great changes in the ozone amount as may be seen from figure 11.

The implication for our present purpose is that any alteration to the radiation budget of the lower stratosphere by only a few tenths of a degree Kelvin per day could have quite large effects on the ozone distribution. This sensitivity to the lower stratospheric heating rates may present difficulties for studies of pollution-induced perturbations, as the total heating is a comparatively small residual of two large numbers. One of these depends to a large extent on the amount of ozone above. It is quite likely, therefore, that the feedback loop, ozone → heating → mean motions → ozone, must be taken into account for reliable computation of pollution effects.

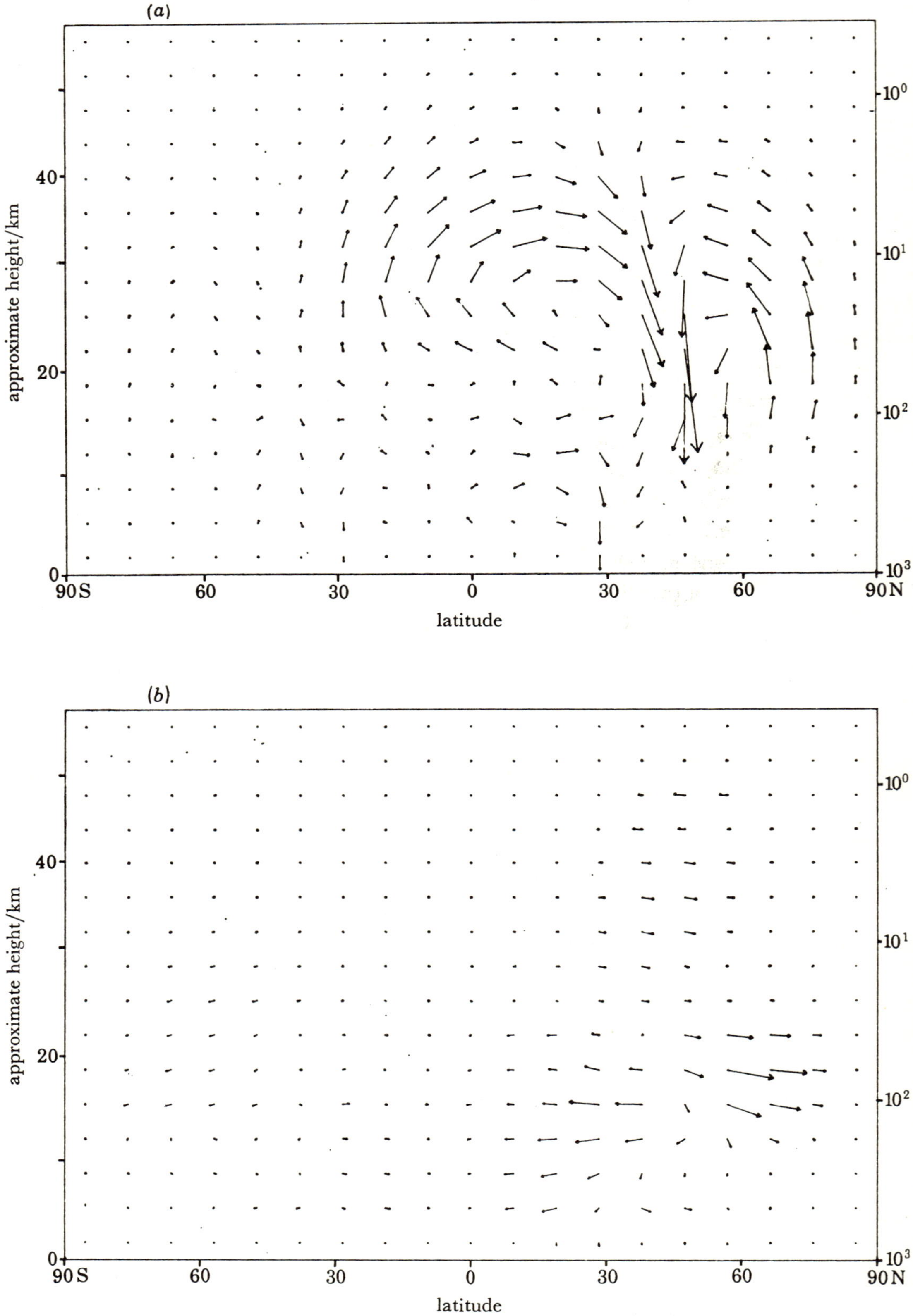

FIGURE 9 *a* and *b*. For description see opposite.

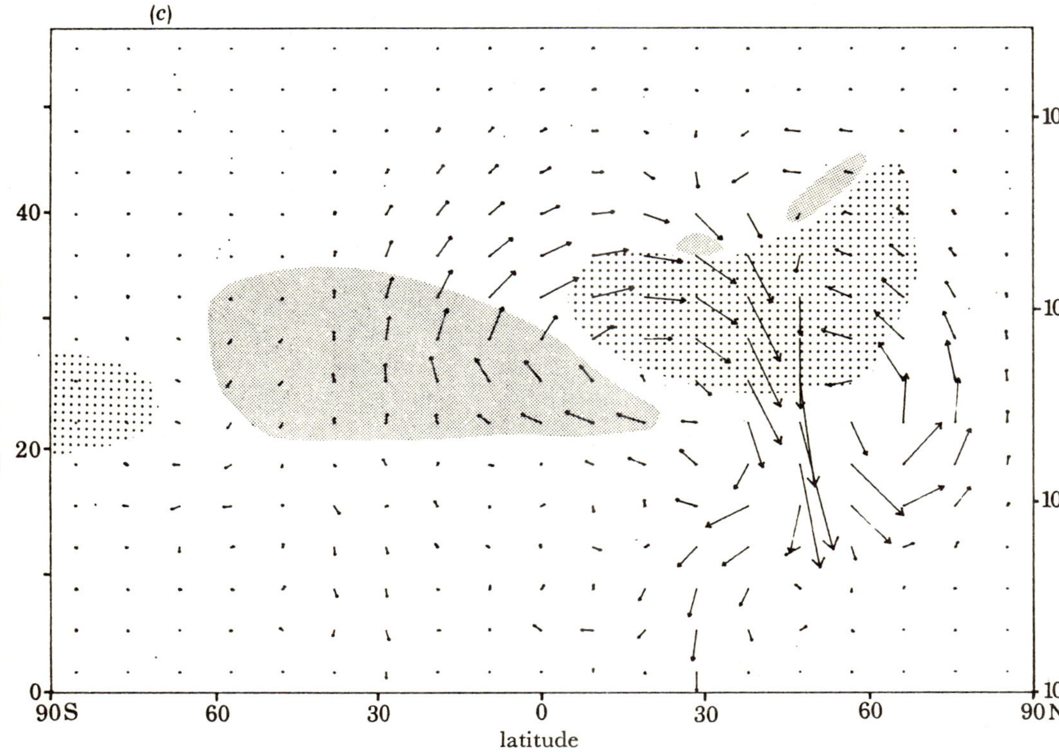

FIGURE 9. Meridional flux of ozone for the same case as figure 8 (*a*) due to mean motions, (*b*) due to eddies, (*c*) total. Fine shading denotes a photochemical source greater than 3×10^9 cm^{-3} day^{-1}. Coarse shading denotes a photochemical sink greater than 3×10^9 cm^{-3} day^{-1} (Harwood & Pyle 1977).

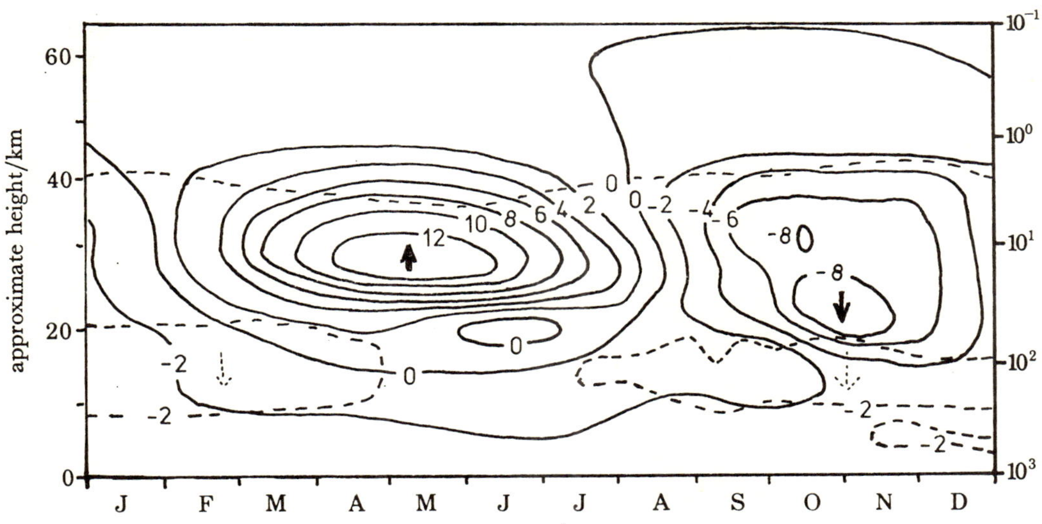

FIGURE 10. Northern Hemisphere average vertical ozone flux (10^{10} mol cm^{-2} s^{-1}) in a model similar to that in figure 7. ——, Mean motions; – – –, eddy motions.

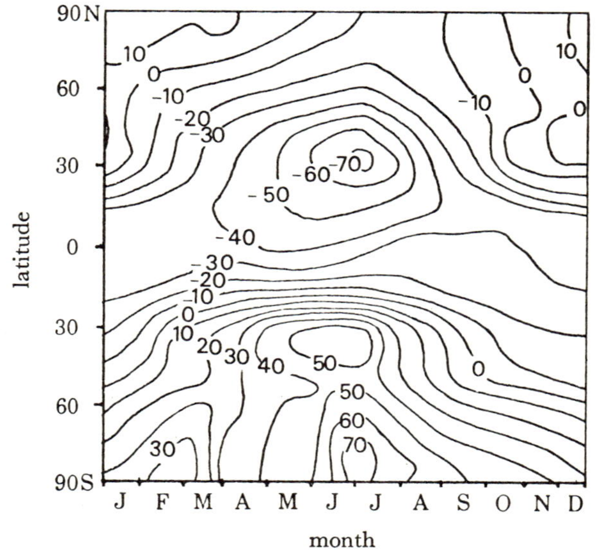

FIGURE 11. Change in total ozone (m atm-cm) caused by changing the lower stratospheric heating rate.

4. MODELLING THE MOMENTUM FLUX

The previous considerations have pointed to the importance of calculating the mean circulation explicitly. To do this it is necessary to determine the eddy momentum flux. Figure 12 provides an example of the dependence of total ozone on the momentum flux. The two curves show the latitudinal distributions of total ozone in two runs of the model described above, which differed only in the value of the eddy momentum fluxes prescribed in the upper stratosphere. These values have been compiled by Dr A. Crane of the Clarendon Laboratory, Oxford, based on temperature observations from the selective chopper radiometer on the Nimbus 5 satellite. The mid-latitude ozone amounts change by about 5 %. The true sensitivity may be underestimated by this experiment as the momentum fluxes in the lower stratosphere were unaltered between the two runs.

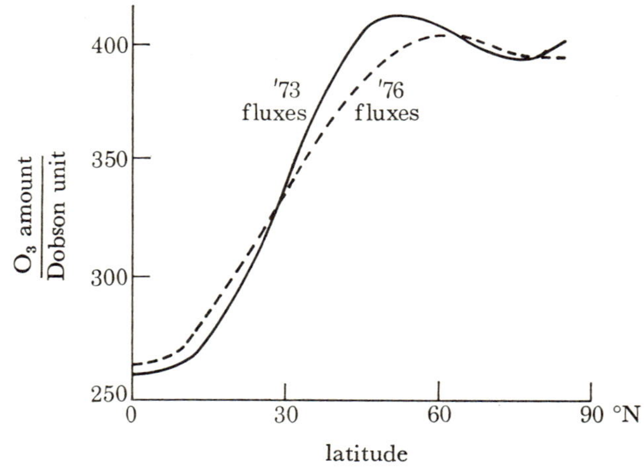

FIGURE 12. Sensitivity of the ozone amount to the momentum fluxes. Latitudinal distribution of total ozone (m atm-cm) on 1 April for two runs differing only in the year of the observed momentum fluxes used in finding the zonal mean circulations.

As yet, no satisfactory method of calculating the momentum fluxes from the model state has been proved in a model, although some attempts have been made. Rao (1973) uses a parametrization of the momentum flux proposed by Williams & Davies (1965) in which the fluxes are proportional to the pole–equator temperature gradient. This parametrization was proposed for tropospheric modelling on the assumption that the biggest contribution to the fluxes is made by the baroclinic eddies. However, even in the troposphere the relevant factors are considerably more complicated than is here implied (Green 1970; White 1977) while the present theories of planetary wave propagation to the stratosphere (Holton 1975) give no encouragement to the belief that the Williams & Davies formulation is realistic above the tropopause.

Another method which has been proposed for the tropospheric eddy momentum fluxes is that of Saltzman & Vernekar (1968). Assuming that the trough ridge systems are produced baroclinically but then tilted in the latitude–longitude direction by barotropic processes, they obtain a parametrization of the flux convergence dependent on the tilting time. While the method appeared to give encouraging results in the troposphere it is unlikely that it provides a suitable scheme for stratospheric eddies.

The success of the theories of planetary-wave propagation (Charney & Drazin 1961; Matsuno 1970; Simmons 1974) offer the possibility that an acceptable configuration of standing planetary eddies could be found every few time steps from which the momentum (also heat fluxes) could be calculated, or even the time evolution of the eddies calculated. In essence this method is used by Matsuno (1971) and Holton (1976) in their studies of the sudden warming phenomenon. Unfortunately, great simplifications were necessary, mainly the linearization of the wave equations. Moreover, the method involves inversion of an elliptic operator, which is computationally expensive, so that a severely truncated spectral model may prove cheaper. Schoeberl & Strobel (1978) have attempted to construct a steady state quasi-geostrophic model based on this method but were forced to abandon it because of numerical instabilities that developed when they used realistic amplitudes for the forcing of the wavenumber 1 wave. Instead, they made use of a Rayleigh friction in which the eddy flux convergence is made proportional to the zonal mean wind. This is simply a mathematical device necessary to prevent the occurrence of strong wind speeds in the absence of a proper eddy formulation; it has no physical justification whatsoever in the stratosphere.

A desirable property of any parametrization of the momentum fluxes is that it should be consistent with the other formulations. Green (1970) has outlined a procedure, which can be adapted to stratospheric use, that has this property. At the heart of the method is a formula relating the eddy momentum flux to the fluxes of potential temperature and potential vorticity. The main adaption needed for the stratosphere is to use an approximate form of the full Ertel potential vorticity rather than the quasi-geostrophic form used by Green. Since potential temperature and potential vorticity are both quasi-conservative properties, at least in the lower stratosphere, it should be possible to treat them by K-theory. Indeed, we have already seen that quite acceptable values can be obtained for heat transport. In conclusion we show some evidence that K-theory is capable of modelling the potential vorticity flux.

Figure 13 shows the observed distribution and flux of potential vorticity at 30 mbar for November 1976–April 1977. In mid-latitudes the flux is directed towards the pole, which is predominantly up the gradient on the constant pressure surface. In low latitudes the flux is towards the equator. Similar results have been obtained for this height for December 1970–January 1971 and for October 1975–January 1976. Hartmann (1977) found rather similar

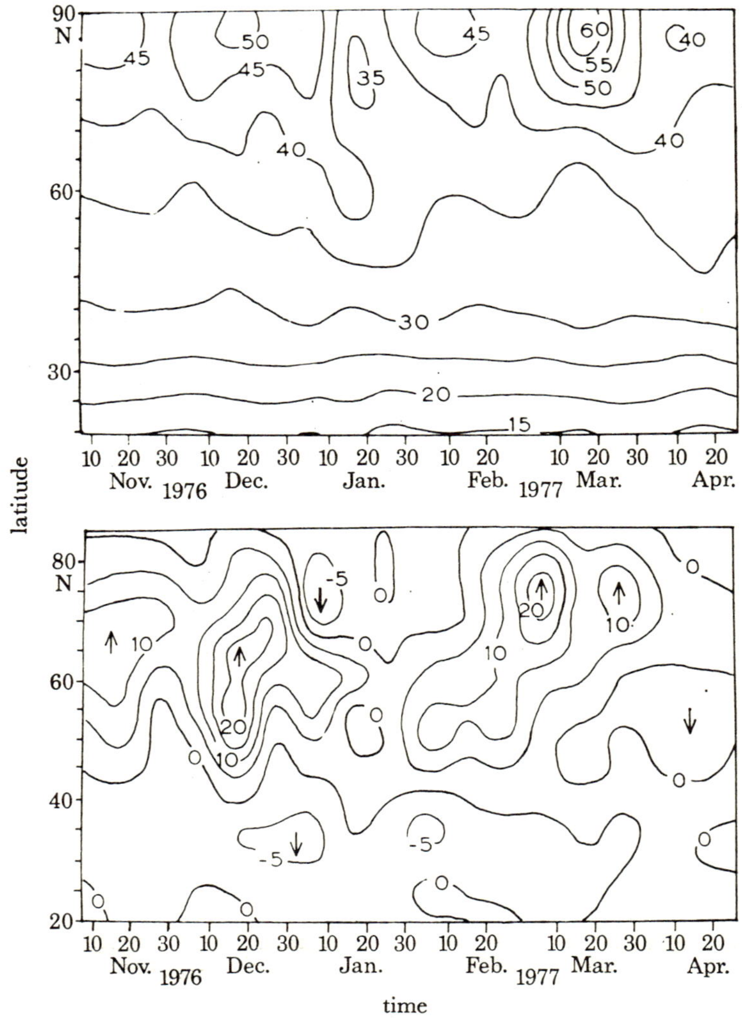

FIGURE 13. Observed time latitude section at 30 mbar of (*a*) potential vorticity, Q, (*b*) northward eddy flux of potential vorticity, $\overline{v'Q'}\cos\phi$, where ϕ = latitude. Note the counter-gradient flux at high latitudes.

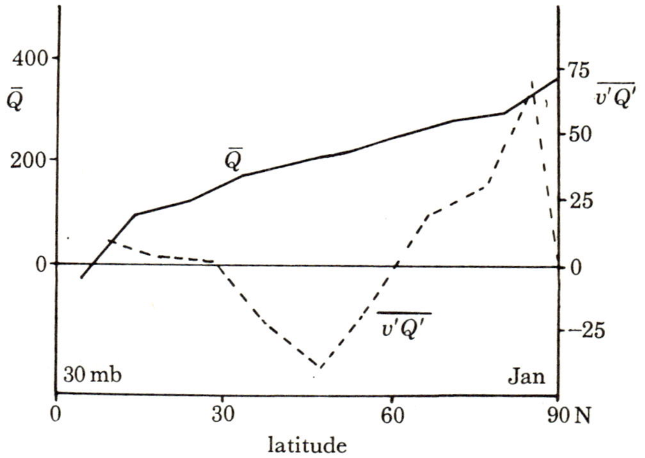

FIGURE 14. Distribution of potential vorticity $\bar{Q}$ and the northward eddy flux $\overline{v'Q'}$, both non-dimensionalized calculated for the C.I.R.A. 65 January atmosphere using the diffusion coefficients of Luther (1974).

transports in the Southern Hemisphere. Note that the 'up-gradient' transport of high latitudes must arise because the mixing paths (assuming the concept is applicable) slope more steeply than the surfaces of constant potential vorticity which slope poleward and downward.

The final figure (14) shows the potential vorticity flux deduced from a January distribution in a standard atmosphere (C.I.R.A. 1965) using the K values of Luther. The qualitative structure is about right but the magnitude of the mid-latitude poleward flux is too small. This suggests that the implied slope of the mixing paths is too shallow in this region. It is significant that Fabian & Libby (1974) found that the main discrepancy between observed and modelled tracer spread using these diffusion coefficients also pointed to an underestimation of this slope.

In view of the large arbitrariness in the usual method of finding diffusion coefficients, it seems probable that a set of coefficients could be found which would give realistic fluxes of both heat and potential vorticity thus allowing the momentum flux and hence the mean circulation to be calculated from the model state. Although such a model would be susceptible to the same criticism as all K-theories, it would have a powerful consistency between the assumptions adopted about the chemical transports and the dynamics.

5. Concluding remarks

The interactions between tracer gases, radiation and dynamics in the stratosphere are clearly too complex for a full and consistent treatment to be achieved with the computing power available in the foreseeable future. Transport processes depend on three dimensional systems with lifetimes of days. As yet, no sound physical basis exists for parametrizing the transports, so the only way to guarantee consistency is to perform a calculation capable of resolving this time and space scale. However, the stratospheric residence times are such that, even if other chemical factors are held constant, a simulation needs to extend for 2 or 3 years for a quasi-steady state to be reached. Such a calculation is at the limit of present computing power. However, the complete treatment of the chemistry requires many species and reactions to be included; it requires runs extending for decades because of species with reaction times of years, and simultaneously requires very short time steps to be taken to accommodate quick reactions or rapid adaptations to sunrise and sunset.

Understanding will grow only by considering a spectrum of models each of which tackles only part of the problem, from sophisticated three dimensional dynamical models with very crude chemistry on the one hand to multi-species chemical models with very crude dynamics on the other. Two dimensional models have an important role to play, lying in the middle of this spectrum. A pressing need here is to 'close' the calculation of the meridional circulation by parametrizing the momentum flux. The remarks in the previous section give some hope that this is possible. Another pressing need in two dimensional models is for a more consistent diffusion theory. Here we may anticipate that the Lagrangian averaging methods currently being explored in wave–mean flow interaction (Andrews & McIntyre 1978) may eventually provide a firmer footing.

I am indebted to Dr J. A. Pyle and Miss J. Haigh for figure 12.

References (Harwood)

Alyea, F. N., Cunnold, D. M. & Prinn, R. G. 1975 *Science, N.Y.* **188**, 117–121.
Andrews, D. G. & McIntyre, M. E. 1978 *J. Fluid. Mech.* **89**, 609–646.
Blake, D. & Lindzen, R. S. 1973 *Mon. Weath. Rev.* **101**, 783–802.
Charney, J. G. & Drazin, P. G. 1961 *J. geophys. Res.* **66**, 83–109.
Chen, T.-S. & Ramanathan, V. 1978 *J. atmos. Sci.* **35**, 615–633.
Clark, J. H. E. 1970 *Mon. Weath. Rev.* **98**, 443–461.
C.O.M.E.S.A. 1975 *The report of the committee on meteorological effects of stratospheric aircraft* (C.O.M.E.S.A.), 1972–1975. Meteorological Office, Bracknell, England.
Corby, G. A., Gilchrist, A. & Newson, R. L. 1972 *Q. Jl R. met. Soc.* **98**, 809–832.
C.O.V.O.S. 1976 *Comité d'Etude sur les conséquences de vols stratosphériques. Activités* 1972–1976: Rapport de synthèse. Société Météorologique de France.
Cunnold, D., Alyea, F., Phillips, N. & Prinn, R. 1975 *J. atmos. Sci.* **32**, 170–194.
Dickinson, R. E. 1968 *Mon. Weath. Rev.* **96**, 405–415.
Dobson, G. M. B., Harrison, D. N. & Lawrence, J. 1928 *Proc. R. Soc. Lond.* A **122**, 456–486.
Dütsch, H. U. 1971 *Adv. Geophys.* **19**, 219–322.
Fabian, P. & Libby, W. F. 1974 *Proc. third conference on the climatic impact assessment program* (ed. A. J. Broderick & T. M. Hard), pp. 103–116. U.S. Department of Transportation, Washington, D.C.
Green, J. S. A. 1970 *Q. Jl R. met. Soc.* **96**, 157–185.
Gudiksen, P. H., Fairhall, A. W. & Reed, R. J. 1968 *J. geophys. Res.* **73**, 4461–4473.
Hartmann, D. L. 1976 *J. atmos. Sci.* **33**, 1141–1154.
Hartmann, D. L. 1977 *J. atmos. Sci.* **34**, 968–977.
Harwood, R. S. 1975 *Q. Jl R. met. Soc.* **101**, 75–91.
Harwood, R. S. 1977 *Proc. scientific seminar on stratospheric monitoring.* Paris 21–23 March 1977. Direction de la Météorologie: Secrétariat d'état auprès du Ministre de l'Equipment, Paris.
Harwood, R. S. & Pyle, J. A. 1975 *Q. Jl R. met Soc.* **101**, 723–748.
Harwood, R. S. & Pyle, J. A. 1977 *Q. Jl R. met. Soc.* **103**, 319–343.
Hesstvedt, E. 1974*a* *Can. J. Chem.* **52**, 1592–1598.
Hesstvedt, E. 1974*b* *Proc. International conference on structure, composition and general circulation of the upper and lower atmospheres and possible anthropogenic perturbations.* I.A.M.A.P./I.A.P.S.O. Special Assembly. Melbourne 1974.
Hirota, I. 1978 *J. atmos. Sci.* **35**, 714–722.
Holton, J. R. 1975 *Met. Mon.* **15**. American Meteorological Society.
Holton, J. R. 1976 *J. atmos. Sci.* **33**, 1639–1649.
Hunt, B. G. 1969 *Mon. Weath. Rev.* **97**, 287–306.
Hunt, B. G. & Manabe, S. 1968 *Mon. Weath. Rev.* **96**, 503–539.
Kats, A. L. 1968 Gidrometeoizdat Leningrad USSR. Reported in *CIAP Monograph*, no. 1, ch. 2. Climatic Impact Assessment Program, U.S. Department of Transportation, Washington, D.C.
Leovy, C. B. & Webster, P. J. 1976 *J. atmos. Sci.* **33**, 1624–1638.
Lorenz, E. N. 1960 *Tellus* **12**, 364–373.
Lindzen, R. S. & Goody, R. 1965 *J. atmos. Sci.* **22**, 341–348.
Luther, F. M. 1974 *Second annual report. DOT-CIAP Program.* URRL-51336-74. Lawrence Livermore Laboratory, California.
Mahlman, J. D. & Moxim, W. J. 1978 *J. atmos. Sci.* **35**, 1340–1374.
Manabe, S. & Mahlman, J. D. 1976 *J. atmos Sci.* **33**, 2185–2217.
Matsuno, T. 1970 *J. atmos. Sci.* **27**, 871–883.
Matsuno, T. 1971 *J. atmos. Sci.* **28**, 1479–1494.
Murgatroyd, R. J. 1969 *The structure and dynamics of the stratosphere – the global circulation of the atmosphere* (ed. G. A. Corby), pp. 159–195. London: Royal Meteorological Society.
National Academy of Sciences 1976 *Halocarbons: Effects on stratospheric ozone.* Washington, D.C.
Newell, R., Vincent, D., Dopplick, T., Ferruzza, D. & Kidson, J. 1969 *The energy balance of the global atmosphere – the global circulation of the atmosphere* (ed. G. A. Corby), pp. 42–90. London: Royal Meteorological Society.
Newson, R. L. 1974 *Proc. 3rd conference on the Climatic Impact Assessment program* (ed. A. J. Broderick & T. M. Hard). Report No. DOT-TSC-OST-74-15. U.S. Department of Transporation, Washington, D.C.
Peng, L. 1965 *Pure appl. Geophys.* **61**, 191–218.
Pyle, J. A. 1976 Ph.D. Thesis. Oxford University.
Pyle, J. A. 1978 *Nature, Lond.* **271**, 42–43.
Ramanathan, V. & Grose, W. L. 1978 *J. atmos. Sci.* **35**, 600–614.
Rao, V. R. K. 1973 *Mon. Weath. Rev.* **101**, 510–527.
Rao, V. R. K. & Christie, A. D. 1973 *J. atmos. Sci.* **30**, 667–676.
Reed, R. J. & German, K. E. 1965 *Mon. Weath. Rev.* **93**, 313–321.
Rodgers, C. D. 1967 *Report no.* A2, Department of Meteorology, Massachusetts Institute of Technology.

Saltzmann, B. & Vernekar, A. D. 1968 *Mon. Weath. Rev.* **96**, 854–857.
Schoeberl, M. R. & Strobel, D. F. 1978 *J. atmos. Sci.* **35**, 577–591.
Simmons, A. J. 1974 *Q. Jl R. met. Soc.* **100**, 76–108.
Telegadas, K. 1971 *Health and Safety Laboratory, Fallout Program Quart. Summary Rep.*, HASL-243, 1–87.
Trenberth, K. E. 1973 *Mon. Weath. Rev.* **101**, 306–322.
Tuck, A. F. 1979 *Phil. Trans. R. Soc. Lond.* A **290**, 477–494.
Vincent, D. G. 1968 *Q. Jl R. met. Soc.* **94**, 333–349.
White, A. A. 1977 *Q. Jl R. met. Soc.* **103**, 93–120.
White, A. A. 1979 *Mon. Weath. Rev.* (to appear).
Widhopf, G. F. & Taylor, T. D. 1975 *Proc. 3rd Conference on the Climatic Impact Assessment program*, (ed. A. J. Broderick & T. M. Hard) *Report no.* DOT-TSC-OST-74-17. U.S. Department of Transportation, Washington, D.C.
Williams, G. P. & Davies, D. R. 1965 *Q. Jl R. met. Soc.* **91**, 471–489.
Wu, M.-F. 1973 *Annual report*, contract DOT-05-2-217, Environmental Research and Technology. Lexington, Mass.

Phil. Trans. R. Soc. Lond. A **296**, 129–148 (1980) [129]
Printed in Great Britain

Towards a Lagrangian-mean description of stratospheric circulations and chemical transports

By M. E. McIntyre

Joint Institute for the Study of the Atmosphere and Ocean, University of Washington, Seattle,†
and Department of Applied Mathematics and Theoretical Physics, University of Cambridge, Silver Street, Cambridge, U.K.

The theory of Lagrangian means and Stokes drifts is reviewed, with particular attention to recent developments and to possibilities for applying the ideas to the stratosphere. The properties of Lagrangian means for finite-amplitude disturbances in spherical geometry pose certain conceptual and technical problems, which appear unavoidable if the notion of Lagrangian mean is to be defined in an exact, coordinate-independent manner compatible with Stokes's classical theory.

1. Motivation

A number of stratospheric phenomena, which seem obscure and complicated within the standard description using Eulerian zonal means and deviations, are better appreciated from a Lagrangian viewpoint (see, for example, Reed 1955; Staley 1960; Danielsen 1968; Mahlman 1969; Shapiro 1978; Dunkerton 1978; Wallace 1978; Matsuno 1979). One example is the well known propensity of Eulerian eddy fluxes of quasi-conservative quantities to have almost any direction relative to their local mean gradients (see, for example, Mahlman 1975, p. 138; Matsuno 1979; Plumb 1979). The mean behaviour of the same quasi-conservative quantities, e.g. the mean motion of certain trace gases, comprises another, related, example (see, for example, Brewer 1949; Dobson 1956; references cited above).

A fictitious stratosphere in which potential temperature θ and tracers are exactly conservative, i.e.

$$\mathrm{D}X/\mathrm{D}t = 0, \tag{1}$$

where $\mathrm{D}/\mathrm{D}t$ is the rate of change following an air parcel, and X stands for θ or for tracer mixing ratio, presents an impenetrable barrier to tracers if statically stable everywhere. Each isentropic surface is a material surface, and no tracer satisfying (1) can be carried across it. This statement is obvious, but the fact that it holds whether or not planetary waves or other disturbances to zonal symmetry are present is significant. It reminds us that the Eulerian eddy fluxes of tracers associated with such disturbances cannot give rise to any net tracer transport across isentropic surfaces, except to the extent that the right-hand side of (1) is not zero. If our 'stratosphere' is frictionless, as well as satisfying (1) with $X = \theta$, then surfaces of constant potential vorticity are similarly impenetrable.

The Eulerian zonal mean of equation (1) is

$$\left(\frac{\partial}{\partial t}+\bar{v}\frac{\partial}{\partial y}+\bar{\omega}\frac{\partial}{\partial p}\right)\bar{X} = -\frac{1}{\cos\varphi}\frac{\partial}{\partial y}(\overline{v'X'}\cos\varphi)-\frac{\partial}{\partial p}(\overline{\omega'X'}), \tag{2}$$

† This paper is JISAO contribution no. 4.

where overbars and primes denote Eulerian zonal means and departures therefrom, v and ω are the usual measures of meridional and vertical velocity in a coordinate system in which the vertical 'coordinate' is pressure p, and y is northward distance, equal to latitude φ times the Earth's radius. If the mean state of our fictitious 'stratosphere' undergoes no systematic changes as time goes on, there must be a long term tendency for the effects of the eddy-flux terms on the right of equation (2) to be cancelled by the effects of the Eulerian-mean meridional circulation $(\bar{v}, \bar{\omega})$ on the left (see, for example, Sawyer 1965, p. 414). As is well known, such cancellations are also observed in the real stratosphere, where for many of the quantities X of interest the right-hand side of (1) is often small even if not zero. These considerations, then, all serve to remind us that the eddy fluxes arising from taking Eulerian means are not very directly related to tracer transport; and they suggest that some kind of Lagrangian averaging (see, for example, Stokes 1847; Riehl & Fultz 1957; Krishnamurti 1961; Mahlman 1969; Kida 1977; Dunkerton 1978; Matsuno & Nakamura 1979; Matsuno 1979) might be a more natural way of defining the mean state of the stratosphere, for use in two dimensional descriptions or models of tracer behaviour.

The simplest concept of Lagrangian averaging is Stokes's classical idea of taking the time mean following a single air parcel (for a good discussion of the classical theory see Longuet-Higgins 1969). This idea is sometimes all that is needed in practice; but it has inherent limitations. If we wish to speak of the Lagrangian-mean velocity at a given point in space, the classical concept cannot apply in any exact sense, because the air parcel we are following will generally wander away from the point under consideration. Thus any exact theory of the Lagrangian-mean velocity *field* must abandon the simple concept of the mean following a single air parcel (Andrews & McIntyre 1978*b* and references, hereafter A.M.). The result will be neither a pure Lagrangian nor a pure Eulerian description but, rather, a *hybrid* Eulerian–Lagrangian description of the motion.

The result of requiring that the concept of 'Lagrangian–mean velocity field' be defined exactly, and that it reduce to the classical theory when the approximations inherent in the latter are applicable, is the generalized Lagrangian-mean (g.l.m.) description of wave, mean-flow interaction developed independently by A.M. and by Bretherton (1979). That theory has provided a powerful analytical key to problems concerning the interaction of waves and mean flows (see, for example, Andrews & McIntyre 1976, 1978*a–c*; Bretherton 1979; Dunkerton 1979; Grimshaw 1979; Leibovich 1979; McIntyre 1979). One reason is that the theory avoids the indirectness of equation (2): when (1) holds the equation for the generalized Lagrangian mean $\bar{X}^{\mathrm{L}}$ of X (to be defined below) can be shown to take the simple form

$$\left(\frac{\partial}{\partial t}+\bar{v}^{\mathrm{L}}\frac{\partial}{\partial y}+\bar{\omega}^{\mathrm{L}}\frac{\partial}{\partial p}\right)\bar{X}^{\mathrm{L}} = 0 \tag{3}$$

(A.M., equation 2.22). That is, it has the same form as (2) except that there are *no eddy-flux terms* on the right. This is not to say that the generalized Lagrangian-mean theory holds all eddy effects to be unimportant; it merely highlights the fact, already mentioned, that the effects of wave or eddy motion on net heat and tracer transport are necessarily bound up with departures from conservative motion, and not necessarily with wave activity *per se*. (Mahlman *et al.* (1979) refer to such statements as 'non-transport theorems'.) Departures from conservative motion entail the appearance of some quantity $Q_X \neq 0$ on the right-hand side of (1), representing effects of small-scale turbulence as well as radiation and photochemistry; the corresponding term then appearing on the right of (3) is $\bar{Q}^{\mathrm{L}}_{X}$. The value of $\bar{Q}^{\mathrm{L}}_{X}$ can be affected by eddy motions (just as can its Eulerian-mean counterpart $\bar{Q}$); such motions, therefore, are certainly important in the equation for $\bar{X}^{\mathrm{L}}$, but they are important *only* in so far as they promote mixing or other departures from conservative motion.

In view of these remarks it is natural to hope that the g.l.m. theory could be applied straightforwardly to the stratosphere, or to numerical models thereof. There are, however, a number of difficulties in the way of using the theory as a practical diagnostic tool, and it is one purpose of this paper to discuss, or at least to state explicitly, what those difficulties seem to be at our present stage of thinking. This is done in § 4 below, after reviewing the basic analytical theory, and its relation with classical concepts, in § 2. In § 3 the important problem of 're-initializing' the theory is discussed briefly; this, too, can have its difficulties.

If blind application of Lagrangian-mean theory is impracticable, this may of course be a good thing in the long run. It can be argued that the very ease with which Eulerian-mean quantities can be calculated, for any complicated fluid flow, has made them something of a two-edged weapon. Elucidation of rates of conversion between mean and eddy energies, for instance, has sometimes been mistaken for a full understanding of the eddy dynamics involved. What the g.l.m. theory does provide at its present stage of development is a precise analytical structure which, first of all, makes clear what is involved in generalizing the classical notions of Lagrangian mean and Stokes drift, and which, secondly, makes available certain analytical tools whose use might prove helpful in the search for a fundamentally improved understanding of the stratosphere. Just how those tools can best be used is not likely to be clear until some detailed applications have been attempted; and it is hoped that the present discussion will help stimulate researchers to tackle some of the intriguing problems involved.

2. Basic theoretical ideas

2.1. *Classical and generalized Lagrangian means*

We start by expressing the classical notion of Lagrangian mean in a form suitable for subsequent generalization, following A.M. For the purposes of exposition we shall forget about pressure coordinates for the present, and use ordinary Cartesian coordinates.† The essential idea is to average over positions displaced by the disturbance motion. In other words, if $\phi(\boldsymbol{x}, t)$ is the quantity to be averaged and $\overline{\phi(\boldsymbol{x}, t)}$ its Eulerian mean at $\boldsymbol{x}$ in any of the usual senses (time, space, ensemble, etc.), then the corresponding Lagrangian mean is defined as

$$\bar{\phi}^{\mathrm{L}}(\boldsymbol{x}, t) \equiv \overline{\phi(\boldsymbol{x}, t)}^{\mathrm{L}} \equiv \overline{\phi\{\boldsymbol{x}+\boldsymbol{\xi}(\boldsymbol{x}, t), t\}}, \tag{4}$$

where $\boldsymbol{\xi}(\boldsymbol{x}, t)$ is the disturbance-induced *displacement* of the fluid parcel, in a sense to be elucidated. (If the disturbance motion is the dominant motion, the meaning of 'displacement' is fairly obvious *ab initio* and is such that if $\overline{(\ )}$ is the time mean at a fixed point, then $\overline{(\ )}^{\mathrm{L}}$ is the time mean following a single fluid parcel, at least approximately.) The Stokes correction to $\bar{\phi}$ is defined as the difference between the Eulerian and the Lagrangian mean:

$$\overline{\phi(\boldsymbol{x}, t)}^{\mathrm{S}} \equiv \overline{\phi(\boldsymbol{x}, t)}^{\mathrm{L}} - \overline{\phi(\boldsymbol{x}, t)}. \tag{5}$$

† This involves no loss of generality, as yet: the use of any particular coordinate system does not in itself restrict the type of mean flow considered, provided averaging is defined in a coordinate-independent way (A.M. § 2.1).

When ϕ is velocity $\boldsymbol{u}$, the Stokes correction $\bar{\boldsymbol{u}}^S$ is sometimes referred to as the Stokes 'drift'. A typical illustration of the Stokes-drift effect is furnished by the parcel trajectory shown in figure 1. This was calculated for a hypothetical, oscillatory fluid motion whose Eulerian-mean velocity $\bar{\boldsymbol{u}}$ is exactly zero: the mean velocity following the parcel is, by contrast, quite different from zero.

To define ξ for small-amplitude waves, it is customary to resort to the simplest possible approximation

$$\frac{\partial}{\partial t}\,\xi(\boldsymbol{x}, t) = \boldsymbol{u}'(\boldsymbol{x}, t), \tag{6}$$

(see, for example, Longuet-Higgins 1969, equation (2)), where we have defined, again bearing subsequent generalization in mind,

$$\boldsymbol{u}'(\boldsymbol{x}, t) \equiv \boldsymbol{u}(\boldsymbol{x}, t) - \overline{\boldsymbol{u}(\boldsymbol{x}, t)} \tag{7}$$

(the familiar Eulerian measure of disturbance velocity), and where

$$\overline{\xi(\boldsymbol{x}, t)} = 0. \tag{8}$$

The last relation formally expresses the idea that ξ should be a *disturbance-associated* quantity.

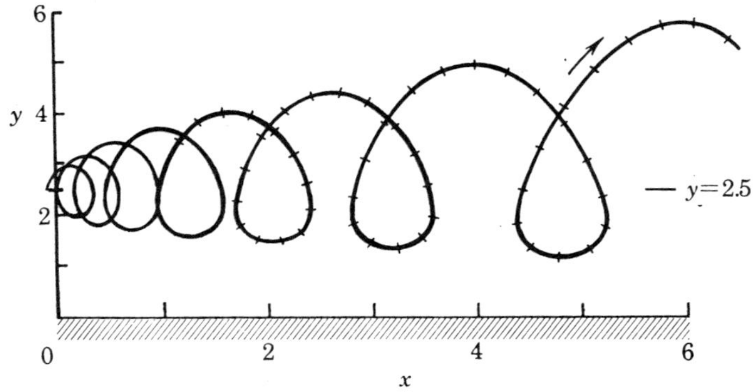

FIGURE 1. Path of a single fluid parcel in the two dimensional velocity field $u = 0.01\,t\cos(x-t)$, $v = 0.01\,yt\sin(x-t)$, satisfying the incompressibility condition $u_x + v_y = 0$. At $t = 0$ the parcel is at $x = 0$, $y = 2.5$; its subsequent trajectory was computed by means of a fourth order Runge–Kutta algorithm. Part of the path is marked at equal time intervals $\Delta t = 0.5$.

Since equation (6) is a differential equation, some kind of initial condition or equivalent information must be prescribed. When $\overline{(\ )}$ is a time average the condition (8) is sufficient to provide that information. But when $\overline{(\ )}$ is a space or ensemble average, (8) is *not* sufficient, because (6) and (8) are still satisfied if we add to ξ any time-independent function $\boldsymbol{f}(\boldsymbol{x})$ with zero mean ($\bar{\boldsymbol{f}} = 0$). This point is not usually discussed, because in classical wave theories there is usually no point in taking $\overline{(\ )}$ as anything other than a time mean, to within the approximations involved in those theories and in equation (6).

It can be shown that the approximations involved in using these ideas in their classical form require, in fact, that

(i) the disturbance amplitude a is small, in the sense that each component of ξ must be much smaller in magnitude than the corresponding smallest length scale of the disturbance; and

(ii) the mean velocities (Eulerian and Lagrangian) are both very small, formally $O(a^2)$ as $a \to 0$. (Thus the last term in (7) is actually irrelevant to the discussion so far.)

Although these requirements are sometimes satisfied in practice, it would seem desirable to have exact definitions applicable to finite-amplitude disturbances on arbitrary mean flows. Such definitions are provided by the generalized Lagrangian-mean theory, which can be seen as a natural generalization of the classical theory if we accept as natural the following three ideas. The first is that we should retain the relations (4), (5) and (8) exactly as they are, for analytical as well as conceptual simplicity. The second is the very reasonable idea that the natural Lagrangian measure of 'disturbance velocity' is not the right-hand side of equation (7) but, rather, a similar expression in which the last term is the *Lagrangian* mean, and the first term on the right is evaluated at the *displaced* position $\boldsymbol{x}+\boldsymbol{\xi}$. The resulting 'Lagrangian disturbance velocity' will be denoted by $\boldsymbol{u}^l$; by definition

$$\boldsymbol{u}^l(\boldsymbol{x}, t) \equiv \boldsymbol{u}\{\boldsymbol{x}+\boldsymbol{\xi}(\boldsymbol{x}, t), t\} - \overline{\boldsymbol{u}(\boldsymbol{x}, t)}^{\mathrm{L}}. \tag{9}$$

Note that this definition together with (4) (with $\boldsymbol{u}$ substituted for ϕ)) implies that $\overline{\boldsymbol{u}^l} = 0$, so that $\boldsymbol{u}^l$ is a disturbance-associated quantity, like $\boldsymbol{u}'$ or $\boldsymbol{\xi}$. Other Lagrangian disturbance quantities ϕ^l are defined similarly.

The third idea, which is obviously necessary if we want to remove the restriction (ii), and obtain a theory which behaves sensibly in a moving frame of reference, is that $\partial/\partial t$ in equation (6) should be replaced by the rate of change following the mean flow; and it is natural in this context to use the *Lagrangian*-mean flow. Putting the last two ideas together, we replace (6) by

$$\left(\frac{\partial}{\partial t} + \bar{\boldsymbol{u}}^{\mathrm{L}}\cdot\nabla\right)\boldsymbol{\xi} = \boldsymbol{u}^l, \tag{10}$$

where ∇ is the three dimensional gradient operator. Equations (4), (8), (9), and (10) are the basic equations of the generalized Lagrangian-mean theory. They assert that for each type of Eulerian-mean operator $\overline{(\ )}$, whether the averaging is done with respect to time, space, or an ensemble of realizations, there is a corresponding Lagrangian-mean operator $\overline{(\ )}^{\mathrm{L}}$. Once again, (10) is a differential equation and so suitable initial (or 'upstream') conditions, actual or hypothetical, are required to complete the definition, as discussed in A.M. (see their postulate (*viii*), p. 617). The simplest such conditions specify an initial state of no disturbance anywhere.

At first sight equations (4), (8), (9), and (10) look frighteningly implicit, but it is not hard to see that in the particular case where $\overline{(\ )}$ is a zonal average their content can be visualized with the aid of the following *mechanical analogy* (A.M. § 2.2). The analogy as we shall first describe it applies to the case where the Eulerian averaging operator $\overline{(\ )}$ is a rectilinear spatial average, parallel to the x axis, say, of some Cartesian coordinate system. We initialize the theory by assuming zero disturbance everywhere at $t = t_0$, say, and fix attention on a row R_0 of marked fluid parcels which are initially spaced at equal distances Δx along a line parallel to the x axis (figure 2*a*). We then watch those parcels as they follow the fluid motion. In the mechanical analogy (whose connection with the fluid motion is only kinematical, not dynamical) we imagine that a massless, rigid rod R initially coincides with R_0, but is subsequently free to move while constrained to remain parallel to the x axis. The position P of a typical parcel of fluid whose initial position was P_0 is joined to the point P_{R} on R, which initially coincided with P_0, by an idealized 'elastic band' $P_{\mathrm{R}}P$. This elastic band is also massless, and is such that P_{R} is pulled towards P (whose position is known from the given fluid motion) with a force just proportional to the distance $P_{\mathrm{R}}P$. Identical elastic bands similarly join all the other parcels to their corresponding points on the rod R, which is imagined to be in static equilibrium under the pull of all the elastic bands (and will therefore follow the parcels if they all happen to have a systematic motion in some direction).

Now if $\boldsymbol{x}$ is the current position of P_{R} at time t, $\boldsymbol{\xi}(\boldsymbol{x}, t)$ is the 'elastic band vector' $\overrightarrow{P_{\mathrm{R}}P}$, in the limit $\Delta x \to 0$, and $\boldsymbol{\bar{u}}^{\mathrm{L}}(\boldsymbol{x}, t)$ is the velocity of the rod R. For, in the first place, equation (8) is satisfied since in the limit $\Delta x \to 0$ it expresses the vanishing of the resultant force on R.

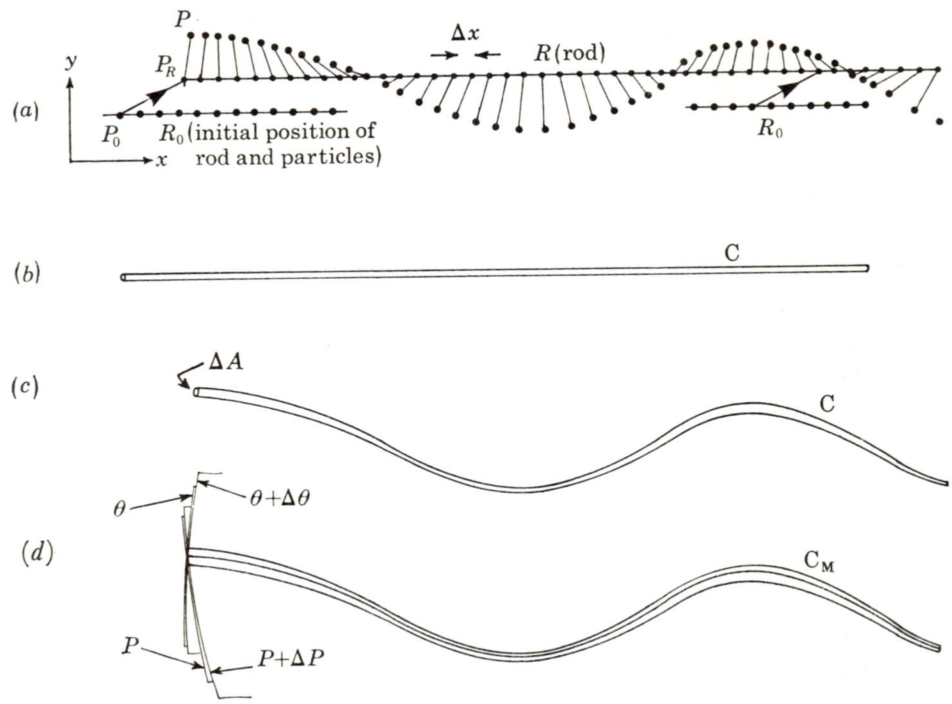

FIGURE 2. (*a*) Mechanical analogy for visualizing both the generalized Lagrangian mean and the finite-amplitude disturbance displacement function $\boldsymbol{\xi}(\boldsymbol{x}, t)$, in the case where averaging is spatial; see text below equation (10). (*b*) Uniform material tube C in a hypothetical initial state of no disturbance, corresponding to R_0 in (*a*). (*c*) Same material tube in the presence of a disturbance, giving domain of averaging for the spatial generalized Lagrangian mean (11). (*d*) Tube $\mathrm{C_M}$ marked out by surfaces of constant potential vorticity P and potential temperature θ.

(This depends on the assumed properties of our idealized elastic bands, and also on the equal spacing along R of the points $\boldsymbol{x}$ typified by P_{R} since the Eulerian averaging operator $\overline{(\ )}$ has uniform weighting in x.) In the second place, the properties of the elastic bands ensure that the rod's velocity will be just the average velocity of all the fluid parcels. But (4) with $\phi = \boldsymbol{u}$ states that the average velocity in this sense is $\boldsymbol{\bar{u}}^{\mathrm{L}}(\boldsymbol{x}, t)$, in the limit $\Delta x \to 0$, again because of the equal spacing of points along R. Finally, the remaining equations (9) and (10) are satisfied because (10) expresses the fact that $\boldsymbol{u}^l$ is equal to the rate of change of an individual elastic-band vector $\boldsymbol{\xi}$ such as $\overline{P_{\mathrm{R}}P}$, while (9) expresses the fact that this rate of change $\boldsymbol{u}^l$ is just the velocity of one end P of the elastic band relative to the other, P_{R}.

It was emphasized that the average given by equation (4) is an average along the line of parcels weighted by the number of parcels per unit length, when the parcels were *equally spaced along* $\mathrm{R_0}$ in the initial state of no disturbance. This is equivalent to an average weighted by mass along a material tube C of fluid which was initially straight and uniform as suggested in figure 2*b*, and which has been distorted by the disturbance into the shape shown in figure 2*c*, with variable cross-sectional area ΔA. That is, the definition (4) is equivalent, in the present case of a Cartesian spatial average, to the alternative definition

$$\overline{(\ )}^{\mathrm{L}} \equiv \int_{\mathrm{C}} (\ )\, \rho \Delta A \,\mathrm{d}s \Big/ \int_{\mathrm{C}} \rho \Delta A \,\mathrm{d}s, \tag{11}$$

where $\mathrm{d}s$ is an element of arc length along C, ρ is density, and the cross-sectional area ΔA of the tube C is taken to be vanishingly small. It is important to realize that the non-uniform weighting of the average along C is essential if $\overline{(\ )}^{\mathrm{L}}$ is to correspond to the classical (unweighted) *time* mean following a single parcel, in situations where the classical approximations apply and where the Eulerian time and space means are equal. One such situation is discussed by Matsuno (1979).

An interesting corollary of the foregoing is that $\bar{\boldsymbol{u}}^{\mathrm{L}}$, besides being the velocity of the rod R in figure 2*a*, is also the velocity of the centre of mass of the tube C in figure 2*c* (A.M. § 4.3). The small-amplitude approximation to this exact result was previously noted by T. Matsuno (personal communication). From the centre-of-mass property we can deduce in turn that the Lagrangian-mean vertical velocity is just the rate of change of potential energy of the material tube C, per unit mass. It follows that the vertical Stokes drift, if defined exactly as in (5), can be regarded as an exact measure of the rate of change of eddy available potential energy; an explicit example will be noted in § 4.

2.2. *Small-amplitude approximations*

For disturbances of small amplitude a, Taylor expansion of (4) with respect to $\boldsymbol{\xi}$ and substitution into (5), making use of (8), shows that

$$\overline{\phi}^{\mathrm{S}} = \overline{\boldsymbol{\xi}\cdot\nabla\phi'} + \overline{\tfrac{1}{2}\boldsymbol{\xi}\cdot(\boldsymbol{\xi}\cdot\nabla)\nabla\overline{\phi}} + O(a^3), \tag{12}$$

where ∇ is the three dimensional gradient operator; for more details see A.M. § 2.4. Written out in Cartesian component form for the case of a rectilinear spatial average with respect to x, the second term becomes

$$\tfrac{1}{2}(\overline{\eta^2}\overline{\phi}_{yy} + 2\overline{\eta\zeta}\overline{\phi}_{yz} + \overline{\zeta^2}\overline{\phi}_{zz}), \tag{13}$$

where $\boldsymbol{x} = \{x, y, z\}$ and $\boldsymbol{\xi} = \{\xi, \eta, \zeta\}$, and where subscripts denote partial differentiation. If in addition the flow is incompressible, the first term on the right of (12) may be rewritten

$$\overline{(\eta\phi')}_y + \overline{(\zeta\phi')}_z + O(a^3), \tag{14}$$

since for incompressible flow it may be shown that

$$\nabla\cdot\boldsymbol{\xi} = O(a^2) \tag{15}$$

(A.M. § 9). In all these small-amplitude formulae, including the general formula (12), we may approximate (10) by

$$(\partial/\partial t + \bar{\boldsymbol{u}}\cdot\nabla)\,\boldsymbol{\xi} = \boldsymbol{u}' + \boldsymbol{\xi}\cdot\nabla\bar{\boldsymbol{u}} + O(a^2). \tag{16}$$

The relation (12) reminds us of the well known fact that the Stokes correction $\overline{\phi}^{\mathrm{S}}$ is $O(a^2)$, that is to say second order in disturbance amplitude a, as $a \to 0$. The first term on the right of (12) represents the result originally derived by Stokes (1847); it expresses the correction to the Eulerian mean arising from displacements through disturbance-associated gradients.

Longuet-Higgins (1969) presents some alternative expressions for this term which hold in cases where all averaged quantities are time-independent. The second term of (12), a correction arising from displacements through non-uniform *mean* gradients of ϕ, is neglected in classical treatments (and is sometimes overlooked in cases where it is not negligible). It is easy to discern examples of meteorological interest where the effects described by this term will not be negligible, and one such example will be mentioned in § 5 below. Another such example is the general problem of the mean zonal acceleration due to equatorial planetary waves (A.M. § 9).

2.3. *Other averages arising in the generalized Lagrangian-mean theory*

The operator $\overline{(\ )}^{\mathrm{L}}$ is by no means the only type of average that arises in the g.l.m. theory; different averages are appropriate to different circumstances. For example the most natural measure of 'mean density' for the purpose of expressing mass conservation analytically turns out to be the quantity

$$\tilde{\rho}(\boldsymbol{x}, t) = \rho(\boldsymbol{x}+\boldsymbol{\xi}, t)\,\frac{\partial(x+\xi, y+\eta, z+\zeta)}{\partial(x, y, z)}, \tag{17}$$

where the last factor is a Jacobian determinant involving the derivatives of $\boldsymbol{\xi}$ with respect to the reference position $\boldsymbol{x}$:

$$\frac{\partial(x+\xi, y+\eta, z+\zeta)}{\partial(x, y, z)} = \begin{vmatrix} 1+\xi_x & \xi_y & \xi_z \\ \eta_x & 1+\eta_y & \eta_z \\ \zeta_x & \zeta_y & 1+\zeta_z \end{vmatrix}. \tag{18}$$

Despite the absence of an overbar on the right of (17), $\tilde{\rho}$ is indeed a mean quantity ($\bar{\tilde{\rho}} = \tilde{\rho}$) when the theory is properly initialized, as is shown in A.M. § 4. It is also shown there that $\tilde{\rho}$ satisfies

$$\partial\tilde{\rho}/\partial t + \nabla\cdot(\tilde{\rho}\bar{\boldsymbol{u}}^{\mathrm{L}}) = 0, \tag{19}$$

which, analytically speaking, is the simplest expression of mean mass conservation in the g.l.m. theory.

Another physically important mean quantity is

$$\begin{aligned} \boldsymbol{U}(\boldsymbol{x}, t) &\equiv \overline{\{\nabla(\boldsymbol{x}+\boldsymbol{\xi})\}\cdot\{\boldsymbol{u}(\boldsymbol{x}+\boldsymbol{\xi}, t)+\boldsymbol{\Omega}\times(\boldsymbol{x}+\boldsymbol{\xi})\}} \\ &= \bar{\boldsymbol{u}}^{\mathrm{L}}+\boldsymbol{\Omega}\times\boldsymbol{x}+\overline{(\nabla\boldsymbol{\xi})\cdot(\boldsymbol{u}^l+\boldsymbol{\Omega}\times\boldsymbol{\xi})}, \end{aligned} \tag{20}$$

where the dot denotes scalar multiplication operating on $\boldsymbol{x}$ and $\boldsymbol{\xi}$, not ∇, and where a frame of reference rotating with constant angular velocity $\boldsymbol{\Omega}$ is assumed. $\boldsymbol{U}$ is the basic measure of 'absolute' mean velocity underlying the 'non-acceleration theorems' and their generalizations of A.M. § 3. When $\overline{(\ )}$ is a rectilinear spatial average with respect to x, the x component U_1 of $\boldsymbol{U}$ is readily expressed in an alternative form paralleling (11),

$$U_1 = \int_{\mathrm{C}} (\boldsymbol{u}+\boldsymbol{\Omega}\times\boldsymbol{x})\cdot\mathrm{d}\boldsymbol{s}\Big/\int_{\mathrm{R}} \mathrm{d}x; \tag{21}$$

i.e. U_1 is Kelvin's *circulation* along C, divided by the length $\int \mathrm{d}x$ of the corresponding rod R in figure 2*a* (with the limit of infinitely long R and C understood if $\overline{(\ )}$ is a non-periodic, rectilinear spatial average). The key role of Kelvin's circulation function in the theory of wave, mean-flow interaction was recognized by Rayleigh (1896, § 352), in the different context of acoustics, and its importance for all kinds of oceanic and atmospheric waves (stemming from *Bjerknes's* circulation theorem) appears to have been first recognized by Bretherton

(1969*b*, 1971). A study of the behaviour of the circulation function itself (the numerator of (21)) in the wintertime stratosphere might be of great interest. Use of the circulation function avoids certain problems in computing Lagrangian-mean angular momentum budgets, the eddy contributions to which involve the pressure or height field (Bretherton 1969*a*, 1971, 1979; A.M. § 8; Uryu 1979; Dunkerton 1979) in such a way as to make accurate computation difficult. The budget of the circulation function, by contrast, does not involve the pressure field directly when C lies in an isentropic surface.

A further measure of mean motion of physical interest, especially for incompressible flow or for hydrostatic flow described in pressure coordinates, is the vector field $\boldsymbol{V}(\boldsymbol{x}, t)$ whose jth cartesian component is given by

$$V_j = \overline{\sum_{i=1}^{3} u_i(\boldsymbol{x}+\boldsymbol{\xi}, t)\, K_{ij}}. \tag{22}$$

Here K_{ij} is the two-by-two determinant consisting of the Jacobian (18) with its ith row and jth column removed. $\boldsymbol{V}$ has a natural interpretation as a measure of mean volume flux. When $\overline{(\ )}$ is a rectilinear spatial average with respect to x, the second or y component V_2 of $\boldsymbol{V}$, when multiplied by a small length Δz, represents the rate at which fluid volume crosses a fixed, ribbonlike surface $\mathrm{R}_{\Delta z}$ whose edges momentarily coincide with the two curves $\mathrm{C}(y, z, t)$ and $\mathrm{C}(y, z+\Delta z, t)$ which correspond to rods R momentarily at (y, z) and $(y, z+\Delta z)$. More precisely,

$$V_2 \Delta z = \int_{\mathrm{R}_{\Delta z}} \boldsymbol{u}\cdot\boldsymbol{n}\,\mathrm{d}S \Big/ \int_{\mathrm{R}} \mathrm{d}x, \tag{23}$$

where $\boldsymbol{n}$ is a unit normal to $R_{\Delta z}$ oriented toward the side corresponding to positive y, $\int_{\mathrm{R}} \mathrm{d}x$ is the length of rod corresponding to $R_{\Delta z}$ and the limit $\int_{\mathrm{R}} \mathrm{d}x \to \infty$ is understood where necessary, as before. The relation (23) is an immediate consequence of equation (A 15) of A.M. A similar relation gives $V_3 \Delta y$ with $R_{\Delta y}$ in place of $R_{\Delta z}$. The relation between $\boldsymbol{V}$ and $\bar{\boldsymbol{u}}^{\mathrm{L}}$ is not very direct, in that the latter does not give the rate at which fluid crosses the surfaces $R_{\Delta z}$, $R_{\Delta y}$ when those surfaces are held fixed in space: rather, the y and z components of $\tilde{\rho}\bar{\boldsymbol{u}}^{\mathrm{L}}$ give the rate at which fluid mass crosses the same surfaces when z, $z+\Delta z$ and y, $y+\Delta y$ are held fixed in the *reference* space of rod positions, in which case the surfaces $R_{\Delta z}$ and $R_{\Delta y}$ will generally be moving. Related to these properties are the facts that for an incompressible flow ($\nabla\cdot\boldsymbol{u} = 0$) we generally have

$$\nabla\cdot\boldsymbol{V} = 0 \quad \text{but} \quad \nabla\cdot(\tilde{\rho}\bar{\boldsymbol{u}}^{\mathrm{L}}) \neq 0. \tag{24}$$

The first relation is easy to verify directly from (22), using equations (A 3) and (A 4) of A.M. [respectively $\Sigma\, \partial K_{ij}/\partial x_j = 0$, and a result equivalent to the chain rule for differentiating a function of $\boldsymbol{x}+\boldsymbol{\xi}(\boldsymbol{x}, t)$]. The second relation of (24), a consequence of (17) and (19), is a property of the classical Lagrangian mean as well as its generalizations (see, for example, McIntyre 1973). It is discussed further in § 4 below.

3. Re-initialisation

We have emphasized the fact that the concept of 'Lagrangian mean' and the definitions of related mean quantities generally depend on a suitable initialization of the theory, implicit or explicit. One cannot normally speak of a unique generalized Lagrangian mean, but only of a generalized Lagrangian mean relative to the initialization used. One important exception, when the averaging is spatial, is strictly conservative motion in the case where the material tubes C can be defined by the intersection of surfaces of constant potential vorticity P and potential temperature θ, as suggested in figure 2*d* and discussed further in A.M. (appendix C). For the real stratosphere, however, P and θ are not precisely conserved, and we must expect the shapes of tubes C to become less and less closely related to the wave motion as time goes on. Such temporally non-uniform behaviour is a basic feature of any description of real, dissipative fluid motion using Lagrangian ideas, and it is clearly an essential part of how the 'non-acceleration' and 'non-transport' constraints are broken by real fluid motion. (In particular, these constraints can be broken catastrophically by large-amplitude wave events, such as those associated with sudden warmings (J. D. Mahlman, C.-P. Hsu, personal communication), which irreversibly coil parts of the tubes C into complicated shapes. We may aptly speak of 'breaking' planetary waves in such cases.)

As a practical matter, therefore, some kind of continuous or repeated *re-initialization* of the theory will be necessary if averages like (11), (21) or (23) are to be useful in numerical or observational diagnostics. Moreover, the most appropriate procedure would depend on the problem at hand. One possibility, when disturbance amplitudes are not too large, is simply to replace (11) by a 'modified Lagrangian mean' operator

$$\overline{(\ )}^{\mathrm{M}} \equiv \int_{C_{\mathrm{M}}} (\ )\,\rho\Delta A_{\mathrm{M}}\,\mathrm{d}s \Big/ \int_{C_{\mathrm{M}}} \rho\Delta A_{\mathrm{M}}\,\mathrm{d}s, \tag{25}$$

where ΔA_{M} is the cross-sectional area of the tube $C_{\mathrm{M}}(P, \theta)$ of figure 2*d*. This tube is bounded by the pair of constant-P surfaces whose potential vorticities are P and $P+\Delta P$, and the pair of isentropes whose potential temperatures are θ and $\theta+\Delta\theta$, where ΔP and $\Delta\theta$ are small. The idea of using C_{M} in this way arose in conversations with T. Dunkerton. $\overline{(\ )}^{\mathrm{M}}$ is the same as $\overline{(\ )}^{\mathrm{L}}$ when P and θ are conserved; but we are at liberty to define and use $\overline{(\ )}^{\mathrm{M}}$ whether or not equation (1) is satisfied, knowing that the whole theoretical structure resulting from its use will be the same as that of the basic generalized Lagrangian-mean theory apart from terms representing departures from frictionless, adiabatic motion. Moreover, the shape of C_{M}, being determined by the fields of potential vorticity and potential temperature, has no systematic tendency to become unrelated to the wave motion as time goes on; therefore $\overline{(\ )}^{\mathrm{M}}$ will not exhibit the non-uniform behaviour in time which frictional and diabatic effects must generally induce in ξ and therefore $\overline{(\ )}^{\mathrm{L}}$. Such a continuously re-initialized theory would apply as long as surfaces of constant P and θ continued to intersect at finite angles. Note that dissipative terms must now appear on the right of equation (10) in particular; the correspondingly modified ξ function will be denoted by ξ^{m}. It is interesting that the operator $\overline{(\ )}^{\mathrm{M}}$ appears implicitly in the quite different but also fundamental context of constructing 'integrals of the motion', in the sense of classical dynamics (Obukhov 1971).

If disturbance amplitudes become too large, then the tube C_{M} may change its topology by splitting into more than one piece; this certainly happens for the breaking planetary waves involved in major warmings. The corresponding ξ^{m} will then behave discontinuously. Examples in the troposphere are also well documented (see, for example, Obukhov 1964; Danielsen 1968). To study isolated events of such large amplitude it may prove more useful to pick an initial time t_0 at which the tubes C_{M} have simple shapes, if that is possible, and then follow the material tubes C which coincided with the tubes C_{M} at time t_0 (Dunkerton 1979). This idea, together with the arguments of A.M. appendix C, could be made the basis of an initialization procedure satisfying the conditions assumed by A.M. (particularly their postulate (viii), p. 617), and avoiding the discontinuous behaviour of ξ^{m}. Even though the initial state of no disturbance envisaged by A.M. may never have existed, the procedure would define a ξ which would coincide with ξ^{m} at time t_0 – i.e. which would correspond to the shapes of the tubes C_M at time t_0 – and which therefore could have evolved, via conservative motion, from a *hypothetical* initial state of no disturbance.

4. Some further properties of generalized Lagrangian means

4.1. *The divergence effect*

In § 2.3 we noted that $\bar{\boldsymbol{u}}^{L}$ is generally divergent even for incompressible flow. Since this fact is sometimes regarded with surprise, it may be worth pointing out (following A.M. § 9) how plausible it becomes as soon as we consider flow near a boundary. Take the case of spatial averaging again, and suppose that the tube C in figure 2*b* is initially parallel to and very close to a wall y = constant. Then if a disturbance grows and C begins to look more like figure 2*c* its centre of mass will usually move away from the wall; a similar effect is of course familiar in problems of turbulent diffusion near boundaries.

It should be recognized that the same phenomenon occurs for the classical Lagrangian time mean following a single fluid parcel, as well as for its various generalizations. This is clear, for instance, from the illustration given in figure 1, depicting the trajectory of a single parcel computed for the incompressible velocity field defined in the figure caption. That velocity field has zero normal component at the boundary shown, and represents an oscillatory disturbance whose amplitude is uniform with distance parallel to the wall, but growing in time. The parcel's position at equal time intervals is marked off along part of the trajectory. It is clear by inspection of these marks that, as time goes on, the parcel spends more time further away from the wall, so that a running time average of the parcel's position over several cycles will show a systematic motion away from the wall. Since in the given velocity field all mean quantities are independent of distance parallel to the wall, *all* fluid parcels initially lying on the plane $y = 2.5$ (from which the computed trajectory begins) will have a mean motion away from the wall. The classical Lagrangian-mean motion is evidently divergent; and it can be shown directly from (12) that the effect is $O(a^2)$ and therefore comparable with all other mean effects of the waves even when their amplitude a is arbitrarily small (e.g. McIntyre 1973).

The divergence effect is further illustrated by a recent calculation (Uryu 1979) of the generalized Lagrangian-mean meridional circulation for the Eady model of baroclinic instability in a channel. In this case the average used is the spatial g.l.m. given by equation (11). Uryu's result is reproduced as figure 3; again, the model flow is strictly incompressible. Nevertheless, $\bar{\boldsymbol{u}}^{L}$ diverges away from the side walls as the disturbance grows; there is a corresponding tendency for convergence in mid-channel. Another feature of interest to note in passing is the fact that the vertical motion is thermally direct, in complete contrast to the Eulerian-mean motion which as is well known has the classical three-cell structure, with a large indirect cell in the middle (Uryu 1979). The thermal directness is reminiscent of results obtained for actual laboratory and atmospheric flows by Riehl & Fultz (1957), Krishnamurti (1961), Danielsen (1968, figure 14), and Mahlman (1969) even though the quasi-Lagrangian averaging procedures of those investigators do not correspond exactly to any of those defined here, if only because the averages were weighted differently. The thermal directness of the Lagrangian-mean vertical motion in figure 3 reminds us that the growing Eady wave is reducing the total potential energy of the system. Note that the potential energy change due to the Lagrangian-mean motion (the motion of the centres of mass of all the tubes C) is the total potential energy change – there is no additional 'eddy' contribution. What is usually called the rate of change of eddy available potential energy appears implicitly as the contribution associated with the Stokes drift, as was mentioned earlier.

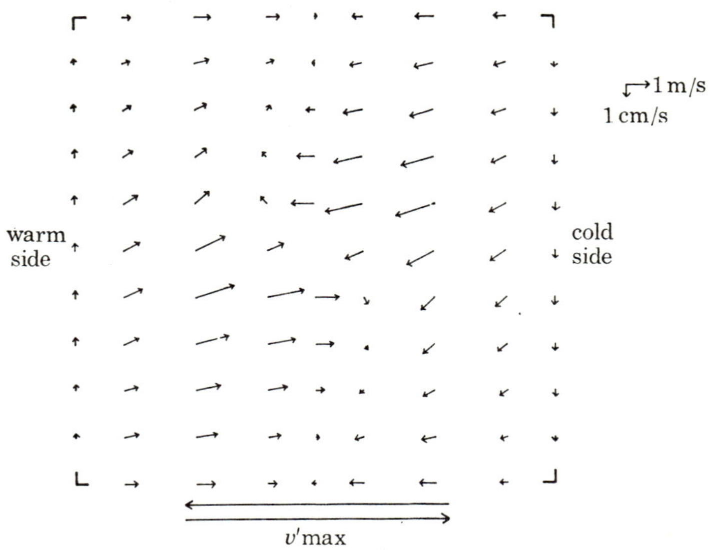

FIGURE 3. Generalized Lagrangian-mean meridional circulation $\bar{v}^L$, $\bar{w}^L$ associated with a growing Eady wave, from Uryu (1979). The fluid motion is incompressible, but the Lagrangian-mean motion is divergent because of disturbance growth (see text). Parameter values are as follows. Disturbance amplitude, as measured by maximum north–south Eulerian disturbance velocity v', 11 m s^{-1} (shown on the same scale as $\bar{v}^L$ by the long arrows at bottom); growth rate 0.7 day^{-1}; wavelength 5000 km; width of channel (meridional *half*-wavelength) 5000 km; height of channel 10 km; vertical shear of basic flow 3 m s^{-1} km^{-1}; buoyancy (Brunt–Väisälä) frequency 10^{-2} s^{-1}; Coriolis parameter 10^{-4} s^{-1}.

As we saw from equation (24) one can define, for incompressible flow (or for hydrostatic flow viewed in pressure coordinates), a mean circulation which is *not* divergent but which is nevertheless closer to Lagrangian concepts than is the standard Eulerian mean. Such a mean motion would not correspond to the classical Lagrangian mean, even in the crudest approximation; and it does not have such a simple relationship with potential energy. Nor does it have such simple analytical properties as $\bar{\boldsymbol{u}}^L$; nevertheless, it might prove useful, and further investigation is warranted.

In all our examples (and the same is true of the example given in McIntyre 1973) the divergent character of $\bar{\boldsymbol{u}}^L$ stems from temporal changes in the amplitude of the disturbance. In the cases considered, $\nabla \cdot \bar{\boldsymbol{u}}^L$ is attributable in the first approximation to the term $\partial\tilde{\rho}/\partial t$ in equation (19). It should not be thought, however, that this term is necessarily zero in all problems for which the wave amplitude is steady. When the fluid behaviour is dissipative, the temporal non-uniformity mentioned earlier can cause any quantity depending on ξ, including $\tilde{\rho}$ in equation (19), to change systematically in time. The effect is formally negligible in certain problems where the wave dissipation is assumed small, but not in general. These remarks are of interest in showing how the present ideas fit in with the results of Kida (1977) on the behaviour of initially zonal lines of air parcels in a general circulation model of the atmosphere. As could have been anticipated from our discussion (see also equation (29) below), Kida found a strong divergence effect in his model troposphere (with a pattern rather like that in figure 3) even though the total eddy field was not far from statistically steady.

One consequence of the divergence effect, for quasi-geostrophic motions such as the Eady wave, is that it can upset the standard quasi-geostrophic scaling for the mean flow. For instance the Lagrangian-mean meridional velocity $\bar{v}^{L}$ in the example of figure 3 exceeds its Eulerian-mean counterpart $\bar{v}$ by a large factor of the order of the inverse Rossby number. This warns us that standard order-of-magnitude arguments may need re-examination when a Lagrangian-mean description is used; further illustrations of this point will be found in A.M. § 9 and in the cited paper by Uryu. An associated phenomenon is that eddy terms involving the disturbance pressure field in the Lagrangian-mean momentum or angular momentum budget (see references cited below equation (21)) tend to be correspondingly large and to oppose the Coriolis force associated with $\bar{v}^{L}$ (Uryu 1979). This is one reason why it was suggested earlier that a direct study of the Kelvin–Bjerknes circulation function (the numerator of equation (21)) might prove more useful in practical diagnostics than the angular momentum itself, important though the latter may be conceptually.

4.2. *Some difficulties in using* $\overline{(\;)}^{L}$ *on the sphere*

For use in connection with realistic models of the stratosphere, the generalized Lagrangian-mean theory will need to be applied on the sphere. Unfortunately this presents some additional difficulties of a rather peculiar kind, for waves of finite amplitude. When $\overline{(\;)}$ is a zonal average on the sphere, the basic prescription of § 3 can be expressed by means of a mechanical analogy similar to that of figure 2*a*; but in place of the rigid rod R constrained to remain parallel to the x-axis, we have to imagine a frictionless, massless 'magic hoop' constrained by marvellous machinery to remain symmetrically disposed about the Earth's axis, while remaining free to expand and contract radially, to move axially, and to rotate about the axis. The positions of air parcels are imagined to be joined to this magic hoop by ideal elastic bands as before. The motion of the hoop under the balanced pull of all the elastic bands then gives the vector $\bar{\boldsymbol{u}}^{L}(\boldsymbol{x}, t)$ at each position $\boldsymbol{x}$ on the hoop. This analogy expresses the content of equations (4), (8), (9) and (10) when $\overline{(\;)}$ is the Eulerian zonal mean on the sphere.† The equivalence between equations (4) and (11) now follows just as before, although the interpretation of equation (11) when a vector quantity is to be averaged requires caution as will be seen shortly. Equation (21) involving the circulation function also holds, provided we interpret U_1 as the zonal component of $\boldsymbol{U}$, and the denominator on the right as the circumference of the magic hoop H.

Now the smallness of the atmosphere's scale height in comparison with the Earth's radius forces planetary-scale parcel excursions to take place along curved, quasi-horizontal paths closely following the shape of the Earth's surface. Consider for example the extreme case sketched in figure 4, in which a hypothetical tube C and its associated magic hoop H are shown. (The elastic bands joining the two are invisible.) Clearly the magic hoop will be pulled *underground* whenever the disturbance displacement attains any substantial amplitude. Now a reference position $\boldsymbol{x}$ located underground does not in itself detract from the power of

† To verify the correctness of the analogy mathematically, it is best to begin with a coordinate-independent statement of what is meant by the zonal average of a vector. A suitable statement is given at the bottom of p. 614 of A.M.; see also the end of p. 622.

the generalized Lagrangian-mean theory to reveal the theoretical generalities of wave, mean-flow interaction – but it is distinctly inconvenient for making mental or actual pictures of specific mean flows. It is, however, an inevitable consequence of requiring that $\boldsymbol{\xi}$ satisfy (8), in a true vector sense independent of particular coordinate systems; and the *analytical* power of the g.l.m. theory seems to depend upon (8).

A related and more serious difficulty is that the tubes C and their sets of reference positions $\boldsymbol{x}$ (lying on the hoops H) are not necessarily in one-to-one correspondence: two C's at different heights in the atmosphere could evidently have the same H. This could happen whenever the disturbance amplitude increased with height at a certain rate, depending on disturbance amplitude. In such cases, which are presumably common enough, the Jacobian (18) and therefore the mean quantity $\tilde{\rho}$ would become infinite somewhere. (So would the mean quantities $\boldsymbol{U}$ and $\boldsymbol{V}$ defined in § 2.3, with the important exception of the zonal component of $\boldsymbol{U}$, in virtue of equation (21); only $\bar{\boldsymbol{u}}^{\mathrm{L}}$ would remain finite in all three components.)

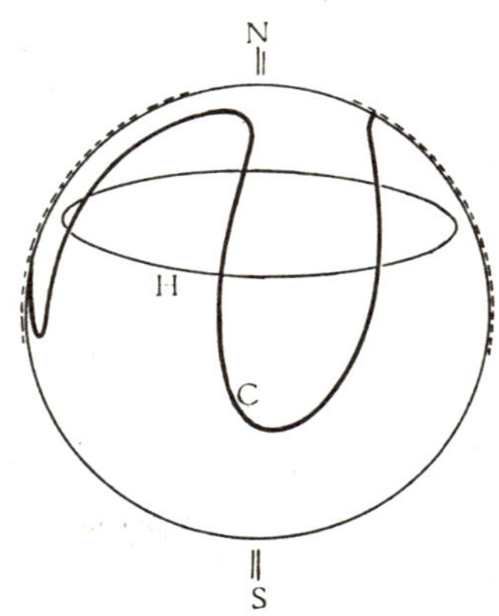

FIGURE 4. Tube C in the presence of a hypothetical, large-amplitude disturbance (not intended to look realistic). The fine dashed lines indicate projections of C onto meridional cross sections. The 'magic hoop' H gives the set of reference positions $\boldsymbol{x}$ associated with C, in terms of which the generalized Lagrangian-mean theory takes the fundamental, coordinate-independent form implied by equations (4), (8), (9), and (10).

One possible way of avoiding this singular behaviour of the reference space would be to use mean latitude ($\bar{y}^{\mathrm{L}}$ or $\bar{\phi}^{\mathrm{L}}$) and height ($\bar{z}^{\mathrm{L}}$) or pressure ($\bar{p}^{\mathrm{L}}$), defined by equation (11), as reference coordinates in place of $\boldsymbol{x}$. The true vector reference position $\boldsymbol{x}$ would still be needed to define the vector $\boldsymbol{\xi}$ (by locating the back end of the corresponding elastic band), but $\boldsymbol{x}$ would no longer be used as the independent variable as well. Part of the price would be loss of the analytical simplicity, as well as the coordinate-independent form, of equation (19). Another alternative which might sometimes be feasible would be simply to use P and θ as coordinates, as suggested by equation (25).

A corollary of the magic-hoop kinematics suggested by figure 4 is that the vertical component of the true vector Lagrangian-mean velocity $\bar{\boldsymbol{u}}^{\mathrm{L}}$ will be large and negative for a rapidly growing disturbance, purely because of the combined effects of disturbance transience and the Earth's curvature. Lest the reader think that such bizarre behaviour is peculiar to the spatial Lagrangian mean under discussion, it should be pointed out that the same thing happens for the classical time mean following a single parcel. If as in figure 1 we consider the behaviour of a single parcel in a growing disturbance, we can see at once that if the disturbance is of the type suggested in figure 4, involving large north–south excursions, then a running time average of the velocity vector for the parcel, taken over several oscillations, will have a large component toward the centre of the earth just as before. The result of applying the operator $\overline{(\ )}^{L}$ to the vertical *component* of velocity will, by contrast, have the smaller order of magnitude which is characteristic of the actual large-scale vertical motion, and may therefore be a more useful quantity to deal with in meteorological practice. It is this quantity, incidentally, rather than the vertical component of the vector $\bar{\boldsymbol{u}}^{L}$, that is relevant to potential energy in a spherical gravitational potential Φ, the potential energy of the tube C being the scalar quantity $\bar{\Phi}^{L}$, per unit mass.

It is evident from what has just been said that, whenever $\overline{(\ )}$ is a curvilinear spatial average, the operation of taking the curvilinear components of a vector does not generally commute with the operation of taking its generalized Lagrangian mean $\overline{(\ )}^{L}$.† It can, however, be shown that when $\overline{(\ )}^{L}$ is applied separately to the latitudinal and vertical components of any vector $\boldsymbol{v}$ in spherical coordinates the result is the 'contravariant components' of $\bar{\boldsymbol{v}}^{L}$ in the sense of general tensor analysis, with respect to a coordinate system in which $\bar{y}^{L}$ and $\bar{z}^{L}$ are regarded as coordinates defining the position of the magic hoop. (The correspondence between $\boldsymbol{x}$ and $(\bar{y}^{L}, \bar{z}^{L})$ is now being looked upon as a coordinate transformation.) It is beyond the scope of this paper to go further into the technicalities of general tensor analysis; the main point to be borne in mind is that *the meaning of such concepts as 'Stokes drift' depends on whether the concept is applied to the velocity vector or only to one of its spherical components.* It is the latter version of the concept that will probably be used most in practice, for reasons already mentioned, and also because it fits in best with the familiar parallelism between spherical formulations and channel or beta-plane formulations. This parallelism allows us to think of spherical coordinates as approximately Cartesian and thus to sidestep the issues raised by figure 4, as is done for example in A.M. § 9, in Matsuno & Nakamura (1979), in Matsuno (1979) and in McIntyre (1979).

Another place where the Jacobian (18) may become singular is near the poles; the region of misbehaviour can be expected to increase in size as disturbance amplitude increases. It is obvious, indeed, that there can be no such thing as a zonally asymmetric set C of air parcels with a mean latitude of 90°. It should be remembered, however, that in the presence of large amplitude disturbances to the zonal flow the operation of taking any kind of zonal average near the pole, while well-behaved mathematically in the Eulerian case, as it happens, is certainly quite meaningless physically. Thus the refusal of the generalized Lagrangian-mean theory to behave itself near the poles, or for that matter to assign any mean values at all to some locations near the poles, could well be regarded as a desirable property of the theory rather than the reverse. A theory which automatically recognizes a domain where zonal averaging becomes physically meaningless should perhaps be regarded as a more 'intelligent' theory than one which does not.

The use of potential vorticity P and potential temperature θ as coordinates may avoid some of the foregoing problems. Equation (25) becomes relevant, in any case, whenever we want to use $\overline{(\ )}^{M}$ instead of $\overline{(\ )}^{L}$ so as to obtain a continuously re-initialized theory. However, breakdown of one-to-one correspondence between pairs of values of (P, θ) and quasi-material tubes C_M might still have to be coped with, since a surface of constant P might intersect a surface of constant θ more than once. Also, a practical disadvantage of P and θ as coordinates is the

† We are taking the fundamental definition of $\overline{(\ )}^{L}$ as equation (4), of course, since equation (11) has no useful coordinate-independent meaning when operating on a vector.

fact that the surfaces of constant P in the extratropical stratosphere tend to be nearly horizontal, and are therefore not usually at a large angle to those of constant θ. This occurs at any rate when P is given its usual definition in pressure coordinates, namely isentropic absolute vorticity times $\partial\theta/\partial p$. The main reason for the strong vertical gradient in P is the roughly exponential fall-off of pressure p with height and the corresponding gradient in $|\partial\theta/\partial p|$. However, whenever P is conserved, so also is the related quantity P_F formed by replacing θ by an arbitrary function $F(\theta)$ in the definition of P; judicious choice of $F(\theta)$ could certainly give a modified potential vorticity P_F with a far weaker vertical gradient and therefore more convenient for use as a coordinate with θ.

Other ways around the various difficulties will no doubt be thought of in due course. Again, it will not be clear how useful the foregoing suggestions will be until more experience is gained, probably in the first place with the results of numerical models. It is hoped, as was said before, that the present, very preliminary, discussion might help stimulate the necessary thinking and experimentation.

5. Concluding remarks

An important topic, as yet little investigated, is how best to express the effects of diabatic heating and other departures from conservative motion on the various mean quantities of interest. These effects are the interesting ones for applications to the real atmosphere, and estimates of diabatic and frictional tendencies in the potential vorticity P from the work of Staley (1960), Danielsen (1968) and others will clearly be essential background, in addition to estimates of radiative and turbulent diabatic heating *per se*.

If equation (1) is replaced by

$$\mathrm{D}X/\mathrm{D}t = Q_X, \tag{26}$$

then (3) becomes

$$\left(\frac{\partial}{\partial t}+\bar{v}^{\mathrm{L}}\frac{\partial}{\partial y}+\bar{\omega}^{\mathrm{L}}\frac{\partial}{\partial p}\right)\bar{X}^{\mathrm{L}} = \bar{Q}_X^{\mathrm{L}}. \tag{27}$$

(A.M., equation 2.21). When $X = \theta$ then Q_X, $= Q$ say, is the total diabatic heating rate expressed as potential temperature change per unit time, and $\bar{Q}^{\mathrm{L}}$ is its generalized Lagrangian mean (as defined for instance by equation (11) above). Dunkerton (1978), using small-amplitude theory for steady waves (and thus avoiding the worst difficulties mentioned in § 4) has argued that for qualitative purposes we may replace $\bar{Q}^{\mathrm{L}}$ on the right-hand side of (27) by $\bar{Q}$, the Eulerian-mean heating (but *not* $\bar{\omega}^{\mathrm{L}}$ by $\bar{\omega}$ on the left). That is, the Stokes correction $\bar{Q}^{\mathrm{S}}$ can be neglected in comparison to $\bar{Q}$ and $\bar{Q}^{\mathrm{L}}$ (although the Stokes correction $\bar{\omega}^{\mathrm{S}}$ cannot). However, the argument must ultimately rely on numerical estimates of net radiative heat-flux divergences $Q_{(r)}$ and turbulent heat-flux divergences $Q_{(t)}$, matters of some uncertainty. It perhaps ought to be remarked that near the bottom of the stratosphere, where turbulent mixing is important at tropopause-folding events (for recent work on this see Shapiro 1978), the correlation between those events and large-scale weather patterns, and therefore with the shapes of material tubes C, could easily make $\overline{(Q_{(t)})}^{\mathrm{S}}$ a significant contribution to $\bar{Q}^{\mathrm{L}}$. (Dunkerton's discussion was, however, concerned less with the bottom of the stratosphere than with the middle atmosphere as a whole.)

At the qualitative level of approximation used by Dunkerton, the temporally non-uniform behaviour of the ξ function (§ 3 above) does not play a significant role in the discussion. For a more quantitative analysis, however, it will be essential to introduce re-initialization procedures (§ 3), and to take account of the fact that dissipative processes make the divergence effect (§ 4.1) significant even for steady waves. For example, in the case of spatial averaging, and in a situation where the modified Lagrangian mean defined by (25) can be used, it is evident from equation (27) that

$$\left(\frac{\partial}{\partial t}+\bar{v}^{\mathrm{M}}\frac{\partial}{\partial y}+\bar{\omega}^{\mathrm{M}}\frac{\partial}{\partial p}\right)\bar{\theta}^{\mathrm{M}} = \bar{Q}^{\mathrm{M}}+D, \tag{28}$$

where
$$D \equiv \partial(\bar{\theta}^{\mathrm{M}}-\bar{\theta}^{\mathrm{L}})/\partial t,$$

and where $\bar{\theta}^{\mathrm{L}}$ in the last expression is defined for material tubes C that coincide with the '$P\theta$ tubes' $\mathrm{C_M}$ at time t, so that $\overline{(\ )}^{\mathrm{M}} = \overline{(\ )}^{\mathrm{L}}$ everywhere at that instant. (Of course $\bar{\theta}^{\mathrm{M}}$ is simply equal to the value of θ for the tube $\mathrm{C_M}$, but $\bar{v}^{\mathrm{M}}$ and $\bar{\omega}^{\mathrm{M}}$ still involve non-trivial averaging.) The term D is written on the right because it is a *dissipative* term, entirely attributable to frictional and diabatic processes; were these zero, C and $\mathrm{C_M}$ would continue to coincide as time progressed and D would vanish.† D also represents an *eddy* process, because D would be zero if all tubes C and $\mathrm{C_M}$ were straight as in figure 2*b* for some time interval around time t. (It would be of some interest to express D explicitly in terms of disturbance-associated quantities, but that has not yet been done.) Similarly, the analogue of equation (19) will take the form

$$\partial\tilde{\rho}^{\mathrm{M}}/\partial t+\nabla\cdot(\tilde{\rho}^{\mathrm{M}}\bar{\boldsymbol{u}}^{\mathrm{M}}) = \mathscr{D}, \tag{29}$$

where $\tilde{\rho}^{\mathrm{M}}$ is given by replacing ξ in (17) by ξ^{m} (see § 3),‡ and $\mathscr{D}$ stands both for 'divergence' *and* 'dissipation' since, like D, it is evidently zero for conservative motion because $\xi^{\mathrm{m}} = \xi$ in that case. $\mathscr{D}$, also, is an eddy term.

It turns out that $\bar{Q}^{\mathrm{M}}$ has a simpler physical interpretation than that associated with (28). It suggests an alternative way of describing the mean meridional circulation, which relates it more directly to $\bar{Q}^{\mathrm{M}}$ and to the analogous quantity $\bar{Q}^{\mathrm{M}}_{P}$ corresponding to $X = P$ in (26). This is obtained by considering the rate at which mass crosses surfaces of constant P and θ (see, for example, Danielsen 1968, p. 518). For instance, the rate at which mass crosses a portion of the isentropic surface marked out by values of the potential vorticity lying between P and $P+\Delta P$ is

$$\Delta\mu(P,\theta) = \bar{Q}^{\mathrm{M}}\sigma(P,\theta)\,\Delta P, \tag{30}$$

where $\sigma(P,\theta)$ is the apparent 'density' of fluid in a meridional cross section in which P and θ are used as coordinates; that is to say $\sigma(P,\theta)$ is the total mass of the tube $\mathrm{C_M}(P,\theta)$ divided by its cross-sectional 'area' $\Delta P\,\Delta\theta$ in (P,θ) space. A derivation of (30) is given in the appendix. ($\Delta\mu$ is closely related to the quantity V defined in § 2.3, in the case of stationary waves and hence immobile isentropic surfaces.) The total mass flux across the finite portion of the isentropic surface marked out by values of P running between P_1 and P_2, say, is given by

$$\mu = \int_{P_1}^{P_2}\bar{Q}^{\mathrm{M}}(P,\theta)\,\sigma(P,\theta)\,\mathrm{d}P. \tag{31}$$

An exactly similar result involving $\bar{Q}^{\mathrm{M}}_{P}$ evidently holds for the rate at which mass crosses a surface of constant potential vorticity P.

It is interesting to consider for a moment what would be involved in calculating the radiative contribution $\overline{(Q_{(r)})}^{\mathrm{M}}$ to $\overline{(Q)}^{\mathrm{M}}$ to a level of approximation beyond that of Dunkerton (1978).

† Note incidentally that in the definition of D, $\partial/\partial t$ may be replaced by $\partial/\partial t+\bar{v}^{\mathrm{M}}\partial/\partial y+\bar{\omega}^{\mathrm{M}}\partial/\partial p$.
‡ To convert to pressure coordinates replace ζ by p^{l}, in the notation of equation (9), as well as z by p.

Since the averaging in (25) involves a weighted mean along the wavy, quasi-material tube C_M one would have to take into account the correlation between tropospheric cloudiness and patterns of troughs and ridges, as well as the fact that some tubes C_M will intersect the edge of the polar night. (Terms like $\frac{1}{2}\overline{\eta^2(Q_{(r)})_{yy}}$ in equation (12) (but involving η^m rather than η) would be important if we were to try to describe the latter effect within the small-amplitude approximation.) Because of the notorious difficulty of the radiative-transfer calculation itself (Houghton 1978), the left-hand side of equation (30) may in actual practice be easier to estimate (by using tracer observations) than the right-hand side. Thus, relations like (30) and (31) may be more useful as constraints on estimates of turbulent mixing and radiative cooling in the lower stratosphere, than as predictors of tracer motion *from* such estimates.

The foregoing discussion has done no more than touch on some of the theoretical facts and difficulties that may have to be taken into account if 'the mean circulation' of the middle atmosphere is to be conceived of in a way more satisfactory than that afforded by the usual Eulerian-mean description. Enough has been said, perhaps, to show that even though Lagrangian-mean theory has furnished some important insights (lucidly brought out in the recent contributions of Dunkerton (1978), Matsuno (1979) and Plumb (1979)), we are still a long way from a quantitative theory of the mean circulation which truly embodies these insights.

I should like to thank D. G. Andrews, S. Death, T. Dunkerton, A. Eliassen, C.-P. Hsu, J. C. R. Hunt, J. D. Mahlman, T. Matsuno, H. Pearson, R. Quiroz, M. Uryu and S. V. Venkateswaran for stimulating discussions and helpful comments on this material. Dr M. Uryu kindly allowed me to quote his work on the Eady problem before publication.

Appendix. Some further relations involving $\overline{(\ )}^{M}$

We first note that (25) can be rewritten as

$$\overline{(\ )}^{M} \equiv \frac{1}{\sigma}\int_{C_M} (\ )\, \rho\alpha\, ds, \tag{A 1}$$

where ρ is density, s is arc length along the tube $C_M(P, \theta)$, σ is defined by

$$\sigma(P, \theta) \equiv \int_{C_M} \rho\alpha\, ds, \tag{A 2}$$

and α is defined by

$$\alpha^{-1} \equiv |\nabla\theta \times \nabla P|,$$
$$= |\nabla\theta|\,|\nabla_\theta P| = |\nabla P|\,|\nabla_P \theta|. \tag{A 3}$$

Here ∇ is the three-dimensional gradient operator as before, ∇_θ denotes the gradient in a surface of constant θ, and ∇_P the gradient in a surface of constant P. In the notation of § 3, $\alpha\Delta P\Delta\theta$ is the cross-sectional area ΔA_M of the tube $C_M(P, \theta)$ and $\sigma\Delta P\Delta\theta$ its mass.

Now consider the rate $\delta\mu$ at which mass crosses an area element δA of an isentropic surface. That rate is $\rho\delta A$ times the component $u^{\perp}$ of fluid velocity normal to the isentropic surface, as seen in a frame of reference fixed in the isentropic surface. If the positive direction is taken towards higher θ, the material rate of change of potential temperature is just $u^{\perp}|\nabla\theta|$. But this is equal to Q by definition, and so

$$u^{\perp} = Q/|\nabla\theta|.$$

Substituting this result into our expression for $\delta\mu$, we get

$$\delta\mu = \rho Q\delta A/|\nabla\theta|. \tag{A 4}$$

Now the area element of the isentropic surface marked out by values P and $P+\Delta P$ of potential vorticity, and values s and $s+\mathrm{d}s$ of arc length along C_M, is

$$\delta A = \Delta P \mathrm{d}s/|\nabla_\theta P|;$$

so the mass flux across this area element is

$$\begin{aligned}\delta\mu &= \rho Q \Delta P \mathrm{d}s/|\nabla\theta|\,|\nabla_\theta P| \\ &= \Delta P Q \rho \alpha \mathrm{d}s, \end{aligned} \qquad \text{(A 5)}$$

by (A 3). Equation (30) now follows, because (A 1) shows that the integral of (A 5) with respect to s is $\sigma\Delta P\bar{Q}^{\mathrm{M}}$. Equation (31) is the result of carrying out a further integration, with respect to P.

It is perhaps worth noting for completeness the corresponding result when the integration is over any arbitrary area A of the isentropic surface, even though the result is hardly novel. The total rate at which mass crosses the area A is

$$\int Q\hat{\sigma}\mathrm{d}A, \qquad \text{(A 6)}$$

where $\hat{\sigma} = \rho/|\nabla\theta|$ is the mass per unit area per unit increment in θ, that is to say $(\Delta\theta)^{-1}$ times the mass per unit area lying between the isentropic surfaces θ and $\theta+\Delta\theta$ at the point under consideration. This result remains useful even in cases where the (P, θ) coordinate system becomes ill-defined. (The expression (A 6) is $\overline{\int\hat{\sigma}\mathrm{d}A}$ times the modified Lagrangian mean of Q corresponding to the Eulerian spatial mean $\overline{(\ )}$ over a horizontal *surface*.)

References (McIntyre)

Andrews, D. G. & McIntyre, M. E. 1976 *J. atmos. Sci.* **33**, 2031–2048.
Andrews, D. G. & McIntyre, M. E. 1978*a* *J. atmos. Sci.* **35**, 175–185.
Andrews, D. G. & McIntyre, M. E. 1978*b* *J. Fluid Mech.* **89**, 609–646.
Andrews, D. G. & McIntyre, M. E. 1978*c* *J. Fluid Mech.* **89**, 647–664.
Bretherton, F. P. 1969*a* *Q. Jl R. met. Soc.* **95**, 213–243.
Bretherton, F. P. 1969*b* *J. Fluid Mech.* **36**, 785–803.
Bretherton, F. P. 1971 *Lectures in applied mathematics* **13**, 61–102. American Mathematical Society.
Bretherton, F. P. 1979 *J. Fluid Mech.* (to appear).
Brewer, A. W. 1949 *Q. Jl R. met. Soc.* **75**, 351–363.
Danielsen, E. F. 1968 *J. atmos. Sci.* **25**, 502–518.
Dobson, G. M. B. 1956 *Proc. R. Soc. Lond.* A **236**, 187–192.
Dunkerton, T. 1978 *J. atmos. Sci.* **35**, 2325–2333.
Dunkerton, T. 1979 *Rev. Geophys. Space Phys.* (submitted).
Grimshaw, R. H. J. 1979 *Phil. Trans. R. Soc. Lond.* A **292**, 391–417.
Houghton, J. T. 1978 *Q. Jl R. met. Soc.* **104**, 1–29.
Kida, H. 1977 *J. met. Soc. Japan* **55**, 71–88.
Krishnamurti, T. N. 1961 *J. Meteorol.* **18**, 172–191.
Leibovich, S. 1979 *J. Fluid Mech.* (to appear).
Longuet-Higgins, M. S. 1969 *Deep Sea Res.* **16**, 431–447.
Mahlman, J. D. 1969 *Mon. Weath. Rev.* **97**, 534–540.
Mahlman, J. D. 1975 *Proc. 4th Conf. Climatic Impact Assessment Prog.* (ed. T. M. Hard & A. J. Broderick), pp. 132–146. U.S. Department of Transportation. DOT-TSC-OST-75-38.
Mahlman, J. D., Levy, H. II & Moxim, W. J. 1979 Three dimensional tracer structure and behaviour as simulated in two ozone precursor experiments. *J. atmos. Sci.* (submitted).
Matsuno, T. 1979 *Pure appl. Geophys.* **118**. (In the press.)
Matsuno, T. & Nakamura, K. 1979 *J. atmos. Sci.* **36**, 640–654.
McIntyre, M. E. 1973 *J. Fluid Mech.* **60**, 801–811.
McIntyre, M. E. 1979 *Pure appl. Geophys.* **118**. (In the press.)

Obukhov, A. M. 1964 *Meteorologiya i gidrologia*, no. 2, pp. 3–9.
Obukhov, A. M. 1971 *Izv. atmos. oceanic Phys.* **7**, 695–704 (English translation pp. 471–475).
Plumb, R. A. 1979 *J. atmos. Sci.* (In the press.)
Rayleigh, Lord 1896 *The theory of sound*, vol. II, § 352, 504 pp. New York: Dover (reprinted 1945).
Reed, R. J. 1955 *J. Meteorol.* **12**, 226–237.
Riehl, H. & Fultz, D. 1957 *Q. Jl R. met. Soc.* **83**, 215–231.
Sawyer, J. S. 1965 *Q. Jl R. met. Soc.* **91**, 407–416 (see p. 414).
Shapiro, M. A. 1978 *Mon. Weath. Rev.* **106**, 1100–1111.
Staley, D. O. 1960 *J. Meteorol.* **17**, 591–620.
Stokes, G. G. 1847 *Trans. Camb. phil. Soc.* **8**, 441–455.
Uryu, M. 1979 *J. met. Soc. Japan* **57**, 1–20.
Wallace, J. M. 1978 *J. atmos. Sci.* **35**, 554–558.

Phil. Trans. R. Soc. Lond. A **296**, 149–160 (1980) [149]
Printed in Great Britain

The chemistry of the stratosphere

BY B. A. THRUSH, F.R.S.
Department of Physical Chemistry, University of Cambridge, Lensfield Road, Cambridge CB2 1EP, U.K.

The chemistry of the stratosphere is dominated by the processes that remove the 'odd oxygen' (atomic oxygen and ozone) generated by the photolysis of molecular oxygen. In recent years it has been established that the Chapman reaction, $O + O_3 = O_2 + O_2$, cannot account for all the observed destruction of odd oxygen. Catalytic cycles involving nitrogen oxides (NO, NO_2), hydrogen radicals (H, HO, HO_2) and chlorine species (Cl, ClO) have been shown to be important and it is now realized that these catalytic cycles are closely linked to each other. Measurements of these trace species, their precursors (e.g. N_2O, H_2O, CH_4) and their reaction products (e.g. HNO_3, HCl) in the stratosphere are essential to understanding its chemistry. The natural variability of the stratosphere places a premium on simultaneous measurements of those trace species that are interconverted by rapid chemical reactions.

INTRODUCTION

The stratosphere is a dry, cloudless region of the Earth's atmosphere which extends from the temperature minimum of the tropopause at an altitude of 10–16 km to the temperature maximum of the stratopause at an altitude of about 50 km (figure 1). It is a region of temperature inversion, this being produced by absorption of visible and ultraviolet light by the few parts per million of ozone that are present in this region (figure 2). The stratosphere contains about one tenth of the Earth's atmosphere by mass, and, apart from the presence of ozone and smaller amounts of other trace species discussed below, its chemical composition is close to that of the troposphere. An exception to this statement is water vapour, which is only present in a few parts per million because the tropopause, where the temperature can be as low as 200 K, acts as a cold trap for water rising from the troposphere, and only small amounts of water are generated in the stratosphere by the photo-oxidation of methane.

Although current concern about pollution problems has led people to speak of an 'ozone layer' as though this were synonymous with the stratosphere where the bulk of the atmospheric ozone resides, the total vertical column of ozone in the Earth's atmosphere corresponds to a layer of pure gaseous ozone only 3 mm thick at ground level.

The stratosphere contains relatively few aerosol particles (about $1/cm^3$) or ions and these play no significant role in its chemistry, which is dominated by the chemical reactions of free atoms and free radicals. Solar radiation with wavelengths below 180 nm is completely absorbed by O_2, N_2 and O, and the photochemical processes that occur in the stratosphere therefore generate atoms and free radicals but not ions (figure 3).

Transport of gases in the stratosphere occurs much more rapidly by movement of air masses than it would by molecular diffusion. These processes are discussed in the papers by Murgatroyd & O'Neill (1979, this symposium) and by Harwood (1979, this symposium) and are not considered further here. However, it should be noted that the amounts of various trace species (e.g. ozone, water vapour or nitrogen oxides) in a sample of air depend on its past history as well as on such parameters as altitude, latitude, intensity of solar irradiation and temperature. In general, chemical conversion processes are more rapid than transport in the upper stratosphere (above 35 km), whereas transport processes can be dominant for some species in the lower stratosphere.

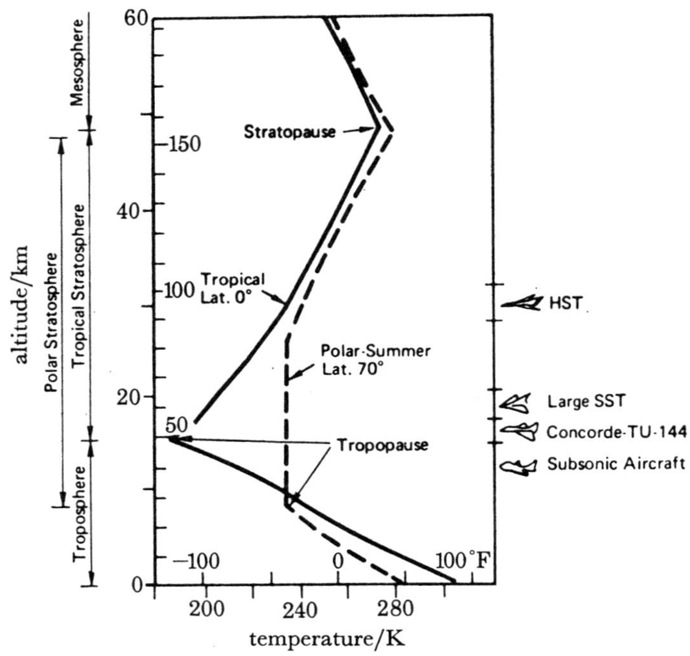

FIGURE 1. Variation of temperature with altitude in the Earth's atmosphere.

OXYGEN ATMOSPHERE

Historically, the basic mechanism for the formation and removal of ozone was proposed by Chapman in 1930. Molecular oxygen is dissociated by light with wavelengths less than 242 nm. This is a slow process because light that is strongly absorbed by O_2 has been filtered out at higher altitudes.

$$O_2 + h\nu \ (\lambda < 242 \text{ nm}) = O + O \quad \text{(slow)}, \tag{1}$$

$$O + O + M = O_2 + M \quad \text{(slow)}, \tag{2}$$

$$O + O_2 + M = O_3 + M \quad \text{(quick)}, \tag{3}$$

$$O_3 + h\nu = O_2 + O \quad \text{(quick)}, \tag{4}$$

$$O + O_3 = O_2 + O_2 \quad \text{(slow)}. \tag{5}$$

Of the subsequent processes, (2) is only significant in the mesosphere (above the stratosphere) where the lower density of oxygen limits the rate of formation of ozone by reaction (3). The ozone formed in reaction (3) absorbs light very strongly below 300 nm and is also dissociated by visible light, the timescale of its decomposition by process (4) being about 1 min by day. However, this process and reaction (3) merely interconvert the 'odd oxygen' species formed by process (1). In contrast, reaction (5) converts the odd oxygen species O and O_3

back to O_2 which can be regarded as even oxygen; this process is unusually slow for a highly exothermic reaction ($\Delta H = -390$ kJ mol^{-1}) involving a free atom. Only in the last decade have laboratory measurements of the rates of these processes become accurate enough to show that reaction (5) accounts for the removal of only about one-fifth of the 'odd oxygen' generated in the atmosphere by the photolysis of molecular oxygen. Rate coefficients for individual reaction processes will not be quoted here, since a compendium has recently been published by the U.S. National Bureau of Standards (1978).

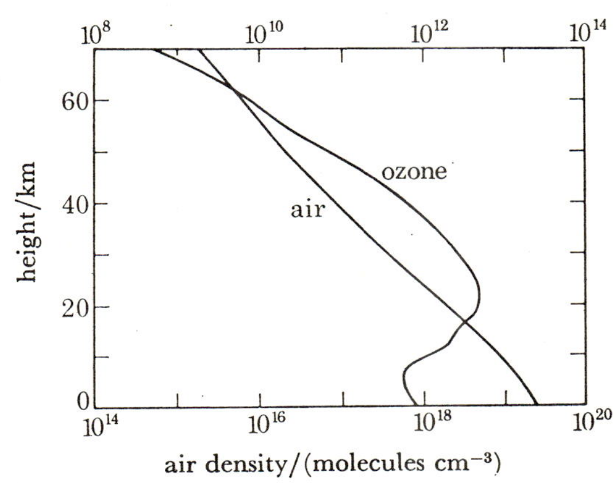

FIGURE 2. The density of air and of ozone in mid-latitudes as a function of altitude. Note that the density scales differ by a factor of 10^6 so that the curves intersect at 1 part/10^6 of ozone.

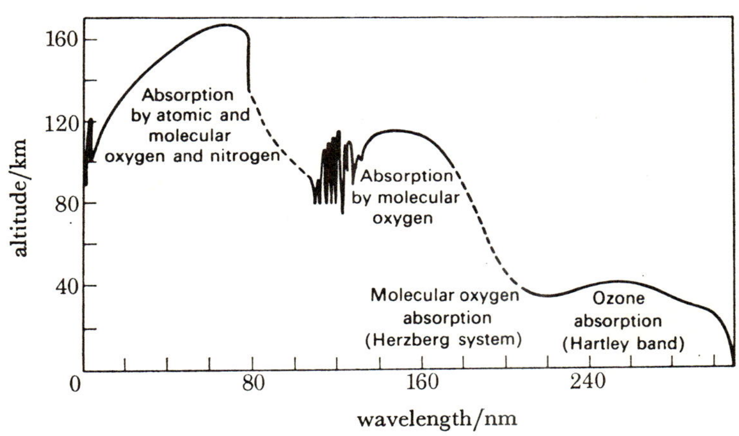

FIGURE 3. Altitudes of maximum light absorption as a function of wavelength for overhead sun.

Other processes removing odd oxygen from the stratosphere can be visualized in terms of a catalytic cycle in which the interconversion of two species X and XO achieves the same effect as reaction (5) and continues to destroy odd oxygen until X or XO is removed by some other process.

$$X + O_3 = XO + O_2, \tag{6}$$

$$XO + O = X + O_2 \tag{7}$$

$$O + O_3 = O_2 + O_2 \qquad (6)+(7)$$

The condition that both these processes should be exothermic is only that the dissociation energy of the X–O bond should lie between the wide limits of 100 and 490 kJ mol^{-1}. There are many free radical species, all of them species with a single unpaired electron for which both reactions (6) and (7) are rapid. The importance of atmospheric cycles involving X = H, HO, NO and Cl has been recognized and these are discussed separately below. The actual position is, however, more complicated than this because these free radical species can undergo rapid reactions with each other as well as with ozone and atomic oxygen, thereby giving cycles that interconvert free radicals X and XO without removing odd oxygen.

A reasonably full description of the chemistry of the stratosphere involves about one hundred chemical and photochemical processes of which nearly half play a significant role. These can be classified into the processes that form the free radicals mentioned above, the balancing removal processes for these radicals and the much more rapid processes by which these radicals are interconverted.

Hydrogen species (HO_x)

Bates & Nicolet (1950) recognized early the importance in the atmosphere of the free radicals H, HO and HO_2 (cumulatively called HO_x). The strongest feature of the airglow at night is the Meinel band emission from the mesosphere, due to vibrationally excited hydroxyl radicals produced in the reaction

$$H + O_3 = HO + O_2. \tag{8}$$

The main source of HO_x in the stratosphere is the rapid reaction of excited oxygen atoms ($O(^1D)$) with water vapour and to some extent with methane transported upwards from the troposphere. $O(^1D)$ is produced by the photolysis of ozone at wavelengths below about 310 nm and is removed mainly by quenching by N_2 and O_2, which occurs with a collisional efficiency of about one quarter.

$$O_3 + h\nu\ (\lambda < 310\ \text{nm}) = O_2 + O(^1D), \tag{9}$$

$$O(^1D) + N_2, O_2 = O(^3P) + N_2, O_2, \tag{10}$$

$$O(^1D) + H_2O = HO + HO, \tag{11}$$

$$O(^1D) + CH_4 = CH_3 + HO. \tag{12}$$

Oxidation of the methyl radicals proceeds by the mechanism

$$CH_3 + O_2(+M) = CH_3O_2\ (+M), \tag{13}$$

$$CH_3O_2 + NO = CH_3O + NO_2, \tag{14}$$

$$CH_3O + O_2 = CH_2O + HO_2, \tag{15}$$

$$HO + CH_2O = H_2O + HCO, \tag{16}$$

$$CH_2O + h\nu = HCO + H, \tag{17}$$

$$HCO + O_2 = CO + HO_2. \tag{18}$$

In the stratosphere, methane is also converted to methyl radicals by the reaction

$$HO + CH_4 = H_2O + CH_3. \tag{19}$$

This is overall a source of HO_x when the formaldehyde produced is decomposed photochemically (reaction (17)) rather than by HO (reaction (16)).

In the upper stratosphere the cycle

$$HO + O_3 = HO_2 + O_2, \quad (20)$$

$$HO_2 + O = HO + O_2, \quad (21)$$

$$O + O_3 = O_2 + O_2, \quad (20)+(21)$$

is an important removal process for odd oxygen, while the corresponding cycle in the lower stratosphere is

$$HO + O_3 = HO_2 + O_2, \quad (20)$$

$$HO_2 + O_3 = HO + O_2 + O_2, \quad (22)$$

$$O_3 + O_3 = O_2 + O_2 + O_2. \quad (20)+(22)$$

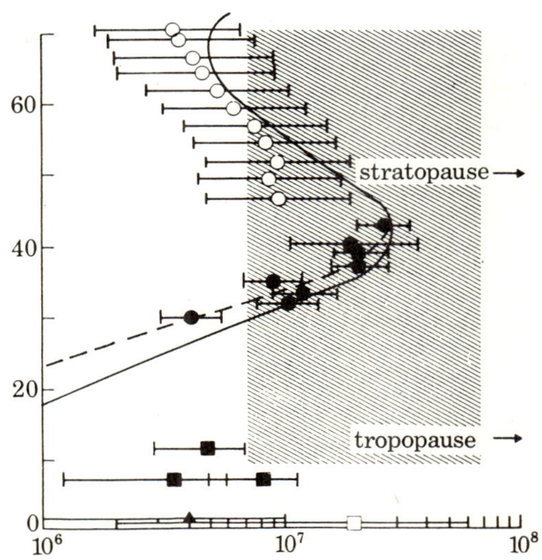

FIGURE 4. Calculated and observed profiles of HO (Hudson 1977).

Although reaction (8) has a rate coefficient many times that of reaction (20) or reaction (22), the cycle

$$H + O_3 = HO + O_2, \quad (8)$$

$$HO + O = H + O_2, \quad (23)$$

is not important in the stratosphere because the reaction

$$H + O_2 + M = HO_2 + M, \quad (24)$$

which has a rate coefficient about one hundred times greater than that of reaction (3), maintains a very high ratio of [HO_2] to [H], particularly in the lower stratosphere where the densities of O_2 and of M are higher.

Recombination of the hydrogen radicals occurs mainly by the processes

$$HO + HO_2 = H_2O + O_2, \quad (25)$$

$$HO_2 + HO_2 = H_2O_2 + O_2. \quad (26)$$

The latter is more important in the lower stratosphere, although the H_2O_2 yielded by the latter has not yet been detected in the atmosphere.

Figure 4 shows that the measured concentrations of HO (and HO_2) in the stratosphere are around 10^7 cm^{-3}, in moderate agreement with the predicted values.

Nitrogen oxides (NO_x)

Nitric oxide (NO) and nitrogen dioxide (NO_2), which are present in the stratosphere at the parts in 10^8 level, are formed predominantly by the reaction of excited oxygen atoms with nitrous oxide (N_2O), a process that has two rapid and equally probable paths:

$$O(^1D)+N_2O = NO+NO, \tag{27}$$

$$O(^1D)+N_2O = N_2+O_2. \tag{28}$$

Nitrous oxide, which is derived from terrestrial denitrification processes, is mainly removed from the stratosphere by photolysis to yield N_2+O and only 1–2 % generates NO by reaction (27). Lesser amounts of nitric oxide are generated in the mesosphere where photoionization processes can occur

$$N_2+h\nu = 2N(^4S, {}^2D), \tag{29}$$

$$N(^2D)+O_2 = NO+O, \tag{30}$$

$$NO+h\nu = NO^+ + e, \tag{31}$$

$$NO^+ + e = N(^2D)+O. \tag{32}$$

By day, the NO_x catalysed odd oxygen removal cycle

$$\begin{aligned} NO+O_3 &= NO_2+O_2, && (33)\\ O+NO_2 &= NO+O_2, && (34)\\ \hline O+O_3 &= O_2+O_2, && (33)+(34) \end{aligned}$$

is partly neutralized by efficient photolysis of NO_2 by ultraviolet light, which initiates the 'no-effect' cycle (Crutzen 1970; Johnston 1971).

$$NO_2+h\nu\ (\lambda < 400\ nm) = NO+O, \tag{35}$$

$$O+O_2+M = O_3+M, \tag{3}$$

$$NO+O_3 = NO_2+O_2. \tag{33}$$

In addition, recent laboratory measurements (Howard & Evenson 1977; Thrush 1979) have shown that reaction (36) is many times faster than had been believed. This means that the mixed cycle

$$HO_2+NO = HO+NO_2, \tag{36}$$

$$NO_2+h\nu = NO+O, \tag{35}$$

$$O+O_2+M = O_3+M, \tag{3}$$

$$HO+O_3 = HO_2+O_2, \tag{20}$$

which does not destroy odd oxygen, is an important process by day in the lower stratosphere.

The main removal process for NO_x also involves an HO_x species:

$$HO + NO_2 + M = HNO_3 + M. \quad (37)$$

The nitric acid formed is transported downwards to the tropopause, where it is rained out. During this process, NO_x can be regenerated by the reactions

$$HNO_3 + h\nu = HO + NO_2, \quad (38)$$

$$HO + HNO_3 = H_2O + NO_3, \quad (39)$$

$$NO_3 + h\nu = NO_2 + O \text{ (predominantly)}, \quad (40)$$

$$NO_2 + h\nu = NO + O. \quad (35)$$

It should be noted that the rapid reaction (36) assists the removal of NO_x by generating both active species for the formation of nitric acid in reaction (37). Thus reaction (36) followed by the above sequence of (37), (39), (40) and (35) removes $HO + HO_2$ from the stratosphere, but unlike their direct reaction (25) generates two oxygen atoms in the process. The recognition of this reaction sequence and of the competition between the odd oxygen destruction cycle (22) + (20) and the no-effect cycle (36), (35), (3) + (20) has led to a drastic reduction in the estimated effects on ozone of the nitrogen oxides released directly into the stratosphere from the engine exhausts of such aircraft as Concorde and the Boeing SP747. The much more uncertain estimates of the effects of increased atmospheric N_2O due to increased use of nitrogenous fertilizers are similarly reduced.

Chlorine compounds (ClX)

The ability of chlorine compounds to catalyse ozone removal was not considered until the besign of the U.S. Space Shuttle, which incorporates solid fuel rockets based on ammonium perchlorate and aluminium powder. Stolarski & Cicerone (1974) showed that the amounts of hydrogen chloride thereby released directly into the stratosphere would have a negligible effect on ozone. Soon afterwards, Molina & Rowland (1974) and Rowland & Molina (1975) pointed out that the transport upwards to the stratosphere of F-11 ($CFCl_3$) and F-12 (CF_2Cl_2), which are accumulating in the atmosphere owing to their extensive use as aerosol propellants, refrigerants and foam-blowing agents, would eventually provide a much larger source of reactive chlorine species in the stratosphere.

F-11 and F-12 have no known tropospheric sinks and are removed from the atmosphere by transport upwards to altitudes of 25–30 km where they are photolysed, to yield chlorine atoms, by light of wavelength between 180 nm and 220 nm in the 'solar window' between the strong absorptions by O_2 and by O_3 (figure 3). This process and the parallel but less important decomposition of F-11 and F-12 by $O(^1D)$ will produce a stratospheric burden of reactive chlorine species (Cl, ClO, HCl) of about 5 parts/10^9 on a timescale of about 50 years. The current burden is 2–3 parts/10^9 and is thought to be partly derived from the attack of HO radicals on CH_3Cl, which is produced by marine algae and by slash-and-burn agriculture and is the main chlorine species in the troposphere (Lovelock 1975). Other contributors are CCl_4 (origin uncertain), F-11, F-12, volcanic activity and sea spray. CH_3CCl_3, which finds increasing use as a cleaning agent, may prove a significant source in the future.

The cycle for the chlorine catalysed decomposition of ozone is

$$Cl + O_3 = ClO + O_2, \qquad (41)$$

$$ClO + O = Cl + O_2, \qquad (42)$$

$$O + O_3 = O_2 + O_2. \qquad (41) + (42)$$

This removal process occurs mainly in the 30–50 km altitude range, and, particularly at lower altitudes, there is a no-effect cycle

$$ClO + NO = Cl + NO_2, \qquad (43)$$

$$NO_2 + h\nu = NO + O, \qquad (35)$$

$$O + O_2 + M = O_3 + M, \qquad (3)$$

$$Cl + O_3 = ClO + O_2. \qquad (41)$$

This is strictly analogous to the cycle (36), (35), (3) + (20) discussed above and it is interesting to note that the parallel behaviour of reactions of Cl and HO, which is familiar in laboratory studies of chemical kinetics, extends to the atmospheric reactions of the analogous species ClO and HO_2. A further example is provided by the similar rate coefficients for the formation of chlorine nitrate and pernitric acid:

$$ClO + NO_2 + M = ClONO_2 + M, \qquad (44)$$

$$HO_2 + NO_2 + M = HO_2NO_2 + M. \qquad (45)$$

Although these molecules have limited thermal stability and are decomposed by sunlight and by free radical attack in a matter of hours, they should be present in the stratosphere in detectable quantities.

The main removal process for ClX species is the reaction of chlorine atoms with methane to yield hydrogen chloride, which is transported downwards to the troposphere where it is rained out:

$$Cl + CH_4 = HCl + CH_3. \qquad (46)$$

Sufficient formaldehyde may be present as an intermediate in the stratospheric oxidation of methane for reaction (47), which has a much higher rate coefficient than reaction (46), to contribute significantly to HCl formation:

$$Cl + CH_2O = HCl + HCO. \qquad (47)$$

Unlike the NO_x system, where HO radicals react to form the stable nitric acid, here the action of HO radicals is to regenerate Cl atoms from hydrogen chloride:

$$HO + HCl = H_2O + Cl. \qquad (48)$$

Recent measurements by three groups have shown that the reaction

$$ClO + HO_2 = HOCl + O_2 \qquad (49)$$

can remove significant amounts of ClO. The importance of this process in stratospheric chemistry depends on the lifetime of the HOCl formed. There is considerable controversy as to the form and strength of the spectrum of HOCl above 280 nm, but present evidence strongly suggests that HOCl is rapidly photolysed to yield the active species HO and Cl.

Figure 5 brings out the similarities and differences in the stratospheric cycles of nitrogen oxides and chlorine compounds. A particular point of interest is the differing proportions of the active species X and XO in the two cases.

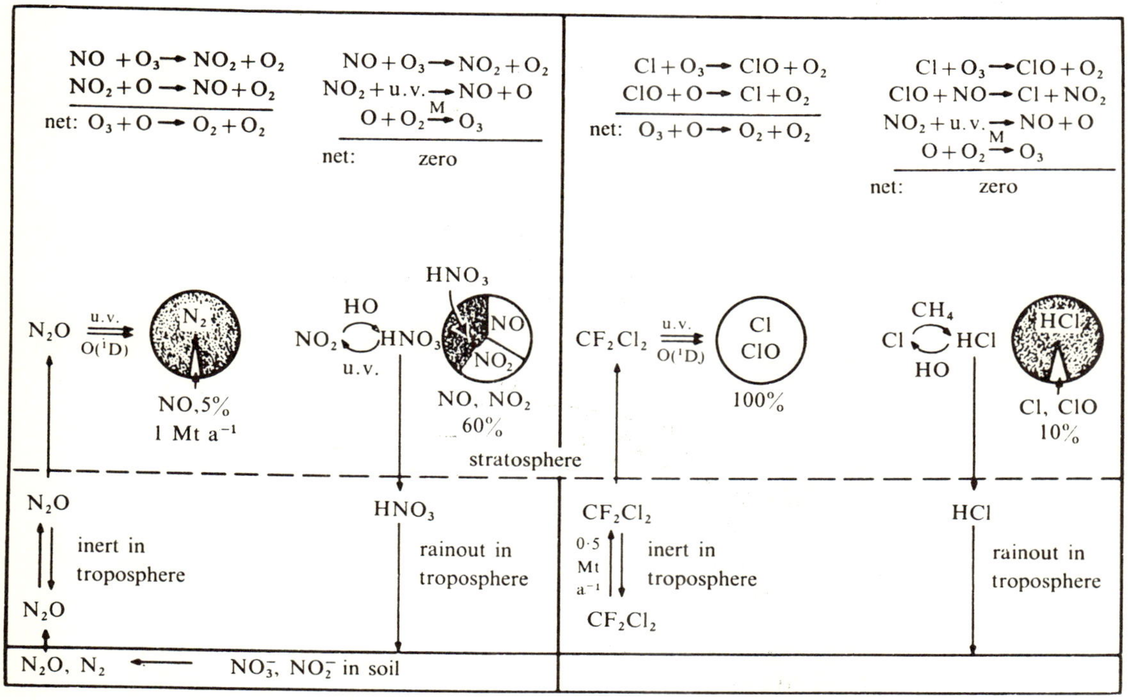

FIGURE 5. The ClX and NO_x cycles in the atmosphere showing the average proportions of the various reactive species (Johnston 1975).

If the atmosphere contained significant amounts of bromine, the bromine catalysed cycle corresponding to reactions (41) and (42) would be very important because the weakness of the H–Br bond makes the reactions corresponding to (46) and (47) endothermic and therefore negligibly slow. Formation of the inactive HBr would then have to depend on the reaction of Br with the much less abundant HO_2 radical

$$Br + HO_2 = HBr + O_2, \tag{50}$$

and the proportion of BrX present as HBr would be much smaller than the corresponding fraction for the chlorine case (figure 5).

In contrast, the great stability of the HF formed by the analogue of reaction (46) ensures that the fluorine catalysed decomposition of stratospheric ozone is negligible.

TRACE SPECIES IN THE ATMOSPHERE

The test of our understanding of atmospheric chemistry is the accuracy with which we can use laboratory values for reaction rates and our knowledge of atmospheric circulation to model the profiles of trace species in the atmosphere.

This is a formidably difficult problem since the concentrations of trace species will vary with latitude, solar elevation, season, etc. Although much progress is being made in the measurement of trace species in the stratosphere (Ehhalt 1979, this symposium), only limited data are available on species other than ozone and any deviations between observed and calculated profiles can arise from the chemical inhomogeneity of the stratosphere as well as from the most obvious causes of inadequacies in modelling in laboratory kinetic data or in the observations themselves.

$$O(^1D) + H_2O, CH_4, H_2 \longrightarrow HO, H$$

$$H_2O + h\nu \longrightarrow HO, H$$

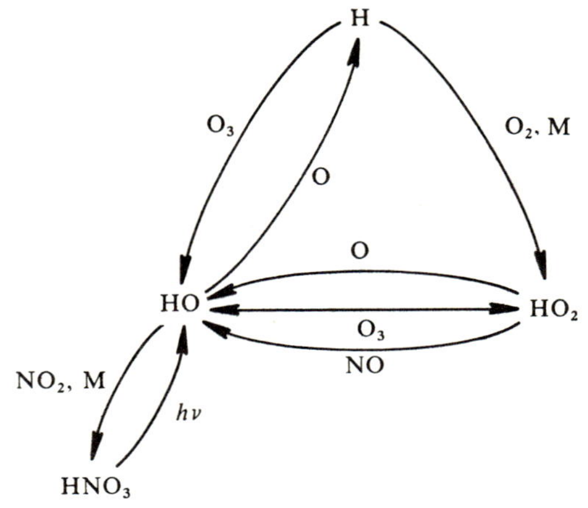

$$HO + HO_2 = H_2O + O_2$$
$$HO_2 + HO_2 = H_2O_2 + O_2$$
$$H + HO_2 = H_2 + O_2$$

FIGURE 6. Diagrammatic representation of the formation, interconversion and removal processes for the hydrogen radicals, H, HO and HO_2.

Currently, it is thought that the NO_x cycle is responsible for about half the rate of removal of odd oxygen from the stratosphere, that the HO_x species and reaction (5) between O and O_3 each remove about 20% and that the ClX cycle removes about 5%. The rapid interconversion of radicals by processes such as (36) and (43) give rise to mixed cycles which make this partition somewhat arbitrary. For this reason it is wrong to assume that increasing the stratosphere burden of NO_x or ClX by a given factor would produce an ozone depletion that could be estimated from the contribution of that species to ozone removal in the unperturbed stratosphere. This linking of the odd oxygen removal cycles and the 'self-healing' effect, by which a lower ozone concentration at higher altitudes allows the penetration to lower altitudes of the ultraviolet light which generates odd oxygen, both make ozone profiles and total column measurements a relatively insensitive test of our understanding of stratospheric chemistry. This is unfortunate because ozone is by far the best studied reactive trace species.

The vertical profiles of species such as CH_4, N_2O, F-11 and F-12, which are liberated at ground level and removed by chemical reaction and/or photolysis in the stratosphere, provide a measure of the competition between these processes and vertical transport. Such profiles are normally used to provide information about transport rather than to verify calculated reaction rates, although the arguments can be somewhat circular since the reaction rates are derived from a model calculation.

With a short-lived species such as $O(^3P)$ or HO, the rate of transport does not govern their transport directly. As figure 6 shows, the concentration of HO is controlled by its rate of formation, by the interconversion reactions of HO_x radicals and by the removal processes for these radicals. Thus, differences between the calculated and observed concentrations (figure 4) can reflect inadequate knowledge of the rates of any of these processes and particularly the rate of the formation reaction (11) since the concentration of water vapour in the stratosphere is somewhat variable. A measurement of the ratios of the concentrations of H, HO and HO_2 would assess our knowledge of the processes for interconversion of these species, which form the centre of figure 6, without being affected by formation, transport and removal processes of the HO_x species as a whole.

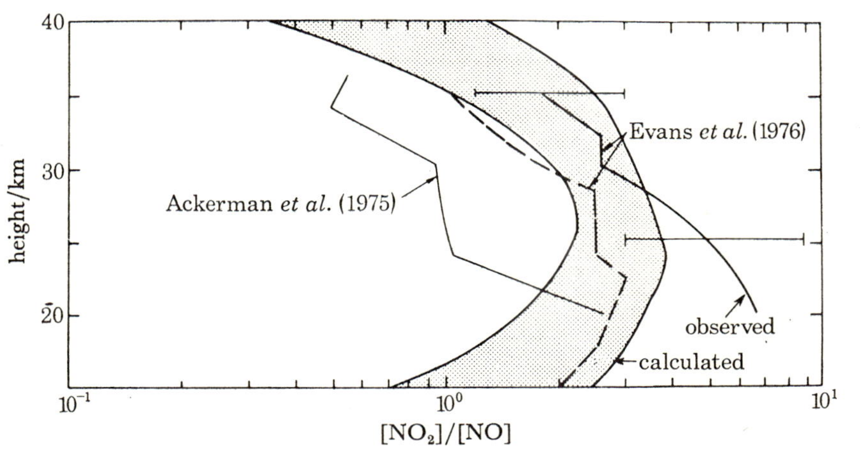

FIGURE 7. Calculated and observed ratios of $[NO_2]$ to $[NO]$ (Hudson 1977).

This approach has been applied most extensively to the nitrogen oxides. Here the ratio of NO to NO_2 at altitudes of 20–25 km is governed by processes (33), (34) and (35)

$$\frac{[NO_2]}{[NO]} = \frac{k_{33}[O_3]}{J_{35}+k_{34}[O]},$$

where the ratio of [O] to $[O_3]$ is determined by processes (3) and (4) (Evans *et al.* 1976). The method can be extended to include HNO_3, which is formed from NO_2 by reaction (37) and converted back to it by process (38) and by reactions (39) and (40). At higher altitudes the ratio of $[NO_2]$ to [NO] is affected by the rapid reactions (36) and (43). Infrared measurements of the ratio of $[NO_2]$ to [NO] agree adequately with the calculated values (figure 7), but extension of this to include HNO_3, which involves the chemistry of HO, produces a significant discrepancy.

Similarly, Anderson *et al.* (1977) find relative values of Cl and ClO in the upper stratosphere that are consistent with the model calculations, but their absolute values do not always agree with the amounts of HCl detected at lower altitudes, which are used to assess total chlorine burdens. Measurement of the ratio of the unreactive HF to the much more reactive HCl as a function of altitude provides a method for estimating the proportion of chlorine present in other forms, e.g. ClO, Cl, $ClONO_2$, HOCl which will vary with altitude (Farmer & Raper 1977).

Thus, measurements of trace species in the atmosphere provide two distinct tests of our understanding of stratospheric chemistry. The neglect or erroneous measurement of the rates of processes interconverting two related trace species will show itself as a disagreement between the observed and calculated ratios of their concentrations. Where species such as HO_2NO_2, N_2O_5 (from $NO_3 + NO_2$), H_2O_2, $ClONO_2$, HOCl, etc. are predicted to survive long enough to constitute an appreciable fraction of the NO_x, HO_x or ClX species at a given altitude, these species should be detectable, particularly by long path infrared spectroscopy. Since the rates of formation of these species are relatively well known, continued failure to detect some of them must imply a lack of knowledge of their decomposition rates.

In this connection it is interesting that unpublished measurements at Cambridge by J. P. Burrows, D. I. Cliff, B. A. Thrush and J. P. T. Wilkinson, and at Harwell by J. P. Burrows, R. A. Cox and G. S. Tyndall, have shown that the rate coefficient of reaction (26), which is the dominant source of H_2O_2 in the stratosphere, is pressure dependent. The 'fall-off' from its high pressure value of 3×10^{-12} $cm^3\ s^{-1}$ begins at pressures around 25 mbar, and at 1 mbar total pressure k_{26} is less than 1×10^{-12} $cm^3\ s^{-1}$. This new work provides a partial explanation of the failure to detect H_2O_2 in the upper stratosphere.

Although Chapman's work around 1930 provided a secure and accurate basis for our understanding of atmospheric chemistry, the subject has only developed rapidly in the last decade. This has stemmed partly from developments in techniques for laboratory and atmospheric measurements, but largely it has been environmental considerations that have shifted the emphasis from the physics to the chemistry of the atmosphere in recent years.

The author thanks the Royal Society, Science Research Council, Manufacturing Chemists Association and Federal Aviation Administration for Research Grants.

References (Thrush)

Anderson, J. G., Margitan, J. J. & Stedman, D. H. 1977 *Science, N.Y.* **198**, 501.
Bates, D. R. & Nicolet, M. 1950 *J. geophys. Res.* **55**, 301.
Chapman, S. 1930 *Mem. R. met. Soc.* **3**, 103.
Crutzen, P. J. 1970 *Q. Jl R. met. Soc.* **96**, 320.
Evans, W. F. J., Kerr, J. B., Wardle, D. I., McConnell, J. C., Ridley, B. A. & Schiff, H. I. 1976 *Atmosphere* **14**, 189.
Farmer, C. B. & Raper, O. F. 1977 *Geophys. Res. Lett.* **4**, 527.
Hampson, R. F. & Garvin, D. 1977 (eds.) Reaction rate and photochemical data for atmospheric chemistry. *U.S. Natn. Bur. Stand. Spec. Publ.* 513.
Howard, C. J. & Evenson, K. M. 1977 *Geophys. Res. Lett.* **4**, 437.
Hudson, R. F. 1977 (ed.) Chlorofluoromethanes and the stratosphere. *N.A.S.A. Ref. Publ.* 1010.
Johnston, H. S. 1971 *Science, N.Y.* **173**, 517.
Johnston, H. S. 1975 *An. Rev. phys. Chem.* **26**, 318.
Lovelock, J. E. 1975 *Nature, Lond.* **256**, 193.
Molina, M. J. & Rowland, F. S. 1974 *Nature, Lond.* **249**, 810.
Rowland, F. S. & Molina, M. J. 1975 *Rev. Geophys. Space Phys.* **13**, 1.
Stolarski, R. S. & Cicerone, R. J. 1974 *Can. J. Chem.* **52**, 1615.
Thrush, B. A. 1979 *Phil. Trans. R. Soc. Lond.* A **290**, 505.

Phil. Trans. R. Soc. Lond. A **296**, 161–173 (1980) [161]
Printed in Great Britain

Spectroscopic observations of middle atmosphere composition

BY J. E. HARRIES
Division of Quantum Metrology, National Physical Laboratory, Teddington, Middlesex, U.K.

The principal advantages and disadvantages of spectroscopic techniques as a means of deducing the composition of the middle atmosphere are reviewed. A brief general survey is given, which includes methods using natural or artificial radiation sources, used in remote-sensing or local sampling configurations, and at spectral wavelengths ranging from the ultraviolet to the microwave. Limitations of space dictate an emphasis on the infrared methods using natural radiation sources employed by most workers in this field.

Composition results obtained by using spectroscopic techniques have often had an important effect on the development of our understanding of the middle atmosphere, and an attempt is made to describe some of the more important measurements of recent years, and the significance of the new results they have produced. This discussion of new results is brought up to date with a description of a recent balloon-borne infrared experiment.

1. INTRODUCTION

The application of spectroscopy to atmospheric studies is a subject with a very long and distinguished history (see, for example, Adel 1949 and references contained therein), so that its use to determine atmospheric composition is not new. The application of spectroscopic methods to study the composition of the middle atmosphere (loosely defined as the layer between 10 and 150 km altitude) is, however, more recent, requiring as it did the development of high altitude balloons and rockets, and more recently spacecraft.

Middle atmosphere spectroscopy, in common with several other disciplines, received a healthy boost during the last decade with the realization that conditions in the middle atmosphere may not be as stable with time as had been imagined and, indeed, might even be susceptible to human influence, albeit unintentional, with possibly detrimental environmental consequences. Spectroscopy is an extremely powerful tool for detecting and monitoring composition even at very low concentrations and, in view of the potential problems, it has been widely used to improve our knowledge, and hopefully our understanding, of the middle atmosphere. The present paper contains a brief description of why and how spectroscopy has been used for this purpose.

We shall first consider (§ 2) the range of instruments and techniques at our disposal for atmospheric spectroscopy; this will orient the readers' thoughts in preparation for later, unqualified, references to a variety of apparatus. Then some consideration will be given, in § 3, to the advantages and disadvantages of spectroscopic methods, and the former will be illustrated by some of the leading work selected from recent years. Finally, in § 4, a description of a recent balloon experiment, which carried several infrared spectrometers into the stratosphere, will be described.

2. Techniques and instruments

A wide variety of instruments and techniques are available to the modern atmospheric spectroscopist, and it is necessary to undertake a brief review of these at this stage.

Table 1 lists some representative instrumental types that have been, or are being, used. These instruments have been employed on a variety of observation platforms, at sea level, on aircraft, balloons, rockets and orbiting spacecraft, and in a variety of observational geometries, looking up, down and sideways depending on the requirements of the experiment.

Table 1. A selection of instrument types

filter radiometer/photometer
grating spectrometer
grille spectrometer
gas correlation radiometer
correlation radiometer
Michelson interferometer
Fabry–Perôt interferometer
microwave spectrometer
i.r. heterodyne spectrometer

The filter radiometer or photometer operates, as is well known, through the use of a simple spectral selection device to isolate a relatively broad spectral range which includes an absorption band of the atmospheric constituent to be measured (see Low & Rieke 1974). Spectral selection may be effected by using the absorption properties of solid-state filters, multilayer interference filters, or other devices. Generally, in such devices, spectral scanning is not attempted.

The grating spectrometer (Schroeder 1974) is perhaps the best known device of those listed in table 1, and effects spectral selection by utilizing the diffraction properties of a ruled grating. A grating may produce a large number of different 'orders' of diffraction and care must be taken in the design of a grating spectrometer to include filters that select only those orders which are required. Spectral scanning can quite easily be implemented by rotation of the grating or, less often, of the input and output slits.

The grille spectrometer, devised by Girard (1963), is essentially a modified grating spectrometer, but deserves special mention because it has been extensively employed to make stratospheric measurements (see below). The input and exit slits of a conventional spectrometer are replaced by an array of slits (a 'grille') which is designed carefully to preserve an acceptable instrumental profile. In this way the energy 'throughput' or 'étendue' of the spectrometer may be increased: such an increase in efficiency is usually important in atmospheric studies where observation time is at a premium.

The use of gas correlation radiometers (Smith & Pigeon 1964) rests upon the novel idea of using, within the instrument, a sample of the gas to be detected in the atmosphere, in order to achieve spectral selection with, in the absence of Doppler shifts of the atmospheric lines, perfect spectral matching, and with the possibility of achieving very high spectral resolution. A number of variations of such a device may be imagined (including ones based on the Michelson interferometer – see Harries & Chamberlain 1976), but the most widely used in atmospheric studies have been the selective chopper radiometer (s.c.r.) and pressure modulator radiometers (p.m.r.) developed by Houghton and coworkers at Oxford; these devices achieve modulation in a spectral line by variation of the pressure of a sample of gas or, equivalently, by switching between cells containing samples at different pressures. A review of these devices is given by Houghton & Taylor (1973).

A correlation radiometer (Barringer & Davies 1973) differs from the latter device in having a *synthetic* version of the spectrum (or interferogram) of the required constituent contained or stored (often electronically) within the instrument. In this case, of course, one avoids the sometimes difficult problem of containing a real sample of a gas; on the other hand a correlation radiometer can never achieve perfect matching with the real spectrum because of imperfections in knowledge, nor, in practice, very high spectral resolution.

The Michelson interferometer and the techniques of Fourier transform spectroscopy (Schnopper & Thompson 1974) have been widely used since the mid-1950s when the Multiplex (or Fellgett) and Étendue (or Jacquinot) advantages were discovered. Because of these advantages, Fourier spectrometers are very efficient instruments for studying weak sources, although care does need to be exercised when designing an experiment that conditions are suitable for the advantages to be effective. Thus, the Multiplex advantage exists only as long as detector noise predominates over other noise terms – in certain circumstances a Multiplex disadvantage can occur (e.g. if noise in the incoming signal dominates). Furthermore, the Étendue advantage is based upon the circular symmetry of the device, due to the absence of slits; however, in certain configurations a slit-like field of view may be necessary (e.g. in limb-sounding – see below), which can more easily be matched to a grating spectrometer slit than the circular field of a Michelson. These points are made to illustrate the fact that the Michelson interferometer may not be the panacea for all weak-source measurement problems. When used properly, however, as we shall see below, it becomes a very powerful, wide spectral bandwidth, high resolution spectrometer.

Fabry–Perôt interferometry (Roesler 1974) is another area of classical spectroscopy that has been applied to atmospheric studies. The Fabry–Perôt is, of course, a plane-parallel interference filter, using a pair of highly reflecting plates and multiple beam interference to limit spectral transmission to certain well-defined frequencies. Energy throughput is good because of the absence of slits; in some cases very high spectral resolution is possible.

Microwave atmospheric spectroscopy has recently been reviewed by Waters (1976). Using monochromatic, tunable (and relatively powerful) local oscillator sources, heterodyne and superheterodyne methods have advanced considerably in the years since 1945. The advantages of such techniques are very high spectral resolution and radiometric sensitivity, coupled with low atmosphere scattering because of the long wavelengths, and measurements of middle atmosphere composition have been made from the ground, aircraft and balloons. Previously the instrumentation required has been extremely heavy and large, but modern microwave spectrometers are light enough to be used on such airborne platforms. Plans are being made for systems that can be operated on orbital spacecraft, and considerable progress can be expected in this field in future years.

Finally, table 1 mentions infrared heterodyne spectroscopy. The development of this method has been hampered until recent years by the lack of suitable local oscillators (lasers) and non-linear mixing elements. Menzies (1976), in a recent review, has shown, however, that many problems have been overcome and that very high spectral resolution infrared heterodyne radiometers can now be developed for atmospheric studies.

To be able to briefly review current activities in atmospheric spectroscopy that use the above mentioned instruments and techniques, table 2 has been compiled. This lists a number of groups active in this field, the instrumental types and techniques used and the types of observational platforms that have been employed. Abbreviations used in the table are identified in the key that follows it. Table 2 is without doubt incomplete but represents a reasonably broad sample of current activity.

The types of instruments have been identified in table 1 and the preceding text, of course. Other points that should be brought out in considering table 2 are:

(i) The use of two distinct remote sensing methods, i.e. solar occultation and atmospheric emission. The former permits much better signal:noise ratios than the latter, though limiting observations (when used in a limb sounding configuration, as is usual) to local sunrise and sunset. Emission measurements can of course be made at all local times and, indeed, in, any direction.

TABLE 2. ATMOSPHERIC SPECTROSCOPY: A SELECTION OF GROUPS/TECHNIQUES

group	*country*	*technique*	*platform*
Ackerman/Girard *et al.*	Belgium, France	grille spectrometer/i.r. solar occ.	balloon, aircraft
Anderson *et al.*	U.S.A.	resonance fluorescence/local measurement	balloon, rocket
Evans *et al.*	Canada	g.s./M.i./u.v. and i.r. solar occ. and atmospheric emission	balloon, rocket
Farmer *et al.*	U.S.A.	M.i./solar occ.	balloon, aircraft
Gille *et al.*	U.S.A.	filter radiometer/atmospheric emission	balloon, satellite
Harries *et al.*	U.K.	f.i.r. M.i./cryogenic g.s./atmospheric emission	balloon, aircraft
Heath/Kreuger	U.S.A.	u.v. spectrometer/solar trans. and backscatter	satellite
Houghton *et al.*	U.K.	s.c.r./p.m.r./atmospheric emission	balloon, satellite, aircraft
Menzies *et al.*	U.S.A.	i.r. laser heterodyne solar occ.	balloon, aircraft
Murcray *et al.*	U.S.A.	g.s. (cooled and uncooled)/solar occ./atmospheric emission	balloon, aircraft
Patel	U.S.A.	Laser absorption using spin-flip Raman laser/local measurement	balloon
Offerman *et al.*	West Germany	cryogenic g.s./atmospheric emission	rocket
Stair *et al.*	U.S.A.	g.s./cryogenic M.i./atmospheric emission	rocket
Waters	U.S.A.	microwave spectrometer/atmospheric emission	balloon, aircraft, ground
Zander	Belgium	g.s./solar occ.	balloon

Key to table 2:

i.r.	infrared	u.v.	ultraviolet
solar occ.	solar occulation	f.i.r.	far infrared
g.s.	grating spectrometer	s.c.r.	selective chopper radiometer
M.i.	Michelson interferometer	p.m.r.	pressure modulator radiometer

(ii) Mention is made of the resonance fluorescence spectrometer measurements of Anderson and coworkers; this technique is a local sampling method, and as such will be discussed in an accompanying paper by Dr Ehhalt. However, it is mentioned in table 2 as a most interesting application of spectroscopy to atmospheric studies. A description may be found in Anderson *et al.* (1977).

(iii) The development of completely cold spectrometric systems – often of quite high degrees of complexity – is being undertaken in several laboratories. Cryogenically cooled devices may achieve improved sensitivity, when looking at very weak emissions, by removing thermal emission from optical components which otherwise could totally swamp the desired atmospheric signal. Of course, cooled detectors have been used in atmospheric spectrometers for many years, and so the use of cryogenic liquids and refrigerators themselves is not new; the cooling of a complete spectrometer does, however, present many new problems including the design of satisfactory moving parts, and the interface between the cold spectrometer and the (relatively) warm atmosphere.

(iv) Even though several new and ingenious devices have been developed in recent years, table 2 illustrates the fact that most spectral devices can find application in one type of investigation or another. It is, indeed, important to carry out investigations with many different types of technique in order to increase the quantity of independent information that is available.

3. Advantages and disadvantages

The widespread use of spectroscopy for atmospheric studies clearly indicates that there are strong advantages implicit in the method; but at the same time certain very important disadvantages exist which must be considered. Table 3 shows a summary of some of the more important aspects. A discussion of table 3 follows.

Table 3. Spectroscopy of the terrestrial atmosphere

some advantages:
- large information content;
- high intrinsic sensitivity;
- possible to integrate signals over long paths to increase sensitivity further;
- local contamination reduced by long optical paths;
- remote sensing possible;
- 'discovery' aspect;
- 'archival' aspect;

some disadvantages:
- complex spectrum to unravel;
- atmospheric structure and spectral structure to deconvolve (e.g. inversion problem);
- instruments can be complicated and/or large and/or expensive.

Clearly, the most attractive feature is the very large information content explicitly displayed in the absorption or emission spectrum of an atmospheric molecule or atom. This may be illustrated by figure 1 which shows, in the lowest frame, the spectrum of solar radiation attenuated by the atmosphere as seen at sea level in the spectral range 1–15 μm; the shading represents regions of absorption and the ordinate is given in units of fractional absorptance (0–100 %). The upper frames show spectra of several atmospheric constituents, recorded in the laboratory, and a comparison of these with the solar spectrum immediately shows that this one spectrum contains information on the concentration of each of these constituents, from the intensity of absorption in each case.

When it is remembered that the spectra shown in figure 1 are recorded at relatively low spectral resolution (*ca.* 5 cm^{-1} or so) it will be appreciated that at the higher resolutions currently achievable (0.1–0.001 cm^{-1}) an enormous quantity of information, on a great number of atmospheric constituents, becomes available to us. Connes & Connes (e.g. 1966) have illustrated on many occasions in their study of planetary spectra with Fourier spectroscopy how increased resolving power produces more and more spectral detail and information.

High sensitivity is mentioned in table 3, and, indeed, in the laboratory spectroscopy is often used as an analytical tool for this reason. In the atmosphere high sensitivity means that spectroscopy may be used to detect rather small quantities of specific constituents.

Sensitivity can, of course, be improved upon, as is mentioned in the table, by integrating absorption and emission signals over very long paths. Indeed, most remote sensing applications of atmospheric spectroscopy necessarily involve long atmosperic paths which in some cases may represent effective path lengths of several hundred kilometers.

In the middle atmosphere, of course, the absolute concentrations of many constituents are very low, and it is not difficult to contaminate severely the locality of a balloon-borne sensor, for example with material outgassing from the balloon itself and/or from the instrument. This problem has been met, for example, in many early attempts to measure stratospheric water vapour density. Spectroscopic methods that integrate over long paths are clearly less sensitive to local contamination because of the large integration path outside the contaminated zone.

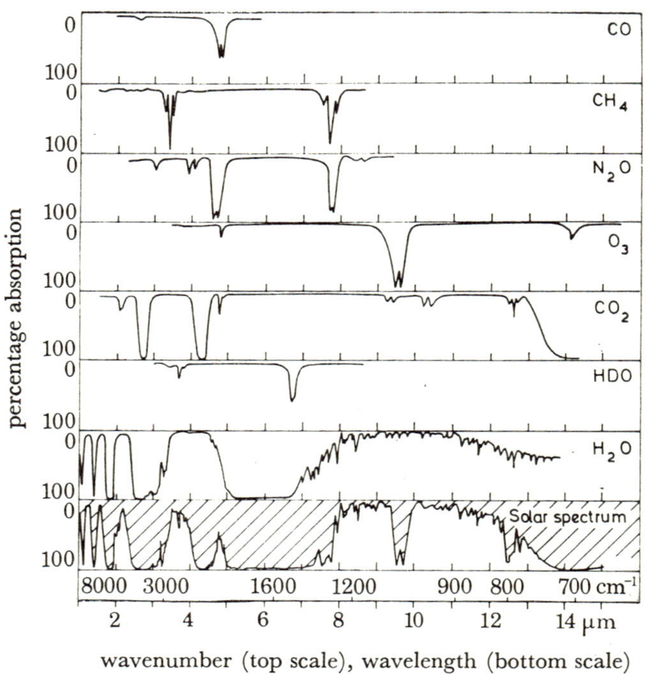

FIGURE 1. Comparison of the transmission of solar radiation through the atmosphere to sea level (bottom curve), compared with the laboratory spectra of atmospheric gases contributing to the atmospheric spectrum. In each curve the abscissa shows wavelength in μm (lower scale) or wavenumber (top scale), and the ordinate is 0–100 % absorptance.

A further aspect listed in table 3 may seem rather obvious but should be mentioned because of its importance: a piece of atmosphere quite remote and distant from the sensor can be investigated or 'remotely sensed'. Ingenious methods have been developed, based on the exponential decrease of atmospheric density with height and 'topside' (or down-looking) viewing, which ensure that most of the radiation detected by a spectroscopic sensor at a given wavelength comes from a rather compact, remote parcel of the atmosphere.

Next, table 3 contains what I have called the 'discovery' aspect of atmospheric spectroscopy. This simply stated means that each time an improved measurement of the atmospheric spectrum is made (e.g. at increased spectral resolution, with a better signal:noise ratio, or in a new spectral region) new bands may be discovered, possibly due to unknown constituents This discovery aspect is not a feature of certain other methods such as chemical sampling, where an experiment is designed specifically for a given measurement (we should point out, of course, that such methods may have other advantages which make them valuable).

A rather beautiful result which illustrates this aspect was the discovery in the late 1960s of the presence of nitric acid, HNO_3, in the stratosphere by Murcray and coworkers at the University of Denver. This group was engaged in studies of the absorption spectrum of the stratosphere with the use of a solar-pointing grating spectrometer mounted on a balloon gondola.

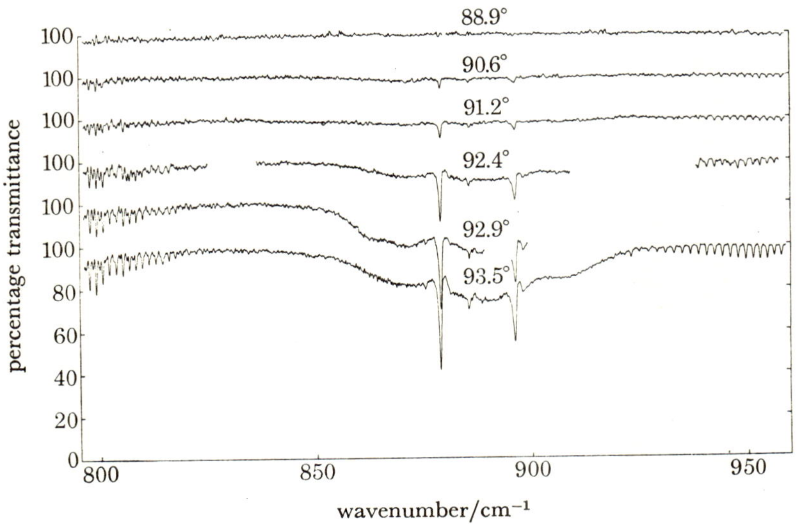

FIGURE 2. Measurements of solar transmission at a number of solar zenith angles (marked on each curve) in the 800–950 cm^{-1} region. Each curve is displaced by 20% for clarity. The observation altitude was 30 km (Murcray *et al.* 1974*a*).

Measurements were made at large zenith angles, in order to maximize the absorption along a very long atmospheric path, to study the detailed spectroscopy of the 5–20 μm region. When measurements were made with solar zenith angles > 90° (i.e. below the local horizontal, looking through atmospheric layers below the flight level) new absorptions were discovered in the 1325 cm^{-1} (7.55 μm) and later in the 890 cm^{-1} (11.2 μm) regions, which were assigned to the HNO_3 molecule. The observations at 11.2 μm were particularly remarkable since prior to this work this region had been known as one of particularly high transmission (a 'window' region). Figure 2 (taken from Murcray *et al.* 1974*a*) shows a series of spectra around 11.2 μm illustrating the rapid growth of the HNO_3 absorption bands as the zenith angle increases. These data were measured from a balloon floating at *ca.* 30 km, with the solar zenith angles θ_z marked on each trace: it can be seen that for $\theta_z < 90°$, the region is one of very high transmission, but that at $\theta_z > 91°$, HNO_3 absorption increases very rapidly.

Table 3 also mentions the 'archival' aspect of atmospheric spectroscopy: thus it has been found in recent experience that high resolution atmospheric spectra recorded previously may be re-examined later either to search for new constituent absorptions or to reanalyse the raw data when improved spectral data or analysis techniques become available.

The most serious drawbacks to the method are, of course, the complexity of the spectrum and the difficulty of extracting unambiguous data from it. Thus a molecular vibration or rotation spectrum is usually extremely complicated and great care and attention needs to be paid to correct assignment of the observed spectral features to the constituent responsible.

A rather beautiful example of spectral analysis is shown in figure 3, again taken from the work of Professor Murcray and his group at Denver University (Murcray *et al.* 1974*b*). The top frame of figure 3 shows an atmospheric transmission spectrum measured by using a grating spectrometer, with the Sun as a source, from an aircraft at an altitude of about 15 km. The lower frames show how theoretical calculations of the transmission spectra of H_2O, CH_4 and NO_2 have been combined (multiplied together) to produce a synthetic spectrum that very

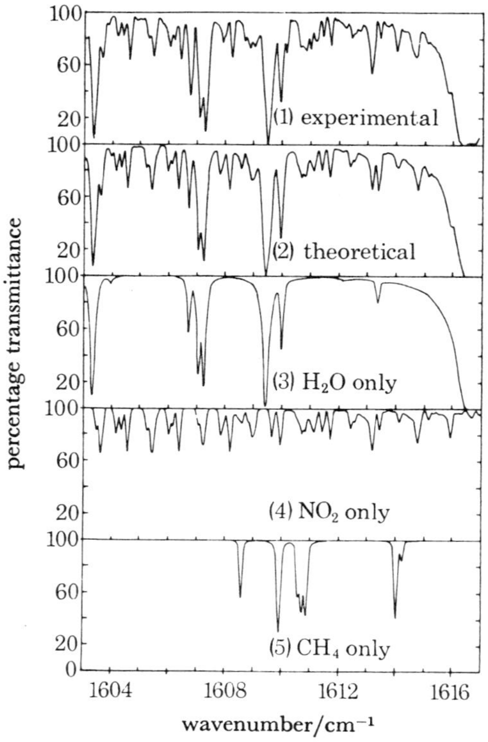

FIGURE 3

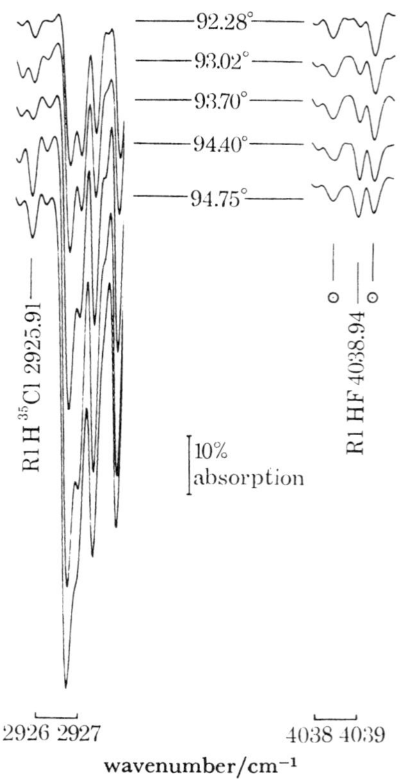

FIGURE 4

FIGURE 3. The atmospheric transmission spectrum measured from an aircraft experiment at *ca.* 15 km altitude is shown in the top frame: the second frame shows a calculated spectrum constructed from the spectra calculated for H_2O (third frame), NO_2 (fourth frame) and CH_4 (fifth frame). The calculated spectra clearly demonstrate that features in the 1604–1606 cm^{-1} and 1611–1614 cm^{-1} regions are due to NO_2 (Murcray *et al.* 1974*b*).

FIGURE 4. Atmospheric transmission measurements made from a balloon at *ca.* 35 km altitude using a Michelson interferometer, in the 2900 cm^{-1} region and 4000 cm^{-1} region. Absorption features due to HF, HCl and solar CO (⊙) are marked (Farmer *et al.* 1977).

closely matches the experimental result. In this case the aim of the exercise was to demonstrate unambiguously the absorption features due to NO_2, and as can be seen by a close inspection of the 1604–1606 cm^{-1} and 1611–1614 cm^{-1} regions in particular, this exercise was very successful. In fact, when faced with the analysis of a very complex line spectrum this process of building up a matching synthetic spectrum species-by-species is the most satisfactory and convincing way of completing such analysis.

This aspect of atmospheric spectroscopy is so close to the heart of the whole topic that it merits further example. Figure 4 shows a result due to Farmer & Raper (1977). This shows a series of atmospheric transmission spectra in the 2900 and 4000 cm^{-1} regions measured by using a rapid-scan Michelson interferometer looking at the Sun from a balloon platform at about 35 km. The simultaneous detection of HCl and HF is observed quite clearly along the line of sight. In this short wavelength infrared region care must be taken to separate absorption features arising within the Earth's atmosphere from those occurring within the solar atmosphere – the spectral features labelled $\odot$ in figure 4 are indeed due to absorption by solar CO.

The problem of deconvolving spectral and spatial structure, also mentioned in table 3, has received considerable attention over the past 20 years. It is a very complicated matter, since the atmospheric structure itself along the line of sight (i.e. the variation of pressure, mixing ratio and temperature) may be highly variable. Furthermore, at wavelengths where the atmosphere is optically thin, photons are received over great distances in the atmosphere, while at optically thick wavelengths photons from great distances are completely absorbed before reaching the observer and so the local atmosphere dominates the signal detected.

The problem may be regarded as an application of information theory, and one of the mistakes it is easy to make is to try to extract too much information from a deconvolution (more commonly called 'inversion' or 'retrieval') exercise, considering the limitations in the real data being used as input. The more advanced methods being used to analyse satellite remote sensing measurements generally provide some extra information to the measurement, in one way or another, in order to constrain the number of possible concentration (or temperature) profiles that may give rise to a measured radiance profile. Valuable reviews of this subject may be found in Rodgers (1976) and Twomey (1977).

An interesting development has been the limb-sounding method (Gille & House 1971) in which the vertical resolution of an atmospheric profile is determined geometrically by scanning the limb (horizon) of the atmosphere with an instrument possessing a narrow vertical field of view. This contrasts with the vertical sounding method originally suggested by Kaplan (1959), which depends on using a variety of optical depths (i.e. different wavelengths across a chosen absorption band) to probe different levels of the atmosphere.

The final point in table 3 is introduced to indicate to the reader not experienced in atmospheric spectroscopy that atmospheric spectrometers may in some cases be large, heavy and complicated and expensive to fly. Indeed, experiments weighing more than 1 ton have been flown on balloons. This is not a general rule, however, and atmospheric spectrometers can range in weight from a few kilograms upwards.

4. A recent balloon-borne spectrometer experiment

At the National Physical Laboratory we have exploited the large information content of the infrared spectrum (particularly the far infrared at wavelengths beyond 100 μm) to explore the possibility of studying many middle atmosphere gases at once. In the far infrared, the pure rotation bands of a large number of atmospheric spectra are overlaid, not separated by differing vibrational fundamental frequencies as in the shorter wave infrared. Measurements have been made from aircraft and balloons of the atmospheric emission spectrum at these very long wavelengths by using Michelson interferometers and Fourier transform techniques.

An example spectrum is shown in figure 5 (Harries *et al.* 1976). Only a small range of the measured spectrum is shown. Atmospheric emission is shown increasing upwards, and the 34–39 cm^{-1} region (λ = 294–256 μm) is displayed. The spectrum was measured by using a liquid helium-cooled InSb detector at an altitude of about 37 km, a zenith angle of about 93°, and a spectral resolution of about 0.05 cm^{-1}. Features can be seen due to $H_2{}^{16}O$, $H_2{}^{18}O$, ${}^{16}O_2$, ${}^{16}O_3$ and NO_2. Elsewhere in the spectrum, lines due to other species, such as HNO_3, occur.

Figure 6 shows profiles of H_2O, NO_2 and HNO_3 calculated from a series of spectra like that shown in figure 5. These profiles were obtained during local daylight, on 20 September 1974, and were measured simultaneously. It is this last point which provides the basic philosophy for our current activities, that is to provide measurements of as many (related) middle atmosphere constituents as possible at the same place and the same time. In this way we hope to avoid the difficulties of combining data measured at different places and different times which may contain important differences due to spatial, temporal and seasonal (i.e. long term temporal) variations.

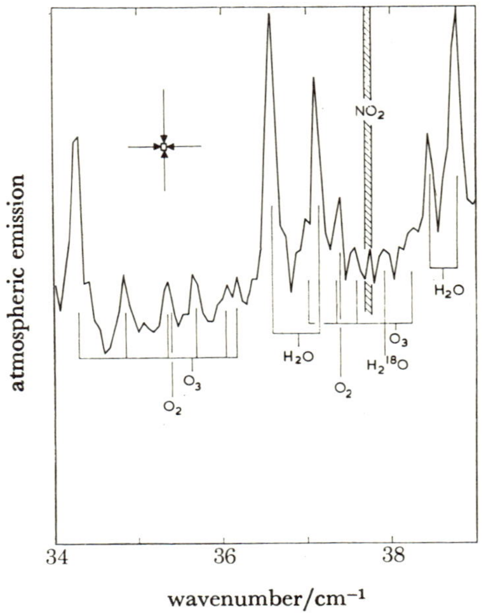

FIGURE 5

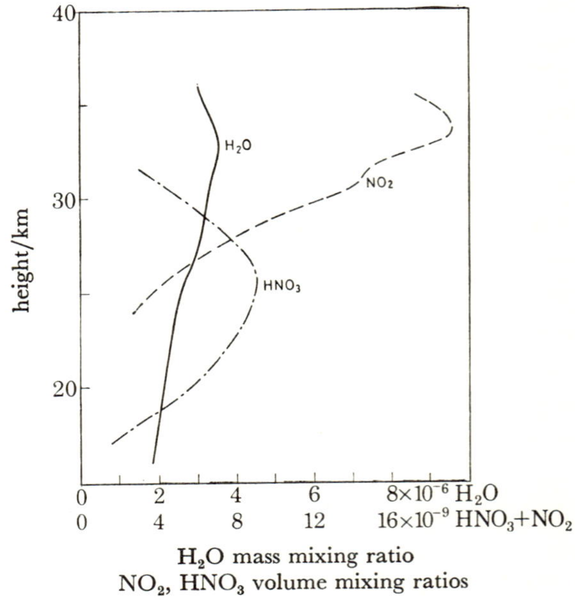

FIGURE 6

FIGURE 5. Atmospheric emission spectrum in the far infrared, measured at *ca.* 37 km altitude and with a zenith angle of 93°. Spectral resolution is 0.05 m^{-1} and atmospheric emission increases upwards. Spectral features due to $H_2{}^{16}O$, $H_2{}^{18}O$, NO_2, O_2 and O_3 may be identified.

FIGURE 6. Profiles of H_2O, HNO_3 and NO_2 derived from spectra including that shown in figure 5 measured simultaneously from a balloon experiment floating at 37 km, during the day on 20 September 1974.

Recently, a balloon flight was made of the latest experiment in our programme. Titled SIBEX (submillimetre infrared balloon experiment) this experiment is being carried out in collaboration with Dr B. Carli and colleagues at the Institute of Research on Electromagnetic Waves (I.R.O.E.) in Florence, Italy. SIBEX comprises two instruments, a liquid nitrogen-cooled grating spectrometer operating in the 8–20 μm region, and a high resolution Michelson interferometer operating in the far infrared (or submillimetre region) between about 100 and 1000 μm. Both instruments view the same parcel of air and detect atmospheric emission by using the limb-sounding technique.

Table 4 illustrates the measurements made by using SIBEX: as can be seen, the combination of the two instruments allows us to measure, simultaneously, the concentrations of a large number of atmospheric species, many of them closely related photochemically.

Sibex was flown on 17 November 1978 from the National Scientific Balloon Facility at Palestine, Texas, U.S.A., The two instruments operated very well, and the analysis of the results is currently under way in Italy and the U.K.

An example of the raw spectral data obtained using the cooled grating spectrometer has been prepared by Dr S. Pollitt and Mr M. J. Bangham, of the N.P.L. group, and is shown in figure 7. The adjacent sections show spectral scans from two of the three channels in the instrument. The spectra have been annotated to indicate emission features due to a number of constituents; the spectra are complicated by the effects of grating blaze efficiency and transmission edges due to order-sorting filters. Each of these effects is, of course, removed during the calibration process which is currently in progress.

Table 4. Middle atmosphere trace constituents to be measured by Sibex

instrument	*spectral region*	*constituent*
far infrared (submillimetre) Michelson interferometer	10–100 cm^{-1} (1000–100 μm)	H_2O O_3 HNO_3 NO_2 N_2O CO HF HCl NH_3 ClO*
mid infrared cooled grating spectrometer	500–1250 cm^{-1} (20–8 μm)	H_2O O_3 CF_2Cl_2 $CFCl_3$ HNO_3 CH_4 CCl_4 $ClNO_3$*

*, measurement uncertain.

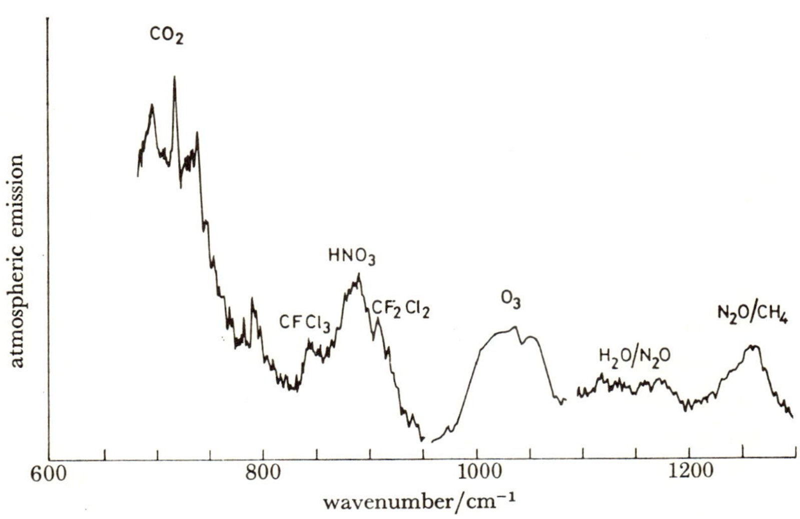

Figure 7. Raw spectral data recorded using the cryogenic grating spectrometer on board the Sibex experiment. The sections of the diagram represent two separate channels of the spectrometer, and in each case the ordinate is signal received at the detectors, and the abscissa is wavenumber. Provisional wavenumber scales and spectral identifications have been superposed.

The far infrared experiment data is still being calibrated and Fourier transformed, and so we can only illustrate the anticipated performance of that device in figure 8. This shows a calculated emission spectrum (Bangham 1978) in the 41–42 cm^{-1} region (λ = 244–238 μm), in which lines due to HF and HCl in the concentrations shown have been added to a spectrum of H_2O, O_2 and O_3. The calculation was carried out for a limb-sounding experiment carried on a platform outside the atmosphere, and a tangent ray height of 30 km, though the result closely approximates that for a balloon floating above 35 km. The simulated spectral resolution is 0.01 cm^{-1}. The estimated experimental noise level is about the magnitude of the weakest O_3 line, and so, as can be seen, both HF and HCl should be clearly detectable in the spectra obtained from this experiment.

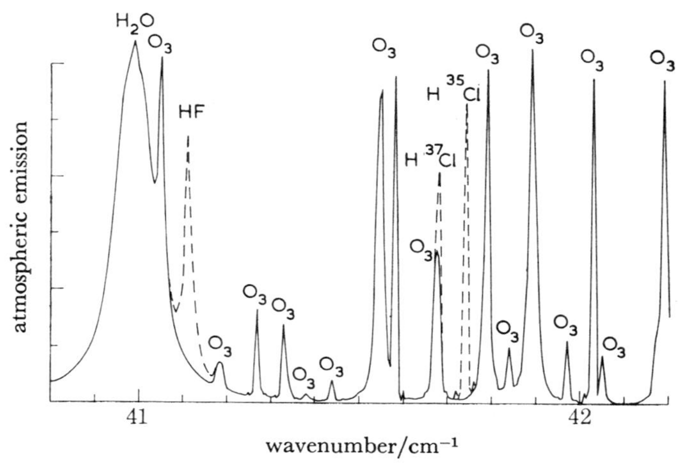

FIGURE 8. A calculated far infrared spectrum of atmospheric emission at high resolution (*ca.* 0.01 cm^{-1}) evaluated for conditions resembling the SIBEX balloon experiment of 17 November 1978. The continuous curve is the emission spectrum due to H_2O, O_2 and O_3. The broken lines indicate the effect of adding HF and HCl with an assumed constant mixing ratio with height of 1×10^{-9}: the zenith angle is *ca.* 91°.

5. CONCLUSION

This paper has considered the justification for applying spectroscopic techniques to studies of middle atmosphere composition, and has attempted to illustrate the intrinsic power of these techniques.

It should be clear from the several examples quoted that the subject of atmospheric spectroscopy is a very active one with many workers applying a variety of ingeniously devised spectroscopic techniques to measurements of composition.

Clearly there is still a strong need for a considerable improvement in the sensitivity and resolving power of spectroscopic instruments, however. As the amount of information to be extracted from a spectrum increases, so must the amount of detail we can resolve. No one instrument will provide all the answers, of course, but a large potential for improvement would seem to exist in the use of i.r. heterodyne radiometers and in microwave spectrometers, in both cases because of their very high resolving power. Gas correlation spectroscopy is currently making great contributions, and I believe we can expect this principle to be used in a variety of applications in the future. One of the other areas in which there is great excitement, because of the extremely high sensitivities which are possible, is in the development of cryogenically cooled high resolution grating spectrometers and Michelson interferometers.

Another area, though somewhat beyond the range of interest defined by the title of this paper, where spectroscopic techniques might be developed considerably is in the remote sensing of atmospheric winds. Currently spectroscopic methods are being tested, which are based on measurements of the Doppler shifts of spectral lines averaged along a given line-of-sight. Sensitivities need to be improved, and consideration needs to be given to the most appropriate instruments for making such measurements.

Finally, I make a plea for the continued support of laboratory spectroscopy in order to obtain the line parameters which are needed for the analysis of all atmospheric spectroscopic measurements. At present, uncertainties in line parameters often represent the largest source of error in a composition measurement, and a significant fraction of available resources should be diverted to support workers active in this field. We should remember that a measurement of a constituent concentration from the best atmospheric spectrum (which may be very difficult and expensive to produce) is only as accurate as the basic spectroscopic data used in the analysis.

References (Harries)

Adel, A. 1949 In *The atmospheres of the Earth and planets* (ed. G. Kuiper), ch. 10. Chicago: University Press.

Anderson, J. G., Margitan, J. J. & Stedman, D. H. 1977 *Science, N.Y.* **198**, 501–503.

Bangham, M. J. 1978 *Infrared Phys.* **18**, 357–369.

Barringer, A. R. & Davies, J. H. 1973 *2nd Joint Conference on sensing of environmental pollutants.* Washington, D.C.

Connes, J. & Connes, P. 1966 *J. opt. Soc. Am.* **56**, 896–910.

Farmer, C. B. & Raper, O. R. 1977 *Geophys. Res. Lett.* **4**, 527–529.

Gille, J. C. & House, F. B. 1971 *J. atmos. Sci.* **28**, 1427–1442.

Girard, A. 1963 *Appl. Opt.* **2**, 79–87.

Harries, J. E. & Chamberlain, J. 1976 *Appl. Opt.* **15**, 2667–2672.

Harries, J. E., Moss, D. G., Swann, N. R. W., Neill, G. F. & Gildwarg, P. 1976 *Nature, Lond.* **259**, 300–302.

Houghton, J. T. & Taylor, F. W. 1973 *Rep. Prog. Phys.* **36**, 827–919.

Kaplan, L. D. 1959 *J. opt. Soc. Am.* **49**, 1004–1008.

Low, F. J. & Rieke, G. H. 1974 In *Methods of experimental physics* (ed. N. Carleton), vol 12A, ch. 9. New York: Academic Press.

Menzies, R. T. 1976 In *Laser monitoring of the atmosphere* (ed. E. D. Hinkley), ch. 7. New York: Springer–Verlag.

Murcray, D. G., Goldman, A., Williams, W. J., Murcray, F. H., Brooks, J. N., Stocker, R. N. & Snider, D. E. 1974*a* *Proceedings of the international conference on structure, composition and general circulation of the upper and lower atmospheres and possible anthropogenic perturbations*, pp. 292–303. U.S. Department of Transportation and Australian Academy of Sciences, Melbourne, Australia.

Murcray, D. G., Goldman, A., Williams, W. J., Murcray, F. H., Brooks, J. N., Van Allen, J., Stocker, R. N., Kosters, J. J. & Barker, D. B. 1974*b* *Proceedings of the third conference on C.I.A.P.*, pp. 184–192. U.S. Department of Transportation, Washington DC.

Roesler, F. L. 1974 In *Methods of experimental physics* (ed. N. Carleton), vol. 12A, ch. 12. New York: Academic Press.

Rodgers, C. D. 1976 *Rev. Geophys. Space Phys.* **14**, 609–624.

Schnopper, H. W. & Thompson, R. I. 1974 In *Methods of experimental physics* (ed. N. Carleton), vol. 12A, ch. 11. New York: Academic Press.

Schroeder, D. J. 1974 In *Methods of experimental physics* (ed. N. Carleton), vol. 12A, ch. 10. New York: Academic Press.

Smith, S. D. & Pigeon, C. R. 1964 *Mem. Soc. R. Liege* **9**, 336–349.

Twomey, S. 1977 *Developments in geomathematics*, vol. 3. New York: Elsevier.

Waters, J. W. 1976 In *Methods of experimental physics* (ed. M. L. Meeks), vol. 12B, ch. 2.3. New York: Academic Press.

Phil. Trans. R. Soc. Lond. A **296**, 175–189 (1980) [175]
Printed in Great Britain

In situ observations

By D. H. Ehhalt
Kernforschungsanlage Jülich GmbH, Institut für Chemie 3, *Atmosphärische Chemie,*
D-5170 *Jülich, Federal Republic of Germany*

In situ observations are defined as those made at the point where the instrument is located. Since a large variety of techniques have been used for *in situ* measurements, the general features, capabilities and problems of these techniques are reviewed. To illustrate capabilities and problems more specifically, examples of recent measurements are presented which have a bearing on the chlorofluoromethane–ozone problem. These include: (1) resonance fluorescence for the measurement of Cl and ClO, (2) grab and cryogenic collection of whole air samples for the measurement of $CFCl_3$ and CF_2Cl_2 (as well as CH_4, H_2, CO and N_2O), (3) impregnated filters for acid chloride, and (4) matrix isolation for HO_2 and NO_2.

Introduction

In situ measurements are those made at the location of the instrument. With this definition a large variety of techniques qualify as *in situ* measurements. Nevertheless, the fact that an instrument has to be carried to the site of the measurement and comes into contact with the sampled air, introduces several common limitations to all those techniques. Measurements can only be made along the flight track of the platform and are thus limited to horizontal or vertical profiles; global data coverage, which requires a large number of such profiles, would be difficult and costly to obtain from *in situ* observations. The fact that the measured air volume is usually inhaled into the instrument raises the possibility of sample modification either by contamination through outgassing or reaction with the surfaces of instrument and platform. Finally, since most *in situ* techniques are designed to measure specific trace constituents, they are usually less suited to exploratory search for unknown compounds.

On the other hand, there are certain advantages. Because they are designed to measure a specific constituent or small group of atmospheric constituents, *in situ* observations generally do not require much compromise and the most sensitive methods available can be employed; for practically any compound of interest an *in situ* technique can be developed. For the same reason, the precision of concentration measurements by *in situ* techniques is usually higher than that of remote measurements. At the same time, because a well defined and often small air volume is sampled, the sampling altitude is better defined, the height resolution in vertical profiles is greater, and the measured signal more easily interpreted than for remote sensing. Generally speaking, *in situ* measurements are better suited to the local instantaneous study of trace constituents, as is needed for detailed validation of chemical models, whereas remote sensing is better suited to providing global averages; in that sense, both approaches complement each other.

General overview of *in situ* techniques

For a number of trace gases, the above mentioned problems of contamination or wall reaction were either small or could be overcome by proper design, and over the years a large number of successful *in situ* measurements have been reported. The most important ones are summarized in table 1. (Some older *in situ* measurements were reviewed by Ehhalt (1974).) *In situ* measurements can be grouped into two classes. In one group, the 'direct measurements', the measuring instrument is carried aloft, and the measurements are truly made *in situ*. In the second group, representative air samples are collected aloft for later analysis in the laboratory. Direct *in situ* techniques have the advantage that they can be adapted to the measurement of most trace constituents. Further, they offer the possibilities of real-time measurement and of taking many data points. However, the development of an instrument that has to function reliably on a balloon, aircraft or rocket requires substantial effort and time. The techniques relying on sample collection on the other hand, have the advantage that the sampling devices can be simple, rugged, and lightweight with small power requirements, i.e. very well suited for fieldwork. Furthermore, the analysis of the samples can be made under ideal conditions in the laboratory with already existing or slightly adapted equipment, which minimizes the cost of development. There are, however, disadvantages too. The air samples have to remain temporarily in a sample container, a requirement that increases the danger of sample modification due to contamination or wall reaction. Thus, collection techniques are limited to the measurements of trace constituents which are stable or can be stabilized. Besides the species measured and the measurement or sampling techniques employed, table 1 also gives an estimate of the precision of each method based on the errors quoted in the literature. Precision is here defined as the long term reproducibility of a measuring

Table 1. List of techniques used for stratospheric *in situ* measurements

('Direct measurement' refers to measurement at altitude; 'sample collection' involves collection at altitude and subsequent analysis in the laboratory.)

species measured	measurement or sampling technique	precision	investigator
Direct measurement			
HO, O, Cl, ClO HO_2, NO	resonance fluorescence	±50%	Anderson 1975, 1976, 1978; Anderson *et al.* 1977
NO	chemiluminescence	±20%	Ridley *et al.* 1975
O_3	chemiluminescence, electrochemical	±10%	
H_2O	frost point hygrometer	±10%	Mastenbrook 1974
$CFCl_3$, CF_2Cl_2† CCl_4, CH_3Cl†	e.c.-gas chromatography	± 5%	Rasmussen 1977
H_2, CO	mercury oxide	± 5%	Schmidt 1974; Seiler & Warneck 1972
Sample collection			
CH_4, H_2, CO	grab sampling	± 5%	Ehhalt *et al.* 1974, 1975
$CFCl_3$, CF_2Cl_2, N_2O	grab sampling	± 5%	Schmeltekopf *et al.* 1975, 1977
CH_4, CO, CO_2, H_2, N_2O	cryogenic sampling	± 5%	Ehhalt *et al.* 1974, 1975
$CFCl_3$, CF_2Cl_2	cryogenic sampling	±10%	Heidt *et al.* 1975, 1976
HNO_3, Cl^-, F^-	impregnated filter	±30%	Lazrus *et al.* 1976
SO_2	impregnated filter	±30%	Jaeschke *et al.* 1976
HO_2, NO_2	matrix isolation	±30%	Mihelcic *et al.* 1978

† Measurements in lower stratosphere only.

method. Many experimenters seem to have underestimated precision and systematic errors of their data. Instrument precision has been improving steadily over the years, and not all the older measurements have the precision quoted here, whereas some of the most recent ones may have better precision than that stated (cf. W.M.O. 1977).

As indicated in table 1, the stratospheric distributions of many important trace constituents were obtained by direct *in situ* measurements: many of the available stratospheric O_3 data were obtained in this way owing to the fact that a lightweight cheap electrochemical O_3 sonde could be developed that allowed balloon measurements on a routine basis. Similarly, most of the stratospheric water vapour profiles were measured by a balloon-borne frost point hygrometer (Mastenbrook 1974). In this case, since the problem of contamination is severe and the experiment is much more difficult, correspondingly fewer H_2O profiles are available. Also in the case of NO, much of the vertical distribution was measured by a direct *in situ* technique based on a chemiluminescent instrument (Ridley *et al.* 1975). The most exciting direct measurements, however, are the experiments of Anderson (1975, 1976) and Anderson *et al.* (1977), who measured stratospheric concentrations of free radicals, using a resonance fluorescence technique. This experiment will be discussed below.

Of the collection techniques grab sampling is the simplest. A previously evacuated container is opened at the sampling site, allowed to fill to ambient pressure, and closed again. The samples are then returned to the laboratory for analysis. Because of its simplicity this technique has been used widely. However, at high altitudes air sample pressure becomes low and the ratio of surface area to sample amount unfavourable. Accordingly preservation of sample integrity becomes more difficult, and the application of grab sampling has been limited to the less reactive and more abundant species, such as CH_4, CO, CO_2, H_2, N_2O, CF_2Cl_2 and $CFCl_3$, and to the lower part of the stratosphere. To overcome the problems of small sample size and to extend measurements to higher altitudes, various methods of sample compression have been tried. The cleanest and most widely applicable of these techniques is cryogenic sampling, which collects large samples by condensing ambient air in sample cylinders cooled by liquid Ne, He, or H_2. Still, even the cryogenic collection of whole air samples is limited to the more stable gases. For certain reactive trace gases, selective sampling techniques have been developed that collect a specific trace constituent or group of trace constituents but discard the bulk of the air In this way, enrichment factors of many orders of magnitude can be achieved, and the method can be quite sensitive. The most widely applied selective sampling technique is the impregnated filter method, which has been used by Lazrus & Gandrud (1974), Lazrus *et al.* (1975, 1976), and Mroz *et al.* (1977) to measure acidic gases involving gaseous nitrate, chloride and fluoride. Another technique of this kind is the matrix isolation technique, which even allows the measurement of free radicals.

Although table 1 is by no means complete, it indicates that a large body of information has been gathered by *in situ* techniques. To illustrate *in situ* techniques, their capabilities and their problems in more detail, examples of different methods and their results are described below. Following the lines of a previous review (Ehhalt 1978), the discussion is essentially limited to measurements of chlorine and fluorine compounds presenting the most recent data. Only in the case of our own measurements are other compounds included.

Direct measurements

Radical measurement by resonance fluorescence

Probably the most advanced *in situ* measurements are the resonance fluorescence experiments for measuring the radical concentrations of O, OH, HO_2, NO, Cl and ClO (Anderson 1975, 1976, 1978; Anderson *et al.* 1977). We will limit our discussion to Cl and ClO, although the instrument used is the same in principle for all these radicals. It is essentially a flow tube suspended on a parachute and dropped from a balloon at altitudes around 45 km. During the fall, ambient air enters from below and flows through the tube at high speeds, between 100 m/s

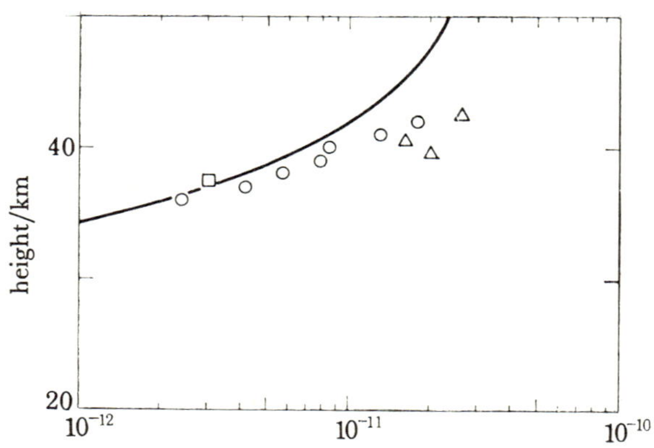

FIGURE 1. Stratospheric measurements of the Cl volume mixing ratio over Palestine, Texas, 32° N (Anderson *et al.* 1977). The curve is a theoretical profile calculated for 30° N latitude and equinox conditions. ○, 28 July 1976; △, 2 October 1976; □, 8 December 1976; ——, K. F. A. model.

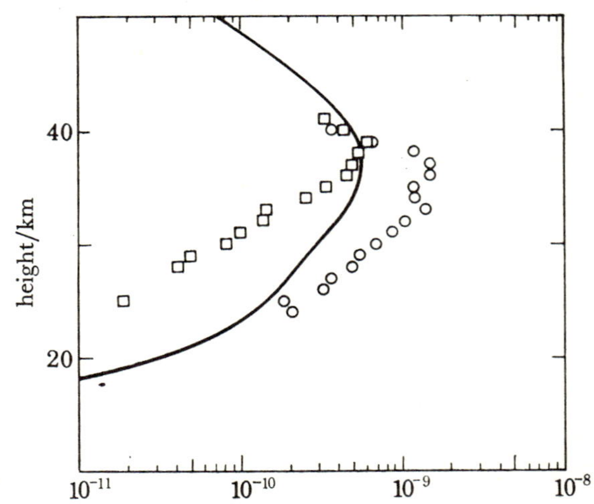

FIGURE 2. Stratospheric profiles of the ClO volume mixing ratio measured over Palestine, Texas, 32° N (Anderson *et al.* 1977; Anderson 1978. The full curve is a theoretical profile calculated for 30° N latitude and equinox conditions. ○, 28 July 1976; □, 8 December 1976; ——, K.F.A. model.

at high altitudes to 30 m/s at low altitudes. This minimizes wall reactions. In order to detect Cl or ClO radicals, a beam of photons at 118.9 nm wavelength from a Cl lamp with a filter of gaseous O_2 is passed perpendicularly through the air stream. The resonant scattered photons are monitored by a detection system mounted at right angles to flow and light beam. The background signal resulting from photons reflected from the walls, photons Rayleigh-scattered by the air molecules and some solar photons reaching the scattering volume, is recorded in flight by eliminating the observed species by a suitable reaction. In the case of ClO which does not fluoresce, the measurement of the background signal does not require a reaction. However, to measure ClO, it is converted to Cl via the reaction $ClO + NO \rightarrow Cl + NO_2$, and the resulting additional Cl atoms are detected by resonance fluorescence. The reactant gas is injected in short alternating intervals through a diffuser loop at the air inlet. Absolute calibration, required to account for geometry and efficiency factors, is performed in the laboratory before and after a stratospheric flight by flowing known amounts of the observed radical through the system. The measured Cl volume mixing ratio are shown in figure 1 and the measured ClO volume mixing ratios are shown in figure 2 (Anderson *et al.* 1977, Anderson 1978). They reflect the newest values given by

Anderson for these flights (Anderson 1978). The measurements were made over Palestine, Texas, 32° N. During 1977 Anderson *et al.* have measured four more ClO profiles which have been discussed at various conferences. Like the ones shown here, they all reveal a strong increase in the lower and middle stratosphere, sometimes reaching a maximum between 35 km and 40 km altitude. Although the data from individual profiles are internally quite consistent, Anderson *et al.* observed large variations between the various profiles, up to one order of magnitude between the maximum and minimum values at a given altitude. There is an indication of a seasonal variation with higher values in July. However, these high values, especially a profile from July 1977 which reached a maximum mixing ratio of $8/10^9$ at 40 km altitude are difficult to reconcile with a number of other observations as well as presently held views on the stratospheric Cl budget. If ClO concentrations were indeed as large as the July 1977 profile seems to indicate, the present understanding of stratospheric O_3 chemistry would demand O_3 concentrations a factor 2–3 lower than the average O_3 profile. However, the O_3 concentrations measured along with that flight showed a reduction of only about 20% (Anderson 1978), which means that such high ClO concentration could have persisted only for a short time and only locally. Furthermore, the present consensus of sources of stratospheric Cl, mainly from CH_3Cl, CCl_4, $CFCl_3$, CF_2Cl_2, limits the sum of all chlorine compounds to $2.3/10^9$, a number in good theoretical agreement with the HCl measurements quoted below. These $2.3/10^9$ for total chlorine are however much less than the maximum mixing ratio of $8/10^9$ measured for ClO alone during the flight in July 1977. In fact, already the ClO profile from July 1976, shown in figure 2 to reach about $1.6/10^9$, would appear fairly high. The four other ClO profiles obtained by resonance fluorescence show reasonable agreement with the mixing ratios predicted from a one dimensional model which are also indicated in figure 2.

There are further, remote measurements of stratospheric ClO, unfortunately also with conflicting results. Measurements from an aircraft-borne microwave spectrometer to detect emission of infrared radiation in the millimetre wavelength range from stratospheric ClO indicated significant amounts of ClO (Waters *et al.* 1978). Comparison of the theoretically calculated lineshapes to the observed ones suggest ClO to be present with peak mixing ratios between $0.5/10^9$ and $1/10^9$. Although these measurements, which were made in June and August 1977, probably still have large errors, they clearly indicate much lower ClO abundances during the summer than those obtained by the resonance fluorescence technique. In addition, there are preliminary data from Menzies previewed in Upper Atmospheric Programs Bulletin (October 1978). Using a balloon-borne laser heterodyne radiometer the following mixing ratios were measured on 20 September 1978, over Palestine, Texas: $(2.5 \pm 0.6)/10^9$ at 37.5 km altitude, $(1.7 \pm 0.4)10^9$ at 33.8 km, $(0.8 \pm 0.2)/10^9$ at 32.8 km, and $(0.22 \pm 0.06)/10^9$ at 29 km. Since the data were taken at sunset (solar zenith angles between 90° and 95°) and ClO has a fairly strong diurnal variation at these altitudes, the corresponding daytime concentrations should be higher by about a factor 2. This would bring the value at 37.5 km into fair agreement with the high values from the July 1977 profile by Anderson (1978). The gradient of the profile by Menzies appears much steeper than Anderson's measurements. This could, however, be due to the fact that the measurements were made in sequence during the rapid setting of the Sun. An explanation will have to await the full publication of these data. Because of the still conflicting measurements it remains difficult to accept the present experimental data as corroboration of the calculated ClO profile as long as the discrepancies are not explained or removed by further measurements.

The Cl measurements shown in figure 1 do not show a particular excess of Cl atoms. For those flights, the Cl/ClO ratio is also in reasonable agreement with the theoretical prediction.

Collection techniques

Grab sampling

The oldest and simplest collection technique is grab sampling. Because the size of the collected air samples becomes small, about 10 cm^3 s.t.p., at 40 km altitude, stratospheric grab sampling places particularly high requirements on the cleanliness of the collection system. These requirements become even more stringent when halocarbons with mixing ratios as low as 10^{-12} are to be measured, and extreme care has to be taken in the choice of components and materials, cleaning and sampling procedures, and the vacuum integrity of the sample containers. A rather sophisticated balloon-borne grab sampler has been developed by the Aeronomy Laboratory of N.O.A.A., in Boulder (Schmeltekopf *et al.* 1976), whose results will be used to illustrate the capabilities of this method. The original sampler consisted of five evacuated 7.5 l stainless steel spheres with specially designed valves. Currently, samplers with up to twelve sample containers are in use. The sphere halves are electropolished inside before welding, passivated after welding by exposing the interior surface to hexamethyldisilazane and vacuum baked at 200 °C. This passivation proved necessary because adsorption sites created at the inner surface during the welding of the spheres chemisorb halocarbons, including $CFCl_3$, from the sample. In fact, some of the early measurements seem to have suffered from a loss of $CFCl_3$, whereas contamination generally turned out to be less of a problem than was anticipated. The present bigger payloads have lost some of their original advantage of being particularly lightweight and compact. Still, they are small enough to be hand launched by a small crew on relatively small balloons. After reaching altitude the payload is cut from the balloon and dropped on a parachute. The samples are collected during rapid descent, which causes any contamination emanating from the package to be flushed away from the inlets. After recovery, the samples are returned to the laboratory and analysed among others for $CFCl_3$, and CF_2Cl_2 by a gas chromatograph equipped with an electron capture detector. Despite the low concentrations encountered, detector sensitivity never presented a serious problem since the electron capture detector is extremely sensitive for halocarbons. However, it took a while before adequate calibration procedures were devised and intercalibration among the various laboratories making such halocarbon measurements became satisfactory. The most recent intercalibrations show that agreement between the laboratories that provided most of the stratospheric $CFCl_3$ and CF_2Cl_2 data is rather good, better than 5 %.

The data obtained at different latitudes by this highly mobile system are shown below in figure 3 and 4, together with the data from cryogenic sampling discussed below.

Cryogenic sampling

Cryogenic sampling is a method for collecting large-volume air samples in the stratosphere that offers the advantages of simultaneous measurement of a number of trace gases and considerable reduction of wall reaction and contamination from the inner surfaces of the sample containers. For example, a 7.5 l grab sample collected at 30 km altitude yields 90 cm^3 s.t.p. of air; a cryogenic sample yields 10 l s.t.p. in one tenth of the volume. Thus the sample amount to surface ratio is improved by a factor of more than 100; the influences of contamination and wall reaction are reduced accordingly. As an example we will discuss the balloon-borne cryogenic sampler developed at the National Center for Atmospheric Research (N.C.A.R.) in Boulder (Ehhalt *et al.* 1974; Lueb *et al.* 1975), which was in a slightly modified version also used by Fabian *et al.* (1978). The cryogenic sampler consists of an array of 8 or 16 electropolished stainless steel cylinders immersed in a dewar filled with liquid neon. Cleaning procedures, vacuum bake-out and passivation are similar to those described for grab sampling. Each cylinder is connected to a common inlet manifold by a bellows-sealed stainless steel high vacuum which is driven by a motor. At liquid neon temperature, about 30 K, most gases in the air are frozen solid with vapour pressures below 10^{-4} Torr. Hydrogen, helium and neon are not condensed and remain gaseous. To prevent fractionation, the sampling cylinders are designed in such a way that the inflowing air effectively prevents the back diffusion of the uncondensable gases (Lueb *et al.* 1975). Thus each tube acts as an efficient cryopump. The samples are also collected during descent. However, sampling times are longer, up to 20 min at maximum altitude and the payload has to descend slowly (at a rate of 50 m/min) while it is suspended on the balloon. To prevent contamination from the balloon, the payload is located 150 m below. To minimize contamination from the payload itself, the inlet line is extended 6 m below the payload and flushed several times before each sampling by a separate cryopump. After being warmed up and returned to the laboratory the samples are analysed for CO_2, CO, CH_4, H_2 H_2O, N_2O, $CFCl_3$, and CF_2Cl_2, mostly by gas chromatography (Ehhalt *et al.* 1974, 1975; Heidt *et al.* 1975; Fabian *et al.* 1978).

The most recent data from both of the above discussed sampling techniques for CF_2Cl_2 and $CFCl_3$ are shown in figures 3 and 4. They include, largely owing to the efforts of the N.O.A.A. group, vertical profiles in tropical and arctic latitudes (Schmeltekopf 1978; Goldan 1978, private communication). To allow identification of an eventual variation with latitude these measurements are shown separately. The tropical data were taken over Panama 8° N and Northern Brazil 8° S. Because of their scarcity, the data in high latitudes include antarctic measurements as well. The profiles were collected in spring or summer. The largest number of profiles is available for mid-northern latitudes from various authors (Schmeltekopf *et al.* 1976; Heidt *et al.* 1976; Fabian *et al.* 1978). These laboratories have intercalibrated their standards and found good agreement. The measurement errors should be less than 5 %. Regardless of latitude, CF_2Cl_2 shows a steep drop with altitude in the stratosphere starting from a value of about 240 parts/10^{12} at the tropopause. It appears, however, that the steepness of the gradient does depend on latitude. At middle and high latitudes the gradients are 20 parts/10^{12} per kilometre in the lower stratosphere, whereas the equatorial gradient amounts to only 7 parts/10^{12} per kilometre. Accordingly the average mixing ratio at 25 km altitude has dropped to 43 parts/10^{12}, nearly a factor of 6, in high latitudes whereas the corresponding value in tropical latitudes is 175 parts/10^{12}, only a factor 1.4 lower.

It appears that the isopleths of constant CF_2Cl_2 mixing ratio in the stratosphere are more steeply inclined with latitude than the contour of the tropopause. This is a surprising finding since, because of the higher photon fluxes in the tropics, one would expect a faster photolysis of CF_2Cl_2 and thus a steeper decrease of the concentration with altitude. The slower decrease measured points to a much faster vertical transport in the equatorial zone at least during the time of sampling. Looking at the scatter of the data at mid-latitudes it becomes clear that a conclusion of this kind is still preliminary. Above 25 km the scatter increases and reaches nearly a factor of 10 between the extreme values at an altitude. Since for the present data the measurement errors are much smaller than that, the fluctuations must be natural, presumably due to large scale transport processes in a concentration field with strong vertical and somewhat weaker horizontal gradients. Like the temperature profiles concentration profiles of longer lived trace gases like CF_2Cl_2 are influenced by advection of air masses of different origin and thus depend on the momentary wind field. To derive a reliable average of the meridional concentration field more profiles have to be measured, particularly at tropical latitudes.

Figure 3 shows also a theoretical profile of CF_2Cl_2 calculated for equinox conditions and 30° N latitude. It falls consistently above the measurements. Some of the discrepancy could be explained by the fact that most mid-latitude data were collected between 40° and 55° N

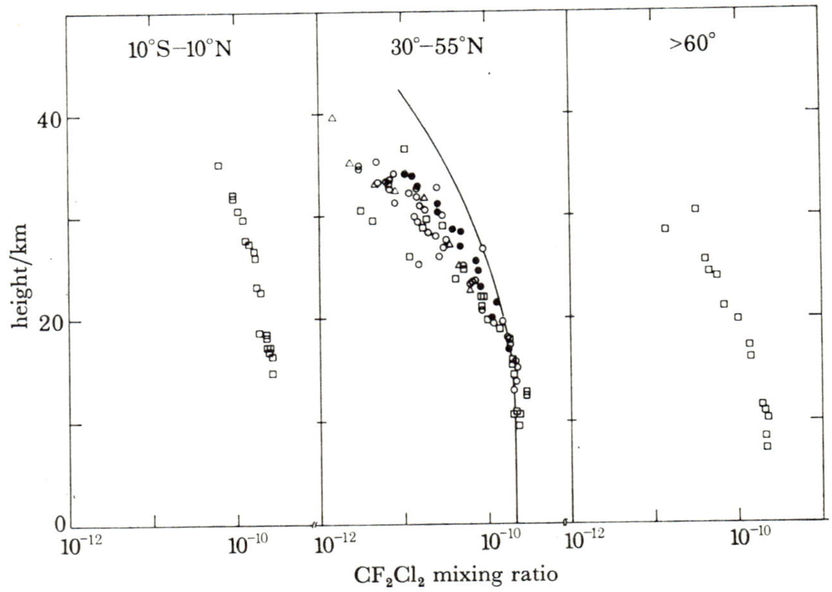

FIGURE 3. Vertical profiles of the CF_2Cl_2 volume mixing ratio at different latitudes. The data for high latitudes include measurements over Antarctica. The theoretical curve shown for mid-latitudes was calculated for 30° N and equinox conditions. △, Yorkton; ●, Heidt *et al.*; □, Schmeltekopf *et al.*; ○, Fabian *et al.*; ——, K.F.A. model.

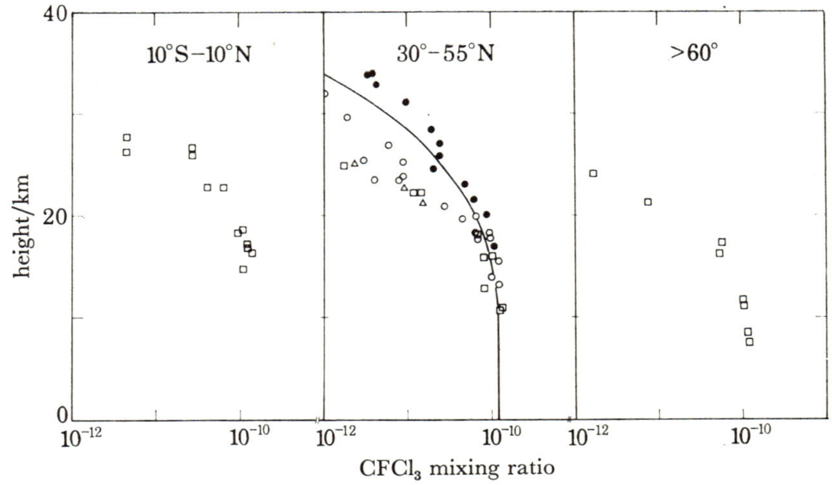

FIGURE 4. Vertical profiles of the $CFCl_3$ volume mixing ratio at different latitudes. The data for high latitudes include measurements over Antarctica. The theoretical curve shown for mid-latitudes was calculated for 30° N and equinox conditions. △, Yorkton; ●, Heidt *et al.*; □, Schmeltekopf *et al.*; ○, Fabian *et al.*; ——, K.F.A. model.

where the latitudinal dependence indicates a steeper gradient. But even the data from 30° N, indicated by the full dots, still fall below the theoretical curve. Most probably this is due to inadequate treatment of the transport in the one dimensional model.

The corresponding $CFCl_3$ data show a much faster fall-off in the stratosphere (figure 4). This is to be expected from its higher photolysis rates. The latitudinal variation of the vertical gradient is less obvious but still present, at least in the lower stratosphere. For $CFCl_3$ the mid-latitude model calculation agrees with the measurements. The photolysis becomes so fast above 30 km that inaccuracies in transport at or above that altitude scarcely influence the calculated profiles.

The samples collected by cryogenic sampling are so large that aliquots can be used for the measurement of other compounds. These include CH_4, H_2, N_2O and CO, CO_2 and H_2O. Results from our own work over Southern France are reviewed briefly in the following. It was done in collaboration with the Max-Planck-Institutes in Lindau and Mainz (cf. Fabian *et al.* 1978). The general flight data are summarized in table 2. The measurements quoted below are from the Institute of Atmospheric Chemistry, K.F.A. Jülich (Ehhalt *et al.* 1977*b*; Fabian *et al.* 1978).

TABLE 2: FLIGHT DATA OF PROFILE MEASUREMENTS IN 1977 OVER SOUTHERN FRANCE, 44° N LATITUDE

flight no.	7701/I	7702/II	7704/I	7705/I
date	7 June	16 June	12 Sept	26 Sept.
maximum altitude/km	35.3	34.8	33.5	32.8
tropopause altitude/km	*ca.* 12.1	*ca.* 12.2	*ca.* 13.2	*ca.* 12.1
number of samples	8 stratosphere 8 troposphere	5 stratosphere	8 stratosphere 16 troposphere	8 stratosphere
date of analysis (K.F.A.)	9–14 June	17–21 June	15–16 Sept.	30 Sept.–2 Oct.

CH_4 AND CO: The CH_4 and CO content of the air samples were measured simultaneously by using a 1 m column of 5 A molecular sieve and a flame ionization detector; CO was converted to CH_4 over a Ni catalyst at 400 °C before detection. The resulting CH_4 profiles are plotted in figure 5. The use of the true altitude in figure 5 introduces some extra scatter in the data of the flights. On the whole, however, the various profiles show a rather good agreement, decreasing steadily from $1.64/10^6$ by volume at the tropopause to $0.6/10^6$ by volume at 34 km altitude. The agreement between flights is much closer than had been observed previously over N.C.A.R's balloon station in Palestine Texas, 32° N, where profiles occasionally showed rather distinct layers (Ehhalt *et al.* 1974). There are two reasons for this finding. The first is that the profiles over Southern France were all obtained during situations of weak stratospheric winds in late spring or late summer, when the probability of long range horizontal advection of air masses with different CH_4 concentrations is relatively small. Secondly, N.C.A.R.'s balloon flights at 32° N were occasionally within the downward branch of the Hadley circulation, which introduces a steepened gradient just above the tropopause. Apart from the latter effect the agreement between the present profiles and N.C.A.R.'s measurements is rather good, in particular both data sets observe about $0.6/10^6$ by volume around 35 km altitude. Figure 5 also shows a CH_4 profile over Southern France obtained by infrared spectroscopy (Ackerman *et al.* 1977). The data points shown here are the average from several lines. Although the agreement between the *in situ* measurements and this profile is much better than with the older infrared measurements (cf. Ackerman *et al.* 1977), the infrared profile still indicates a somewhat steeper gradient than those obtained by us. The tropospheric measurements lie uniformly close to $1.64/10^6$ by volume. This agrees also well with N.C.A.R.'s most recent tropospheric profiles over Eastern Texas (Heidt *et al.* 1976). These concentrations, however, are higher than N.C.A.R.'s earlier measurements from 1965–1967 over Scotts Bluff, Nebraska and the Eastern Pacific which averaged $1.41/10^6$ by volume (Ehhalt & Heidt 1973). Since N.C.A.R. has a continuous sequence of intercalibrated working standards this increase should be real (L. Heidt, private communication).

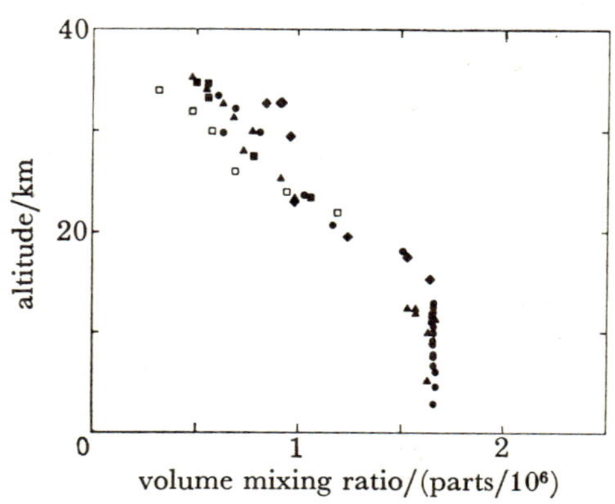

FIGURE 5. Vertical profiles of CH_4. The relative accuracy of analysis was $\pm 2\%$. The open squares represent the results from a balloon-borne infrared experiment (Ackerman *et al.* 1977. ▲, 7 June 1977; ■, 16 June 1977; ●, 12 September 1977; ◆, 26 September 1977.

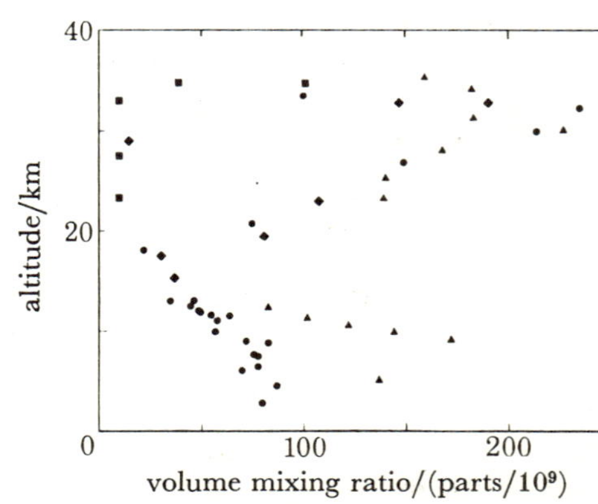

FIGURE 6. Vertical profiles of CO. The results above 20 km altitude are influenced by contamination. The relative accuracy of analysis was $\pm 3\%$ or $\pm 5/10^9$ by volume. ▲, 7 June 1977; ■, 16 June 1977; ●, 12 September 1977; ◆, 26 September 1977.

The corresponding CO measurements are shown in figure 6. The increase of the CO concentration shown above 20 km altitude and the large scatter of the data at these altitudes are most likely due to contamination. Unfortunately, at the time of writing the laboratory tests to identify the cause of this contamination are not conclusive. The speculation is that organic exhaust vapours from the launch vehicle, which find their way into the inlet line react when exposed to ozone-rich air during sample collection at altitude to form CO. It appears that by the time the balloon reaches 20 km altitude the inlet line is flushed free from contamination and data below that altitude should be correct. The tropospheric data showing a decrease with altitude fall well within the range previously observed.

H_2 : H_2 was measured by using a 3 m column of 5 A molecular sieve and a He ionization detector. The H_2 profile obtained is shown in figure 7. The individual profiles show a very similar behaviour and the scatter of the data points is small. Nearly all of it could be explained by the relative accuracy of the analysis, which is about 5%. The stratospheric part shows a weak, linear decrease in the mixing ratio from $0.55/10^6$ by volume at the tropopause to about $0.45/10^6$ by volume at 35 km altitude. There is no hint of a slight maximum of about $0.6/10^6$ by volume at 28 km altitude which was occasionally indicated in some of the N.C.A.R. flights over Palestine, Texas. The troposphere does not show any vertical gradient at all. The average of $0.55/10^6$ by volume differs somewhat from the value of $0.5/10^6$ by volume obtained during aircraft flights in 1966, 1967, 1971 and 1972 due to a difference in absolute calibration (Ehhalt *et al.* 1977*a*).

N_2O: N_2O was measured together with $CFCl_3$ and CF_2Cl_2 on a gas-chromatograph equipped with a 3 m Porasil C column and an electron capture detector. To ensure better resolution of N_2O and a faster elution of the heavier halocarbons the column was temperature-programmed, starting at −10 °C and rising to 150 °C in 8 min. The N_2O data are shown in figure 8. Similarly to CH_4, the N_2O measurements from the four flights showed quite consistent results.

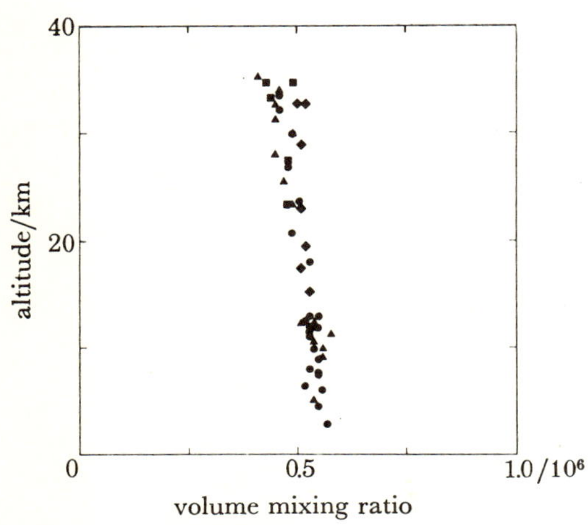

FIGURE 7. Vertical profiles of H_2. The relative accuracy of analysis was ±5%. ▲, 7 June 1977; ■, 16 June 1977; ●, 12 September 1977; ◆, 26 September 1977.

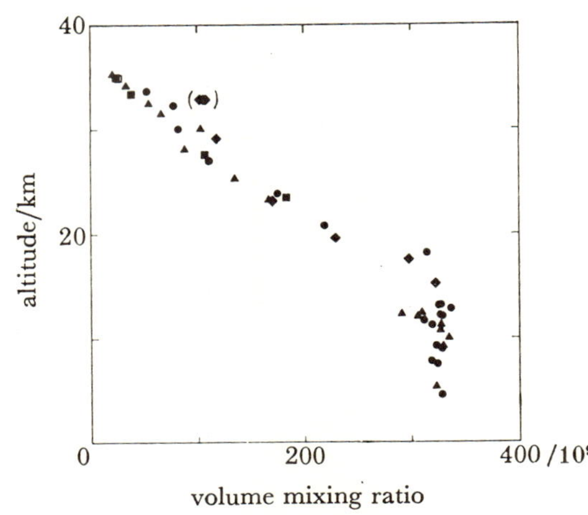

FIGURE 8. Vertical profiles of N_2O. The relative accuracy of analysis was ±5% or $\pm 3/10^9$ by volume. ( ): Collected during floating period. ▲, 7 June 1977; ■, 16 June 1977; ●, 12 September 1977; ◆, 26 September 1977.

Aside from the uppermost data from 26 September 1977, which were collected during float, the observed average fluctuations above 20 km are less than 5%, presumably for the same reasons as discussed for CH_4. The resulting average profile shows a strong, nearly linear decrease from $330/10^9$ by volume at the tropopause to about $20/10^9$ by volume at 35 km, although in flight 7704 nearly tropospheric N_2O mixing ratios persist to 18 km altitude, well above the tropopause. The profile agrees well with the stratospheric N_2O data of Schmeltekopf *et al.* (1977) over Wyoming, 41° N. The troposphere shows a rather uniform N_2O concentration with values between 325 and $335/10^9$ by volume and an average of $327/10^9$ by volume in good agreement with the tropospheric averages reported by Rasmussen (1977) and Goldan (1977).

SELECTIVE SAMPLING

Impregnated filter method

An interesting selective sampling technique used in the stratosphere is the measurement of acidic gases by means of a base-impregnated filter. The technique was developed by Lazrus *et al.* (1975), using an IPC-1478 aerosol filter. Before collection of gaseous acidic chloride, essentially the sum of HCl, $ClONO_2$, and if present HOCL, the filters are washed and purified to minimize the chloride background; they are then impregnated with tetrabutylammonium hydroxide, a strong organic base, which mixes with the organic oil already present on the filter and remains 'sticky' even at the low stratospheric sampling temperatures. To obtain the blank values, the filters are subdivided into four sections. Two of the sections are impregnated, and two remain untreated. One impregnated section and one unimpregnated section are shielded from the airstream by a mask, so that they do not collect and remain as a blank. The filters were flown on aircraft as well as on balloons. In the first case the ram pressure provides the airflow; on the balloons, airflow is maintained by a blower. In either case the airflow is monitored. Collec-

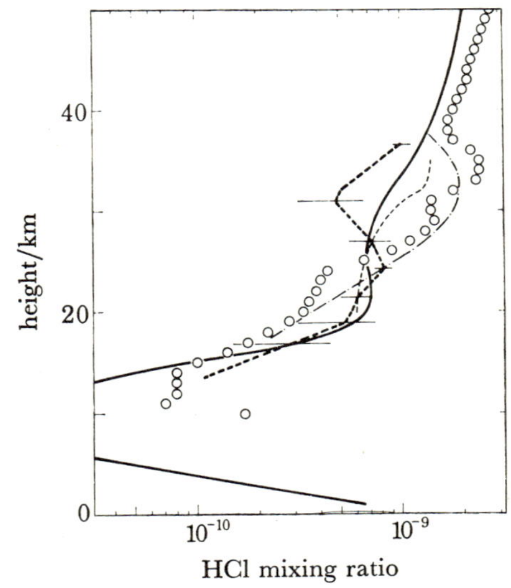

FIGURE 9. Average stratospheric volume mixing ratio profile of gaseous acidic chloride over Holloman Air Force Base, N.M. 33° N (curve labelled Lazrus *et al.*). The other data represent individual profiles of the HCl volume mixing ratio measured by infrared spectrometry. The full curve represents a theoretical profile calculated for 30° N latitude and equinox conditions. - - -, Lazrus *et al.* 1977; ○, Evans, 1 December 1977; – – –, Raper *et al.* 1977; –·–·–, Eyre & Roscoe 1977; ——, K.F.A. model.

tion efficiency for HCl is reported to be now 100 %; for $ClONO_2$ to be at least 80 % (Lazrus *et al.* 1977). The possible collection of HOCl has not been considered by the authors. Measurement of the collection efficiency of ClO is in progress. The effect of CH_3Cl and other halocarbons should be small. In the laboratory the filter sections are separated and washed. The extracted solutions are analysed for the chloride ion by a colorimetric method. The concentrations of the gaseous compounds are obtained after subtracting the contributions from particulate matter and the blank, which are determined from the unimpregnated and masked sections, respectively. An error of ± 30 % was quoted for the acidic chloride measurement by Gutowsky (1976, p. 309). Figure 9 shows the annual average data for acidic chloride, which were obtained over Holloman Air Force Base, New Mexico 33° N. It also contains some of the more recent HCl data obtained by remote infrared techniques. Lazrus *et al.* did not find a seasonal variation but they found rather regularly a minimum in the acidic chloride at around 30 km altitude which interrupts a general increase with altitude in the stratosphere. There is no evidence for such a minimum in the infrared data for HCl. On the contrary some indicate an intermediate maximum at about the same altitude which causes a relatively large discrepancy between the *in situ* and remote measurements at the heights around 30 km. Otherwise the agreement is reasonable. If for the moment we accept the *in situ* measurements of acidic chloride as reliable, their agreement with the optical HCl data and even more their tendency to be smaller around 30 km altitude and would indicate that the other acidic chloride gases, $ClONO_2$ and HOCl have very low concentrations, certainly lower than presently predicted by models. This is somewhat difficult to accept because the calculated HCl profile agrees reasonably well with the ones measured by the infrared techniques and the total chlorine, essentially the sum of HCl, $ClONO_2$ and HOCl mixing ratio, has to be at least $2/10^9$ to which HCl contributes only $1/10^9$.

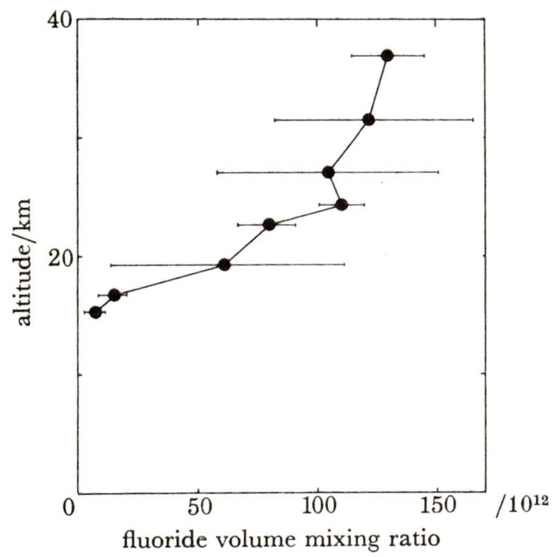

FIGURE 10. Vertical profile of stratospheric gaseous acidic fluoride (Mroz *et al.* 1977).

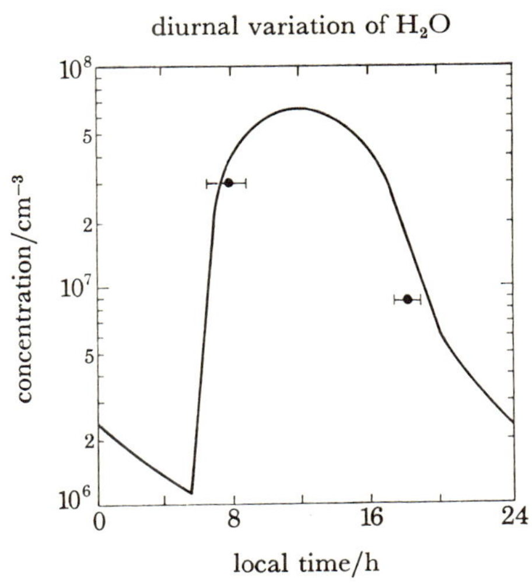

FIGURE 11. Comparison of the experimental HO_2 concentrations (Mihelcic *et al.* 1978) with a theoretical diurnal variation calculated for 30 km altitude, 30° N latitude and equinox conditions (Logan *et al.* 1978).

Similar measurements were made for the total stratospheric acidic fluoride concentrations (Mroz *et al.* 1977). Besides HF, gaseous acidic fluoride includes COF_2, another degradation product of chlorofluoromethanes. The data are summarized in figure 10. They agree reasonably well with a determination of the HF/HCl ratio of 0.1 in the height range from 14 to 30 km by Farmer & Raper (1977) and the HCl mixing ratios in figure 9.

Matrix isolation

There are several ways in which cryogenic techniques can be used to selectively sample trace substances and to reject the bulk of the air sample. One of the most interesting is the so-called matrix isolation, a well known laboratory technique. In its application to stratospheric measurements, this technique employs the condensation of a solid, inert matrix of CO_2 on a cold surface for sample collection (Mihelcic *et al.* 1978). In the process of matrix deposition, free radicals and other condensable trace gases are trapped and embedded in the matrix structure. As long as there is a large surplus of matrix molecules and the matrix is kept sufficiently cold, the free radicals remain isolated and cannot react with each other. The use of atmospheric CO_2 for matrix deposition offers two critical advantages. First, large enrichment factors can be obtained. Second, liquid nitrogen temperature, 77 K, is sufficient for sample collection. This greatly reduces logistic problems in the field and allows one to build a rugged, lightweight sampler suitable for balloon-borne measurements. The sampler consists essentially of a liquid nitrogen cooled cold finger on which the CO_2 matrix is deposited, a liquid nitrogen cooled cryopump which absorbs the non-condensable gases, essentially N_2 and O_2, and maintains the airflow, and a nozzle which allows the air to enter the sampling device in a narrow, well defined beam enveloping the cold finger. The nozzle can be opened and closed by a valve. After sample collection and recovery of the payload, the cold sample is returned to the laboratory and its radical content is analysed by electron paramagnetic resonance (e.p.r.) Matrix isolation combined with e.p.r. detection can be applied to detect nearly all radicals. However, a CO_2 matrix at 77 K is not yet completely rigid and light radicals such as OH can still diffuse through the lattice. Thus the method is currently limited to measuring HO_2 and NO_2. The spectra of these two major radicals are sufficiently different to allow unscrambling of the composite spectrum one expects from stratospheric samples (cf. Mihelcic *et al.* 1978). Two measurements of stratospheric HO_2 by this method have been reported so far. One was made over North Germany, 52° N, on a sample collected at 32 km altitude between 6.40 and 9.00 a.m. on 8 August. It gave a concentration of 3×10^7 cm^{-3}. The other was obtained from a balloon flight over Holloman Air Force Base, New Mexico, 33° N, on 4 March 1977, 4.45–5.48 p.m. at 29.8 km altitude. The HO_2 concentration was 8.5×10^6 cm^{-3}. Because of difficulties in the laboratory calibration, both measurements have an error of about a factor of 3. The data are compared to a theoretical curve of the diurnal HO_2 variation (Logan *et al.* 1978) in figure 11. The curve was calculated for 30° N. and equinox conditions. To match the experimental points to the proper solar zenith angle, they were plotted at a slightly shifted daytime. Despite the different sampling locations the agreement between theoretical prediction and measured points is surprisingly good. In principle it should also be possible to measure ClO with this technique. However, so far no measurements have been reported.

Conclusion

Although limited to a few examples, the foregoing discussion shows that *in situ* techniques have provided numerous interesting measurements. In particular, they provide the best and, in some cases, the only way to measure rare and reactive species such as radicals. Such measurements will present a much more stringent test of our understanding of stratospheric chemistry than was previously possible.

References (Ehhalt)

Ackerman, M., Frimout, D. & Muller, C. 1977 *Nature, Lond.* **269**, 226–227.

Anderson, J. G. 1975 *Geophys. Res. Lett.* **2**, 231–234.

Anderson, J. G. 1976 *Geophys. Res. Lett.* **3**, 165–168.

Anderson, J. G. 1978 *W.M.O. Symposium on the Geophysical Aspects and Consequences in the Composition of the Stratosphere*, Toronto, 26–30 June.

Anderson, J. G., Margitan, J. J. & Stedman, D. H. 1977 *Science, N.Y.* **198**, 501–503.

Ehhalt, D. H. 1974 *Can. J. Chem.* **52**, 1510–1518.

Ehhalt, D. H. 1978 *Rev. Geophys. Space Phys.* **16**, 217–224.

Ehhalt, D. H. & Heidt, L. E. 1973 *J. geophys. Res.* **78**, 5265–5271.

Ehhalt, D. H., Heidt, L. E., Lueb, R. H. & Pollock, W. 1975 *Pure appl. Geophys.* **113**, 389–402.

Ehhalt, D. H., Heidt, L. E., Lueb, R. H. & Roper, N. 1974 *Proc. 3rd Conference on the Climatic Impact Assessment Program.* Cambridge, Mass., pp. 153–160. February 26–March 1.

Ehhalt, D. H., Schmidt, U. & Heidt, L. E. 1977*a* *J. Geophys. Res.* **82**, 5907–5911.

Ehhalt, D. H., Schmidt, U., Volz, A., Khedim, A. & Johnen, F. J. 1977*b* *Proc. 1st C.E.C. Meeting of the Coordination and Study Group on the Ozone Shield Depletion Problem*, Brussels, 24–25 October (In the press).

Evans, W. F. J., Fast, H., Kerr, J. B., McElroy, C. T., O'Brien, R. S., Wardle, D. I., McConnell, J. C. & Ridley, B. A. 1978 *W.M.O.-Symposium*, Toronto, 26–30 June.
Eyre, J. R., Roscoe, H. K. 1977 *Nature, Lond.* **226**, 243–244.
Fabian, P., Borchers, R., Weiler, K. H., Schmidt, U., Volz, A., Ehhalt, D. H., Seiler, W. & Müller, F. 1979 *J. geophys. Res.* **84**, 3149–3154.
Farmer, C. B. & Raper, O. F. 1977 *Geophys. Res. Lett.* **4**, 527–529.
Goldan, P. D., Bush, J. A., Fehsenfeld, F. C., Albritton, D. L., Crutzen, P. J., Schmeltekopf, A. L. & Ferguson, E. E. 1978 *J. geophys. Res.* **83**, 935–939.
Gutowsky, H. S. 1976 *Report of the Panel on Atmospheric Chemistry*, National Academy of Sciences, Washington D.C.
Heidt, L. E. & Pollock, W. H. 1976 *Proc. Symp. on the Non-Urban Tropospheric Composition* 10–12 November, Hollywood, Flo., 6.01–6.11.
Heidt, L. E., Lueb, R., Pollock, W. & Ehhalt, D. H. 1975 *Geophys. Res. Lett.* **2**, 445–447.
Heidt, L. E., Pollock, W. H. & Lueb, R. A. 1976 *Joint Symposium on Atmospheric Ozone International Commission on Atmospheric Ozone/Commission on Atmospheric Chemistry and Global Pollution*, Dresden, East Germany, August.
Jaeschke, W., Schmitt, R. & Georgii, H. W. 1976 *Geophys. Res. Lett.* **3**, 517–519.
Lazrus, A. L. & Gandrud, B. W. 1974 *J. Atmos. Sci.* **31**, 1102–1108.
Lazrus, A. L., Gandrud, B. W., Greenberg, J., Bonelli, J., Mroz, E. & Sedlaeck, W. A. 1977 *Geophys. Res. Lett.* **4**, 587–589.
Lazrus, A. L., Gandrud, B. W., Woodard, R. N. & Sedlaeck, W. A. 1975 *Geophys. Res. Lett.* **2**, 439–441.
Lazrus, A. L., Gandrud, B. W., Woodard, R. N. & Sedlaeck, W. A. 1976 *J. geophys. Res.* **81**, 1067–1070.
Logan, J. A., Prather, M. J., Wofsy, S. C. & McElroy, M. B. 1978 *Phil. Trans. R. Soc. Lond.* A **290**, 187–234.
Lueb, R. A., Ehhalt, D. H. & Heidt, L. E. 1975 *Rev. Sci. Instrum.* **41**, 702–705.
Mastenbrook, H. J. 1974 *Can. J. Chem.* **52**, 1527–1531.
Mihelcic, D., Ehhalt, D. H., Klomfass, J., Kulessa, G. F., Schmidt, U. & Trainer, M. 1978 *Pure appl. Geophys.* **116**, 530–536.
Mroz, E. J., Lazrus, A. L. & Bonelli, J. E. 1977 *Geophys. Res. Lett.* **4**, 149–150.
Raper, O. F., Farmer, C. B., Toth, R. A. & Robbins, D. B. 1977 *Geophys. Res. Lett.* **4**, 531–534.
Rasmussen, R. 1977 *International Symposium on the Influence of the Biosphere upon the Atmosphere*, Mainz, Germany, July 11–14.
Ridley, B. A., Schiff, H. I., Shaw, A. W. & Megill, L. R. 1975 *J. geophys. Res.* **80**, 1925–1929.
Seiler, W. & Warneck, P. 1972 *J. geophys. Res.* **77**, 3204–3214.
Schmeltekopf, A. L., Albritton, D. L., Crutzen, P. J., Goldan, P. D., Harrop, W. J., Henderson, W. R., McAffee, J. R., McFarland, M., Schiff, H. I. & Thompson, T. L. 1977 *J. atmos. Sci.* **34**, 729–736.
Schmeltekopf, A. L., Goldan, P. D., Harrop, W. J., Thompson, T. L., Albritton, D. L., McFarland, M., Sapp, A. E. & Henderson, W. R. 1976 *Rev. Sci. Instrum.* **47**, 1479–1485.
Schmeltekopf, A. L., Goldan, P. D., Henderson, W. R., Harrop, W. J., Thompson, T. L., Fehsenfeld, F. C., Schiff, H. I., Crutzen, P. J., Isaksen, I. S. A. & Ferguson, F. E. 1975 *Geophys. Res. Lett.* **2**, 393–396.
Schmidt, U. 1974 *Tellus* **26**, 78–90.
Waters, J. W., Gustincic, J. J., Kakar, R. K., Roscoe, H. K., Swanson, P. N., Phillips, T. G., de Graauw, T. & Kerr, A. R. 1978 *J. geophys. Res.*
World Metorological Organization 1977 *Global Ozone Research and Monitoring Project, Report of the Meeting of Experts on Measurements on Rare Species Relevant to the Ozone Budget, Seattle*, 18–21 August.

Phil. Trans. R. Soc. Lond. A **296**, 191–204 (1980) [191]
Printed in Great Britain

Satellite ozone measurements

By A. J. Krueger, B. Guenther, A. J. Fleig, D. F. Heath, E. Hilsenrath, R. McPeters and C. Prabhakara
Laboratory for Atmospheric Sciences, Goddard Space Flight Center, Greenbelt, Maryland 20771 *U.S.A.*

Three classes of ozone sounders have been developed since the first Echo Satellite measurements in 1960. They are the (1) backscatter ultraviolet (b.u.v.), (2) infrared limb and nadir radiance, and (3) stellar and solar occultation methods. With these techniques, ozone has been measured from 20 to 100 km. Tropospheric ozone measurements are beyond present technology, but total ozone is determined with the b.u.v. and nadir infrared methods.

Results from the occultation methods are excessively variable, indicating a need for refinement of the technique. Evidence from this technique for a mesospheric ozone maximum is strong. The most extensive set of ozone observations has come from the Nimbus 4 b.u.v. between 1970 and 1977. From this and earlier b.u.v. experiments, it has been determined that the ozone tends toward photochemical steady state densities for levels above 5 mbar and that the temperature coefficient near 1 mbar is approximately 1000 K, in accordance with odd hydrogen catalysis of ozone. At lower altitudes, odd nitrogen catalytic chemistry has been verified from b.u.v. ozone observations during a solar proton event. Ozone profile results from the limb infrared radiance method are discussed in a companion paper by Dr J. C. Gille.

Total ozone soundings from the b.u.v. agree with Dobson results and details of seasonal and latitudinal variations are now available. The total ozone field has also been inferred from nadir infrared radiance data. Monthly mean total ozone maps from this method and the b.u.v. each indicate standing waves at all latitudes.

Introduction

In the last 15 years, interest in stratospheric ozone has changed from scientific curiosity to the present need for monitoring for environmental reasons. Fortunately, a variety of remote ozone sounding techniques were proposed and tested with satellites during the 1960s and early 1970s so that the current needs for research and climatological data can be met with a combination of satellite and conventional sounding devices. This paper is a review of satellite methods and, in conjunction with Dr Gille's paper, a summary of principal findings.

Ozone molecules absorb radiation strongly in the middle ultraviolet (u.v.) Hartley–Huggins bands (220–320 nm), moderately in the infrared (9.6 μm), and weakly in the visible Chappuis bands (0.6 μm). Perhaps the first measurement of ozone to use artificial Earth satellites in connection with the selective absorption of light employed the Echo balloon satellite. Sunlight in the Chappuis band wavelengths which passed through the Earth's limb was measured with a ground based telescope after reflexion from the satellite (Venkateswaran *et al.* 1961). A useful remote sounding technique, namely *limb occultation,* was demonstrated, although, because of geometric problems, the resulting ozone profiles were incorrect. This technique, as with most others, had a pre-satellite origin; that of lunar eclipse photometry. A chronology of developments in limb occultation methods and in the other major satellite techniques for ozone measurement is given in table 1.

The second major remote sensing method to be proposed was the *backscatter ultraviolet* (b.u.v.) technique. Sunlight in the Hartley–Huggins bands penetrates to various stratospheric altitudes (which depend on the wavelength-dependent ozone absorption cross section) where the photons are either absorbed or Rayleigh scattered. Measurements of the Earth albedo at several u.v. wavelengths can yield an ozone profile. The method was proposed by Singer & Wentworth (1957) and newer methods for inversion of the albedo measurements have been developed and reviewed by Mateer (1972). An important extension of the b.u.v. method was to measure total ozone as suggested by Dave & Mateer (1967).

TABLE 1. SATELLITE EXPERIMENTS TO MEASURE OZONE

type	satellite	wavelengths nm	latitude coverage	comments	references
occultation					
solar	Echo 1	590, 529.5	17 N	Dec. 1960	Venkateswaran *et al.* (1961)
	U.S.A.F. 1962	260	33 S–13 S	July 1962	Rawcliffe *et al.* (1963)
	Ariel 2	200–400	50 S–50 N	April, May, Aug. 1964	Miller & Stewart (1965)
	AE-5	255.5	5 N	Dec. 1976	Guenther *et al.* (1977)
stellar	OAO-2	250	16 S–43 N	Jan. 1970, Aug. 1971	Hays & Roble (1973)
	OAO-3	258–343	12 S–3 N	July 1975	Riegler *et al.* (1976)
backscatter u.v.					
profile	U.S.A.F. 1965	284	60 S–60 N	Feb.–Mar. 1965	Rawcliffe & Elliott (1966)
	U.S.S.R.	225–307	60 S–60 N	Apr. 1965	Iozenas *et al.* (1969)
		250–330	60 S–60 N	June 1966	Iozenas *et al.* (1969)
	1966-111 B	175–310	80 S–80 N	1966	Elliott *et al.* (1967)
	OGO-4	110–340	80 S–80 N	Sept. 1967– Jan. 1969	Anderson *et al.* (1969)
	Nimbus 4 b.u.v.	255.5–305.8	80 S–80 N	Apr. 1970– July 1977	Heath *et al.* (1973)
	AE-5 b.u.v.	255.5–305.8	20 S–20 N	Nov. 1975– Apr. 1977	Frederick *et al.* (1977*a*)
	Nimbus 7 s.b.u.v.	255.5–305.8	80 S–80 N	Nov. 1978*	Heath *et al.* (1975)
total	Nimbus 4 b.u.v.	312.5–339.8	80 S–80 N	Apr. 1970– July 1977	Mateer *et al.* (1971)
	AE-5 b.u.v.	312.5–339.8	20 S–20 N	Nov. 1974– Apr. 1976	
	Nimbus 7 t.o.m.s.	312.5–339.8	global	Nov. 1978*	Heath *et al.* (1975)
infrared emission		μm			
profile	Nimbus 6 l.r.i.r.	9.6	65 S–90 N	June 1975– Jan. 1976	Gille (1979, this symposium)
	Nimbus 7 l.i.m.s.	9.6	65 S–90 N	Oct. 1978– May 1979	Nimbus project (1978)
total	Nimbus 3 i.r.i.s.	9–10 spectral scan	80 S–80 N		Hanel *et al.* (1970)
	Nimbus 4 i.r.i.s.	9–10 spectral scan	80 S–80 N	Apr. 1970– Jan. 1971	Prabhakara *et al.* (1976)
	Block 5 m.f.r. (4 flights)		global	Mar. 1977*	Lovill *et al.* (1978)
	Tiros *N* h.i.r.s.	9.71	global	Nov. 1978*	

* Currently in operation (Nov. 1979).

Satellite measurements of ozone profiles by the backscatter method were first made by Iozenas *et al.* (1969) and Elliott *et al.* (1967), who used wavelength scanning monochromators viewing the nadir atmosphere. These were followed by a similar experiment on the OGO-4 satellite in 1967 (Anderson *et al.* 1969), in which data were collected over a 17-month period. The first major backscatter u.v. experiment with an instrument dedicated to ozone measurements was the b.u.v. on Nimbus 4 in a polar Sun-synchronous orbit. Launched in April 1970, this sensor continued to collect data until 1977. The instrument was a double monochromator which sampled 12 discrete wavelengths bands (Heath *et al.* 1973).

Residual Nimbus 4 b.u.v. hardware was later flown on the Atmospheric Explorer-5 spacecraft in November 1975, with an orbital inclination of 20°. Tropical data were collected and local time ozone changes were determined.

The most recent b.u.v. experiment is the Nimbus 7 s.b.u.v./t.o.m.s., launched in October 1978, in a Sun-synchronous polar orbit. This dual instrument continues the Nimbus 4 nadir ozone data set with the solar backscatter u.v. sensor (s.b.u.v.) (an improved version of the Nimbus 4 sensor) and adds spatial scanning of total ozone with the total ozone mapping spectrometer (t.o.m.s.).

Infrared emission methods for remote measurement of ozone are of two types (see table 1). Ozone profiles are determined from measurements of the altitude variation of the atmospheric radiance by viewing the Earth's limb (the limb radiance technique) and total ozone can be inferred from measurements of the atmospheric radiance in the nadir direction. The nadir sounding method was first tested on the Nimbus 3 satellite with an infrared interferometer spectrometer (i.r.i.s.) (Hanel *et al.* 1970). This was followed by a repeat flight on Nimbus 4. The retrieval methods and results were reported by Prabhakara *et al.* (1976).

The limb emission method is a more recent development, which has the advantages of good vertical resolution and simultaneous information about air temperature. Retrieval methods have been developed by Gille & House (1971) and the system has flown on the Nimbus 6 and 7 satellites.

Occultation techniques and results

Limb absorption or occultation measurements have been made with either the Sun or stars as the light source. Ozone profiles have been determined between about 40 and 100 km, thus giving the technique the distinction of working to higher altitudes than any other current sounding technique. Atmospheric attenutation is measured and subsequently interpreted in terms of Rayleigh scattering, atmospheric refraction and ozone slant path optical depth. Stellar occultations can, in principal, be obtained at any local time, although only night observations have been made with existing instrumentation. They are capable of high vertical resolution but are limited by the availability of suitable stars.

Solar occultation measurements are possible from a spacecraft sensor within an hour of sunrise or sunset at the ground twice during each orbit. The geographic coverage is rather severely limited by the orbit geometry. The vertical resolution is about 15 km unless the field of view is limited with a pointed telescope. The useful altitude range is determined by the number of wavelength channels, and the accuracy does not depend on absolute system calibration, but only on an electronic linearity and noise, and errors in effective ozone absorption coefficients.

The published results from occultation experiments are listed with references in table 1 and, since the quantity of soundings is small, the principal results are shown in figure 1 where the profiles are identified by the satellite name.

The K–M mid-latitude ozone model, (Krueger & Minzner 1976) which, for altitudes above 60 km, is derived primarily from twilight rocket flights, is also shown for comparison. The satellite profiles apply to a variety of latitudes, seasons and phases of the solar cycle and, as noted above, were obtained under near twilight or night conditions.

Night-time profiles from the orbiting astronomical observatories, OAO-2 and -3 show strong secondary maxima centered near 85 km and 95 km respectively. The OAO-3 results have been somewhat enigmatic because, at least below 80 km, the ozone densities are more than a factor of 2 greater than the results from the twilight experiments. If true, they would imply a night-time doubling in ozone density even at 50 km. The OAO-2 results exhibit the opposite effect below 75 km, namely a night-time ozone decrease relative to twilight measurements. The band of observed profiles from OAO-2 lies about a factor of 4 less than the K–M mid-latitude model at 50 km and appears to diverge at lower altitudes. The differences between the two OAO experiment results, and between them and the twilight results, are much greater than has been predicted for diurnal variations in the stratosphere and mesosphere.

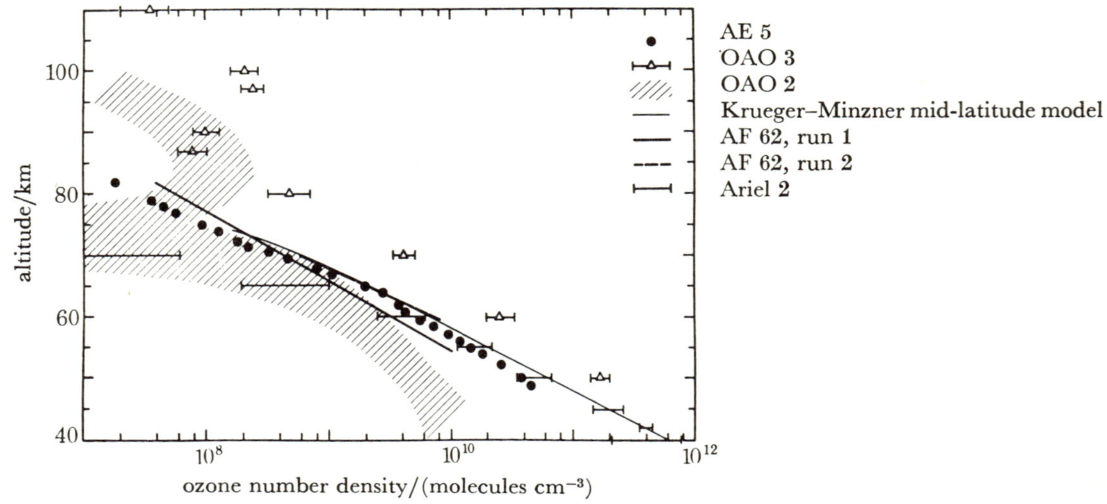

FIGURE 1. A comparison of ozone density distributions obtained with solar and stellar occultation methods.

The AE-5, AF 62, and Ariel 2 experiments are of the solar occultation type and, except for Ariel 2 at upper levels, the results tend to agree with the K–M mid-latitude model. The Ariel 2 data diverge toward lower values at higher altitudes and are consistent with the OAO-2 data at 65 and 70 km. The AE-5 results are 20–50 % less than the K-M mid-latitude model below 65 km. At those levels the K–M model becomes progressively more weighted toward daytime values. The AF 62 experiment determined only a two parameter profile, but two separate analysis techniques yielded different profiles (run 1 and run 2 in figure 1). The ozone densities from the AF 62 run 2 are quite close to AE-5 and K–M model results.

In summary, the solar occultation experiments differ significantly between themselves and from the rocket-derived K–M mid-latitude model. The ozone distribution shows a monotonic decrease with altitude with a scale height between 4 and 5 km for altitudes between 50 and 80 km. The ozone densities decrease from perhaps 5×10^{10} cm^{-3} at 50 km to about 1×10^{8} cm^{-3} at 75 km. The differences from the K–M model at 50 km may be due to latitude gradients, since most of the occultation results come from near the equator.

The stellar occultation results, taken at night, would require an extraordinarily variable mesosphere to account for the 10- to 100-fold differences in their profiles. These profiles are particularly hard to accept near 50 km, where a relative wealth of other data exists.

Clearly, the occultation techniques are still in a developmental stage and mesospheric ozone profiles and fields have not been adequately determined.

The backscatter ultraviolet (b.u.v.) method

The b.u.v. method has proved to be a useful technique for observing global characteristics of the high altitude ozone field. The technique is successful owing to the intense ozone absorption continuum in the middle ultraviolet, the apparent lack of contamination of this band by interfering species and the demonstrated lifetime of the satellite hardware. The principal shortcoming is the inability to resolve-structure in the ozone profile less than about an atmospheric scale height. This, incidentally, is not a severe drawback due to the general 'well behaved' nature of the ozone distribution in the upper stratosphere. However, no information is returned about the ozone distribution below the ozone maximum, although total ozone can be determined.

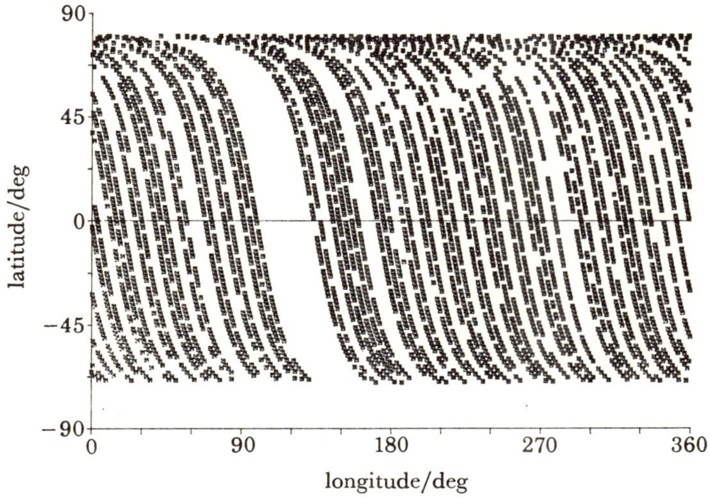

FIGURE 2. Data coverage for a typical week of b.u.v. data. Each square represents a b.u.v. sounding.

The inference of total ozone amounts and its vertical distribution is possible by solution of an integral equation. The fundamental observed parameters are the incident u.v. solar spectral irradiance and the atmospheric spectral radiance in the middle u.v. (250–340 nm). In addition, the surface reflectance must be determined by measurements at a wavelength that is free of ozone absorption. The wavelength dependence of the ozone absorption coefficient and the exponential decrease of air density with height leads to a series of contribution functions peaking at different levels corresponding to different wavelengths between 250 and 300 nm. The ozone profile can be inferred by using this altitude separation. At wavelengths greater than 300 nm, the contribution functions peak below the stratosphere and a total ozone measurement is possible.

Design of a spectrometer to measure the atmospheric radiance is difficult because the Earth's albedo changes drastically with wavelength between 250 and 340 nm, owing to ozone absorption. Furthermore, the solar flux is a factor of 10 less at 250 nm that it is at 340 nm. As a result, in a spectral scan of the earth radiance, the flux changes by 10^4. Further, observational complications come from polarization of the backscattered light. To meet these stringent observational requirements, the Nimbus b.u.v. instruments are double (tandem) Ebert–Fastie monochrometers with collinear narrow-band filter photometers. The instrument views the terrestrial atmosphere at the nadir. At the northern terminator, a diffuser plate is deployed to compare the solar spectral irradiance to the atmospheric radiance.

The Nimbus 4 Backscatter Ultraviolet (b.u.v.) experiment was launched in April of 1970. The spacecraft is in a near polar Sun synchronous orbit and data are available over almost all of the sunlit portion of the earth with the observations taken near local noon. Coverage for a typical week is shown in figure 2 where each 200 km by 200 km sounding is represented by ×. There are an average of 800 data points per day for the first year of operations. Increasingly severe spacecraft power problems starting in July 1972 have caused a substantial reduction in operating time for the instrument although data are still occasionally collected from it.

Vertical profile results from b.u.v. observations

The first experiment in the development of the backscatter u.v. methods used a single channel photometer. Rawcliffe & Elliott (1966) inferred from the radiance data that seasonal differences were present in upper stratospheric ozone. The first findings from an ozone profile experiment using an ultraviolet spectrometer were that the upper stratospheric ozone scale height was constant at sub-polar latitudes and that detailed horizontal structure existed near 30 mbar† (Iozenas *et al.* 1969). These results indicated that photochemistry prevailed over dynamics at the upper levels. In succeeding experiments with spectrometers, it was suggested that instrument calibration uncertainties or possible aerosol scattering might present real problems for accurate estimation of ozone profiles (Elliott 1971). However, Anderson *et al.* (1969) found that profiles could be recovered after calibration with rocket ozone data and that latitudinal and seasonal ozone variations could nevertheless be determined (London *et al.* 1977). In the two data periods (Sept.–Nov. 1967 and Jan.–Feb. 1968) that have been analysed from the OGO-4 experiments, the average ozone fields at 5 mbar were found to have small geographic variations. A broad maximum appears at mid-latitudes and, in the tropics, some longitudinal asymmetry is present with higher ozone mixing ratios over the Atlantic and Africa. These geographical variations diminish with altitude in agreement with the earlier observations. The mixing ratios at higher latitudes were found to increase from autumn to winter.

Initial analysis of the Nimbus 4 ozone profile data (Krueger *et al.* 1973) revealed that the highest mixing ratios were near 10 mbar at lower latitudes. During the equinoxes, the mixing ratios decrease rather uniformly toward the poles. However, during the autumn and winter secondary maxima form near the 2 mbar level at Southern Hemisphere mid-latitudes and at Northern Hemisphere high latitudes. Perhaps the simplest explanation of these upper stratospheric features lies in the temperature dependence of the ozone chemistry. The southern maximum, in particular, was found to vary in amplitude at different longitudes. Barnett *et al.* (1975) used these b.u.v. results, together with s.c.r. temperature data, to estimate the temperature coefficient for steady state ozone mixing ratios. They concluded that the exponential coefficient was of the order of 900–1100 K, a value consistent with oxygen and hydroxyl chemistry near 1 mbar.

The Nimbus 4 b.u.v. was also taking data during the major solar proton event of August 1972. Ozone decreases of 20% were found at 2 mbar in the upper stratosphere at geomagnetic

† 1 mbar = 10^2 Pa.

latitudes irradiated by protons (Heath *et al.* 1977). This was the first direct evidence for catalysis of ozone by odd nitrogen compounds in the stratosphere.

Methods for improvement of retrievals of high level ozone profiles from reprocessed b.u.v. Earth albedo data are currently under study. This new data set incorporates corrections for instrument sensitivity changes and spacecraft attitude errors. The aerosol effects found earlier by Elliott (1971) do not appear in Nimbus 4 b.u.v. data, although anomalous radiances appear to exist at 255 nm. Recently, the global behaviour of 1 mbar ozone from April 1970 to May 1972 has been studied at Goddard Space Flight Center by using an interim inversion product that is valid for the upper stratosphere – Mateer's formulation of the Laplace transform method (Twomey 1961). The variation of the zonal mean of the 1 mbar ozone mixing ratio at 60 N for the period Nov. 1970–Nov. 1971 is shown in figure 3. The 2 mbar air temperatures for the same latitude band and time period derived from s.c.r. observations (Barnett 1974) are shown in the same figure. Both b.u.v. and s.c.r. observe the radiation from rather broad height regions so the excellent inverse relationship of temperature and ozone is quite pleasing.

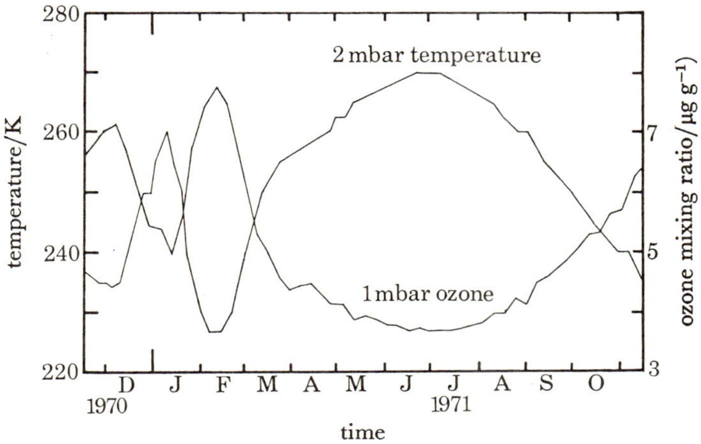

FIGURE 3. Temporal variation of the 2 mbar temperature from s.c.r. and the 1 mbar ozone mass mixing ratio in a latitude band centred at 60° N from 1 year of Nimbus 4 data.

The overall behaviour observed is a summer temperature maximum (270 K) and ozone minimum (3.71 μg/g) and a winter temperature minimum (227 K) and ozone maximum (7.62 μg/g). The large sudden stratospheric warming from mid-December to mid-January is mirrored exactly as a decrease in 1 mbar ozone. Even a small scale feature at the beginning of September is observed in both temperature and ozone. Temperature (T) and the ozone mixing ratio (r_3) are found to be related by the equation

$$r_3 = 0.072 \exp(1062/T)$$

with a correlation coefficient of 0.991 for the fit. This relation holds for the entire year except during the stratospheric warming, when the pre-exponential coefficient 0.072 μg/g increases to 0.082 μg/g. The temperature coefficient of 1062 K compares reasonably well with the value of 928 K reported by Barnett *et al.* (1975) at 50 S, but the correlation coefficient they report of 0.817 is substantially lower. This is due in part to our using much better calibrated ozone data and in part to using zonally averaged data. The increase in the pre-exponential coefficient during the stratospheric warming period is evidence for modification of the chemistry and dynamics from that normally present. This could happen if significant vertical motions arise.

Data from the AE-5 b.u.v. have been partially analysed. Principal findings to date are:

(1) the tropical profiles are consistent with current photochemical model results below 2 mbar but not higher altitudes (Frederick *et al.* 1978).

(2) the transition from dynamic to photochemical control of the ozone takes place between 5 and 10 mbar in the tropics (Frederick *et al.* 1977*b*) and

(3) negligible spatial structure exists in the integral ozone above 1 mbar at tropical latitudes when measured with 25 km horizontal resolution (Guenther & Dasgupta 1979).

TABLE 2. B.U.V./DOBSON OVERALL COMPARISON STATISTICS CHART

Year 1 (April 1970–April 1971)	number of coincidences	$\bar{\Omega}_N$ (m atm-cm)	$\bar{\Omega}_D$ (m atm-cm)	bias ($\bar{\Omega}_N - \bar{\Omega}_D$)	s.d.
Dobson 00 code, 2° sep, all R;	2024	315.9	326.4	−10.6	18.44
1° sep, $R \leqslant 0.2$	328	304.9	316.6	−11.8	16.28
all codes	3860	325.6	336.4	−10.9	21.29
all codes exc. 00 code	1841	336.4	347.7	−11.3	24.02
M-83 Network†	1270	349.7	358.8	−9.1	59.50
Year 2 (May 1971–April 1972)					
Dobson 00 code, 2° sep, all R;	1944	304.0	317.9	−13.9	18.83
1° sep, $R \leqslant 0.2$	301	292.0	304.4	−12.4	16.26
all codes	3392	313.9	329.2	−15.4	20.57
all codes exc. 00 code	1487	327.3	344.2	−17.0	23.65
M-83 network†	1117	342.3	368.9	−26.6	51.51

† Non-Dobson Russian filter photometer.

Total ozone from b.u.v. observations

The first application of the Dave & Mateer (1967) method for total ozone estimation from u.v. earth albedo measurements was on the Nimbus 4 b.u.v. Samples of the data were processed to establish the feasibility of the technique (Mateer *et al.* 1971) and to indicate the nature of time variations in the global ozone field (Krueger 1974). In 1976, as a result of the increasing importance of accurate monitoring of global ozone, a special effort was organized to process all the available data with the best possible accuracy. This effort included development of improved processing algorithms, compensation for instrument changes with ageing, careful review of the spacecraft ephemeris and attitude data and quality control of the output data. All the total ozone data have been reprocessed and archived in the National Space Sciences Data Center. Specific information as to how to acquire the processed Nimbus 4 total ozone data may be obtained by writing to World Data Center for Rockets and Satellites, Code 601, Goddard Space Flight Center, Greenbelt, Md. 20771, United States of America. There has been an extensive effort to assess the quality of the resulting data set. The primary comparison has been with the ozone measurements obtained from *Ozone data for the world* as provided by the Atmospheric Environment Services of Canada. All available spacecraft and ground data collocations have been identified and a variety of statistics computed. Results from several cases for the first 2 years are shown in table 2. All possible cases meeting the collocation criteria are included in each line with no preselection of either the ground or spacecraft data. The first line for each year is for AD-direct Sun wavelength pair (00 code) observations taken on the same day, with the centre of the spacecraft field-of-view within 2° of the ground station. A scatter plot of this set of data is shown in figure 4. The second line in the table 2 categories is a subset of the first, with a colocation limit of 1° and only cases where the spacecraft scene is cloud free are included. The third line includes all types of Dobson data and the fourth line is a subset of the third in which all except the 00 code data are included. A similar comparison of b.u.v. data and M-83 data is shown in the last line for each year. There are four major types of errors that contribute to the overall variance between b.u.v. and ground data shown in this table: random errors in the b.u.v. data; random errors in the ground data; noise introduced by imperfect colocation in time, space and area sampled; and noise resulting from the apparent biases of individual ground stations as shown in the Dobson intercomparisons performed at Belsk and Boulder. It is not possible to establish how much of the bias between a specific ground station and the b.u.v. data originates in the ground station; however, is can be shown that roughly half the overall variance can be removed by nulling these biases out on a station by station basis. When reasonable estimates are made for each of the types of errors mentioned above, it appears that the b.u.v. data is similar in accuracy to a well run Dobson station.

Seasonal variations of total ozone have now been recomputed from the current b.u.v. data sets (Hilsenrath *et al.* 1978). A time–latitude cross section of zonally averaged total ozone was generated from average daily means for 10° wide latitude zones. In figure 5, the total ozone has been contoured with 20 m atm-cm increments for the 26-month period from April 1970

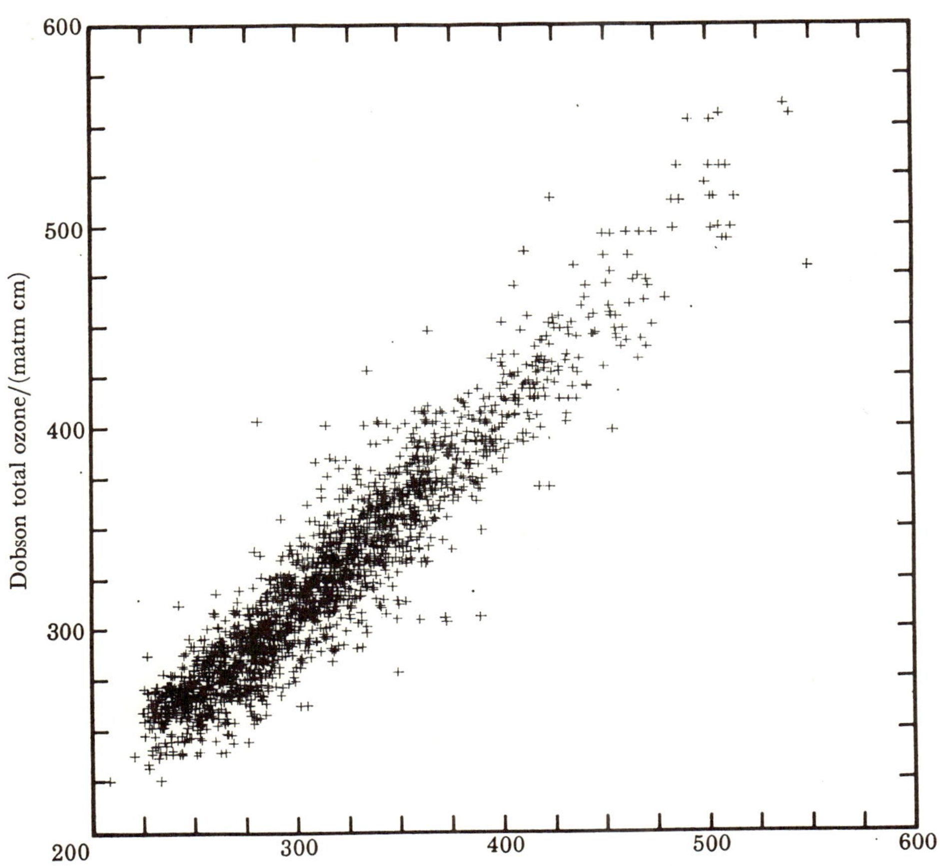

FIGURE 4. Comparison of b.u.v. total ozone estimations with coincident Dobson spectrophotometer total ozone measurements during the period 10 April 1970 to 30 April 1971. Only ADDS Dobson observations are included.

to May 1972 (contours are missing in the polar night because there are no observations during that period). The seasonal march of total ozone in the two hemispheres is clearly seen. A comparison with analyses of Dobson data (see, for example, London *et al.* 1976) reveals close correspondence with the major features in the ozone climatology. The b.u.v. results point up the distinctive differences in the seasonal trends for each hemisphere. These differences are as follows:

(1) In the Northern Hemisphere the spring maximum occurs nearly simultaneously at mid and high latitudes. The maximum value occurs near the pole, with total ozone amounts of about 500 m atm-cm in 1970 and decreasing to 460 and 440 m atm-cm in the spring of 1971 and 1972.

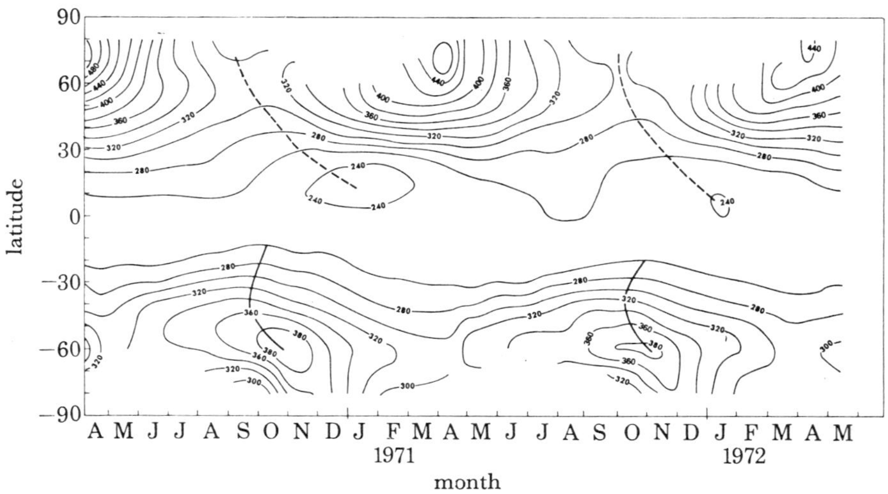

FIGURE 5. Time–latitude cross section of zonally averaged total ozone obtained from the Nimbus 4 b.u.v. from April 1970 to May 1972. The values are given in m atm-cm of ozone.

(2) In the Southern Hemisphere the spring build-up begins at 50 S in September with ozone values substantially lower than those in the Northern Hemisphere spring. The maximum values occur about 1 month later at polar latitudes. The asymmetry in the winter build-up in the two hemispheres is indicative of differences in the circulation features in the two hemispheres.

(3) In the tropical regions, the average total ozone amount is about 250 m atm-cm and a small seasonal variation is present, showing more rapid changes than in the long term Dobson averages. The ozone minimum, which is centred below the equator in April in both years moves northward after August. The lowest ozone amounts, which are found in the northern tropics, appear to follow after Southern Hemisphere spring maximum and are coincident with the Northern Hemisphere winter build-up.

A harmonic analysis has been made of the zonal mean data for the 2-year period (Hilsenrath *et al.* 1978). The results of this analysis are the following:

(1) More than 90 % of the variance of the zonal mean values at most latitudes is found in the two harmonics that represent the annual and semi-annual waves. The search for periodicities shorter than 1 month found no amplitudes greater than 5 m atm-cm at mid to high latitudes and 2 m atm-cm at low latitudes.

(2) At mid to high latitudes the annual wave dominates in both hemispheres. In the Northern Hemisphere the amplitude of the annual wave increases with latitude while in the Southern

Hemisphere the peak amplitude occurs at 50 S in the first year and at 60 S in the second year. The amplitude of the annual wave is a minimum at the equator in the first year and at 10 S in the second year. This shift in the minimum of the annual wave could be a result of the year to year change in the strength and location of the Hadley Cell circulation in the hemisphere.

(3) The semi-annual wave was observed at the higher latitudes in both hemispheres for the 2 years and in the tropics in the first year. This pattern is very similar to the analysis by London *et al.* (1976) of the Dobson data base. However, compared with the annual wave, the fractional variance in the semiannual wave is significant only in the tropics and at high southern latitudes. The tropical amplitude is small and probably is only an expression of the annual waves in the two hemispheres and, indeed, is in phase with these waves. The amplitude of the polar semi-annual wave is about 20 m atm-cm. This is comparable to the amplitude of the annual wave at 80 S but only a quarter of that amplitude at 80 N.

Infrared remote sensing of total ozone

The Nimbus 4 satellite carried an infrared interferometer spectrometer (i.r.i.s.), which provided measurements of the thermal emission spectrum of the Earth and atmosphere between about 400 and 1600 cm^{-1} with an apodized spectral resolution of 2.8 cm^{-1} (Hanel *et al.* 1971). The field of view of the instrument projected on the surface of the Earth was a circle with a diameter of approximately 95 km and all data were acquired near either local noon or local midnight between 80 N and 80 S. Measurements were essentially continuous from April to December 1970 and were available for a few days in January 1971.

The spectral measurements of i.r.i.s. in the 15 μm CO_2 band were inverted to derive the vertical distribution of temperature in the atmosphere (Conrath 1972). Using these temperature data, together with i.r.i.s. spectral measurements in the 9.6 μm ozone band, Prabhakara *et al.* (1976) deduced the global distribution of total ozone for about 10 months.

Though the i.r.i.s. was not a dedicated experiment for ozone soundings, the 9.6 μm band basically yields one parameter about ozone. This parameter depends both on the total ozone and its vertical distribution. Therefore, additional information must be provided to obtain total ozone explicitly. This additional information is provided by an empirical relation, based on balloon observations, which implies that increases in total ozone occur mainly because of increases in ozone concentration in the layers of the atmosphere below the ozone maximum. This empiricism and the crude vertical resolution, approximately 1 scale height, of the remotely sensed temperature profile in the lower stratosphere lead to a basic limitation on the infrared ozone remote sensing technique described by Prabhakara *et al.* (1976).

Since the time of the i.r.i.s flights, several temperature sounding instruments have included a channel to sense the 9.6 μm ozone band. Beginning in March 1977, the multichannel filter radiometer instruments on four of the Defense Meteorological Satellite Program series of satellites has included such a channel. Lovill *et al.* (1978) have reported preliminary results. The most recent 9.6 μm ozone sensor is part of the high resolution infrared radiometer (h.i.r.s.) sensor on the Tiros N Satellite, launched in November 1978.

Despite the limitations of the nadir infrared method, good coverage of the satellite data over the globe permit us to get some gross information on the geographic and seasonal variation of total ozone. The sensing capability at night and in the polar night could be very useful.

In figure 6, a map of the global distribution of ozone derived from i.r.i.s. for the month of December 1970 is shown to highlight the ozone information over the northern polar latitudes in winter. Longitudinal changes of about 100 m atm-cm can be seen between the maximum values east of Siberia and the minimum values in the North Atlantic. In the tropics the lowest values appear in the Indonesian sector, perhaps in association with the Siberian ozone high. The tropical ozone tends to have a wavenumber 1 characteristic throughout the year with the maximum over the Atlantic as shown in the December chart. The maximum amplitude of this wave is in July and August.

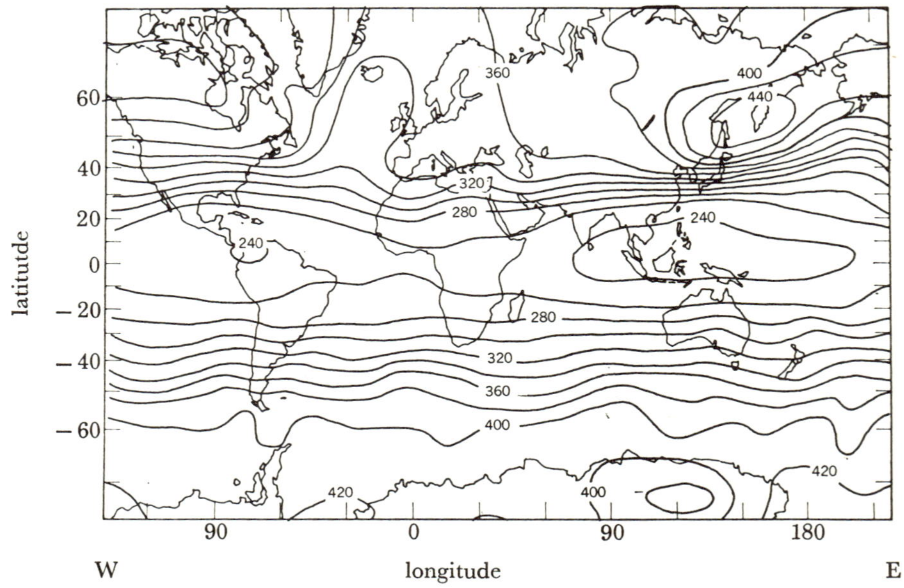

FIGURE 6. Mean global distribution of total ozone for December 1970 obtained from the Nimbus 4 i.r.i.s. instrument. The values are given in m atm-cm of ozone.

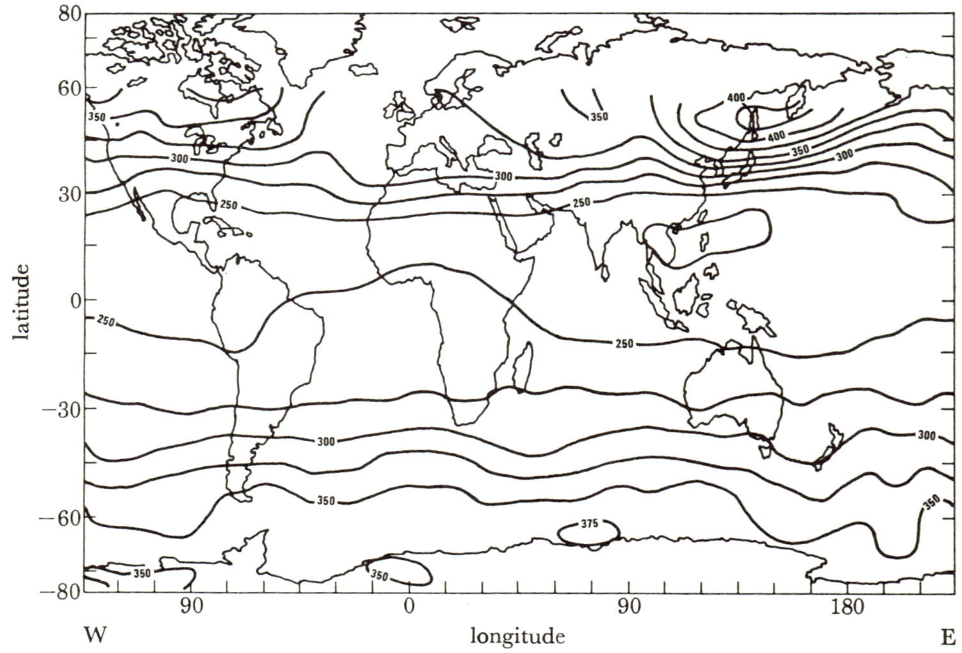

FIGURE 7. Mean global distribution of total ozone for December 1970 obtained from the Nimbus 4 b.u.v. instrument. The values are given in m atm-cm ozone. This chart may be compared with figure 6.

In the Southern Hemisphere, the December (summer) ozone field is fairly uniform at all latitudes except for the Antarctic regions. The i.r.i.s. December map exhibits a 40 m atm-cm zonal variation with the minimum located at the same longitude as the tropical minimum.

Comparative results from b.u.v. and i.r.i.s.

A comparison of the ozone results from i.r.i.s. and b.u.v. is of interest. Figure 7 is a December map of total ozone derived from the Nimbus 4 b.u.v. instrument. This can be compared directly with the map derived from i.r.i.s. (figure 6). In general the observed ozone features from the two instruments are the same. Taking into account the different contour intervals in the two maps, however, some differences in the absolute values can be found. This is expected from the study performed by Prabhakara *et al.* (1976), who demonstrate a reduction in the seasonal amplitude of the i.r.i.s. total ozone amounts. The two satellite measurements compare well in the tropics. The absolute amounts are very close and the minima occurring over Central America and Indonesia are observed by both instruments. In the Southern Hemisphere both measurements show a uniform latitude gradient from low to high latitudes; however, the b.u.v. values are about 15 % lower than the i.r.i.s. values. In the Northern Hemisphere the ozone high east of Siberia and the features in the North Atlantic are detected by both instruments. The absolute values agree here a little better than in the Southern Hemisphere.

Although the two instruments were on the same satellite, the data bases for these maps are not identical. B.u.v. observations are possible only in daytime, while i.r.i.s. soundings are taken day and night and, in addition, during the polar night. Furthermore, the instrument duty cycles were not identical and the data recovery varied between i.r.i.s. and b.u.v. However, it is encouraging that the features are reproduced since the measurement techniques are vastly different.

Conclusions

During the 1960s and into the 1970s, several methods for remote measurement of ozone were devised and tested. The most promising techniques for stratospheric soundings in the 1980s seem to be the backscatter u.v. system for ozone profile and total ozone monitoring and the limb infrared emission system for investigations requiring good vertical resolution of the ozone profile and complementary information on air temperatures and trace constituents. For mesospheric ozone measurements, the occultation methods have shown promise, although the unreasonable differences existing between experimental results must be resolved.

At present, several data sets have been collected and climatological descriptions are being assembled. The interpretation of these results in terms of atmospheric processes is only beginning. Already, photochemical modelling has been guided and theories tested by the satellite data. The ultimate utility will perhaps be found after the chemistry is well understood. At that point the global ozone data can be used for tracing air motions as a complementary source of information for dynamic models of the atmosphere.

REFERENCES (Krueger *et al.*)

Anderson, G. P., Barth, C. A., Cayla, F. & London, J. 1969 *Ann. Geophys.* **25**, 239–243.
Barnett, J. J. 1974 *Qt. Jl R. met. Soc* **100**, 505–530.
Barnett, J. J., Houghton, J. T. & Pyle, J. A. 1975 *Jl R. met. Soc* **101**, 245–257.
Conrath, B. J. 1972 *J. atmos. Sci.* **29**, 1262–1271.
Dave, J. V. & Mateer, C. L. 1967 *J. atmos. Sci* **24**, 414–427.
Elliott, D. D., Clark, M. A. & Hudson, R. D. 1967 *Aerospace Techn. Report no.* TR-0158 (3260–10), 2 September 1967.
Elliott, D. D. 1971 *Space Research* XI, pp. 857–861. Berlin: Akademie–Verlag.
Frederick, J. E., Hays, P. B., Guenther, B. W. & Heath, D. F. 1977*a* *J. atmos. Sci.* **34**, 1987–1994.
Frederick, J. E., Guenther, B. W. & Heath, D. F. 1977*b* *Beitr. Phys. Atmos.* **50**, 496–507.
Frederick, J. E., Guenther, B., Hays, P. B. & Heath, D. F. 1978 *J. geophys. Res.* **83**, 953–958.
Gille, J. C. & House, F. B. 1971 *J. atmos. Sci.* **28**, 1427–1442.
Guenther, B. & Dasgupta, R. 1979 In preparation.
Guenther, B., Dasgupta, R. & Heath, D. F. 1977 *Geophys. Res. Lett.* **4**, 434–437.
Hanel, R. A., Schlachman, B., Clark, F. D., Prokesh, C. H., Taylor, J. B., Wilson, W. M. & Chaney, L. 1970 *Appl. Opt.* **9**, 1767–1774.
Hanel, R. A., Shlachman, B., Rogers, D. & Vanouse, D. 1971 *Appl. Opt.* **10**, 1376–1382.
Hays, P. B. & Roble, R. G. 1973 *Planet. Space Sci.* **21**, 273–279.
Heath, D. F., Mateer, C. L. & Krueger, A. J. 1973 *Pure appl. Geophys.* **106–108**, 1239–1253.
Heath, D. F., Krueger, A. J., Roeder, H. A. & Henderson, B. D. 1975 *Opt. Engng* **14**, 323–331.
Heath, D. F., Krueger, A. J. & Crutzen, P. J. 1977 *Science N.Y.* **197**, 323–331.
Hilsenrath, E., Heath, D. F. & Schlesinger, B. M. 1979 *J. geophys. Res.* **84**. (In the press.)
Iozenas, V. A., Krasnopol'skiy, V. A., Kuznetsov, A. P. & Lebedinsky, A. I. 1969 *Atmos. oceanic Phys.* **5**, 149–159.
Krueger, A. J. 1974 *Proc. Int. Conf. on Structure, Composition and General Circulation of the Upper and Lower Atmosphere and Possible Anthropogenic Perturbations*, vol. 1, pp. 467–477.
Krueger, A. J., Heath, D. F. & Mateer, C. L. 1973 *Pure appl. Geophys.* **106–108**, 1254–1263.
Krueger, A. J. & Minzer, R. A. 1976 *J. geophys. Res.* **81**, 4477–4481.
London, J., Bojkov, R. D., Oltmans, S. & Kelley, J. I. 1976 *N.C.A.R. Tech. Note* TN/113+STR. National Center for Atmospheric Research, Boulder, Colorado.
London, J., Frederick, J. E. & Anderson, G. P. 1977 *J. geophys. Res.* **82**, 2543–2556.
Lovill, J. E., Sullivan, T. J., Weichel, R. L., Ellis, J. S., Huebel, J. G., Korver, J. A., Werdhaas, P. P. & Phelps, F. A. 1978 *Lawrence Livermore Laboratory* UCRL-52473, May 25.
Mateer, C. L. 1972 In *Mathematics of profile inversion* (ed. L. Colin), pp. 2–25. N.A.S.A.TMX-62,150.
Mateer, C. L., Heath, D. F. & Krueger, A. J. 1971 *J. atmos. Sci.* **28**, 1307–1311.
Miller, D. E. & Steward, K. H. 1965 *Proc. R. Soc. Lond.* A**288**, 540–544.
Nimbus Project 1978 *The Nimbus* 7 *Users Guide*, NASA, Goddard Space Flight Center, Greenbelt, Md.
Prabhakara, C., Rodgers, E. B., Conrath, B. J., Hanel, R. A. & Kunde, V. G. 1976 *J. geophys. Res.* **81**, 6391–6399.
Rawcliffe, R. D., Meloy, G. E., Friedman, R. M. & Rogers, E. H. 1963 *J. geophys. Res.* **68**, 6425–6429.
Rawcliffe, R. D. & Elliott, D. D. 1966 *J. geophys. Res.* **71**, 5077–5089.
Riegler, G. R., Drake, J. F., Liu, S. C. & Cicerone, R. J. 1976 *J. geophys. Res.* **81**, 4997–5001.
Singer, S. F. & Wentworth, R. C. 1957 *J. geophys. Res.* **62**, 229–308.
Twomey, S. 1961 *J. geophys. Res.* **66**, 2153–2162.
Venkateswaran, S. V., Moore, J. G. & Krueger, A. J. 1961 *J. geophys. Res.* **66**, 1751–1771.

Phil. Trans. R. Soc. Lond. A **296**, 205–218 (1980) [205]
Printed in Great Britain

Temperature and composition measurements from the l.r.i.r. and l.i.m.s. experiments on Nimbus 6 and 7

By J. C. Gille,† P. L. Bailey† and J. M. Russell III‡§

† *National Center for Atmospheric Research, Boulder, Colorado* 80307, *U.S.A.*
‡ *N.A.S.A. Langley Research Center, National Aeronautics and Space Administration, Hampton, Virginia* 23665, *U.S.A.*

The limb radiance inversion radiometer (l.r.i.r.) on Nimbus 6 was the first orbiting infrared limb scanner. It had four channels with which to determine temperature, O_3 and H_2O in the stratosphere and low mesosphere. The limb infrared monitor of the stratosphere (l.i.m.s.) is a similar six-channel instrument launched on Nimbus 7 in October 1978 to measure temperature, O_3, H_2O, NO_2 and HNO_3. The instrumentation and inversion techniques are briefly described. In this method, the outwelling radiance in the 15 μm bands of CO_2 is inverted to yield temperatures as a function of pressure; the temperature is then used with the radiance emitted by a trace gas to determine its concentration. L.r.i.r. temperature and ozone results show high precision and good agreement with rocket measurements from the tropopause into the mesosphere. Preliminary l.i.m.s. results show that temperatures may be retrieved into the troposphere, and the capability to determine constituent concentrations in the part/10^9 range from a satellite for the first time.

The application of such data for photochemical, dynamical and transport problems is discussed.

1. Introduction

Understanding the many processes – radiative, chemical and dynamical – that take place in the middle atmosphere requires global observations not only of temperature but also composition. The retrieval of temperature and composition information from measurements of infrared radiation emitted by the atmosphere, observed as a function of position on the Earth's limb, has advantages over other techniques for obtaining these data.

The limb radiance inversion radiometer (l.r.i.r.) on Nimbus 6 was the first satellite-borne limb scanner. It obtained its measurements in four channels, in order to infer the vertical distribution of temperature (T), ozone (O_3), and water vapour (H_2O). Subsequently, the limb infrared monitor of the stratosphere (l.i.m.s.) instrument was developed for flight on Nimbus 7. It is similar to the l.r.i.r., but incorporates additional channels for the measurement of nitrogen dioxide (NO_2) and nitric acid (HNO_3), as well as changing the H_2O channel and other improvements.

In this paper we shall review the instrumentation and principles of these limb scanners, present examples of the results, and discuss the applications of the data.

The features of limb scanning, discussed by Gille & House (1971, hereafter referred to as G.H.), follow from the geometry. The atmospheric emission stands out against cold space, and does not require the removal of a background signal. It is enhanced by the large amount

§ Associated only with l.i.m.s. experiment.

of emitting material in the long tangential paths through the atmosphere. Both of these features are very useful for determining the distribution of trace species. The measurement of emission, of course, allows measurements at all local times and latitudes which may be viewed from the spacecraft.

Another feature is the inherent potential for high vertical resolution. The path through a shell 1 km thick at the tangent point is nearly 4.5 times longer than that through the shell 1 km higher. (Thereafter, the geometric path decreases more slowly, and the decrease of opacity is generally dominated by the decrease of constituent concentration with altitude.)

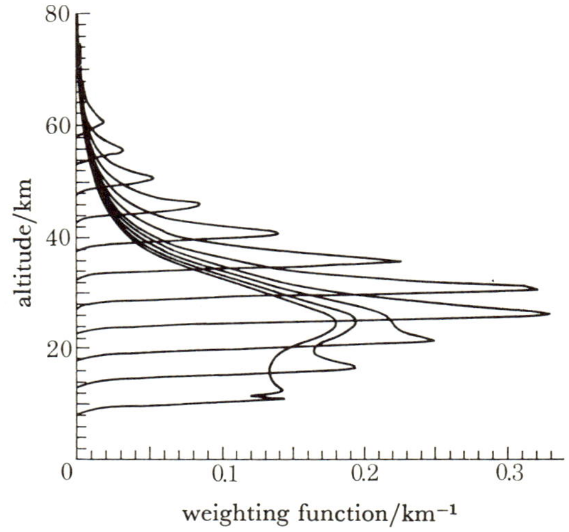

FIGURE 1. Representative weighting functions for the l.r.i.r. wide 15 μm CO_2 channel (WCO_2) used for temperature retrievals. These were calculated by convolving infinitesimal weighting functions (Gille & House 1971) with a nominal 2 km field of view. Actual weighting functions were spaced every 1.5 km in the vertical.

The altitude distributions of opacity or weighting functions (see below) for an optically thin ray path through the atmosphere have half-power points which are about 3 km apart in altitude, roughly a quarter of the width of weighting functions for downward viewing. These, of course, must be convolved with a finite field of view for a real instrument. The instruments described here have employed narrow fields of view in order to take advantage of this capability. A set of representative weighting functions for temperature sounding are shown in figure 1. Note that their vertical spacing depends upon scan rate and sampling frequency, while the width and shape depend upon field of view and atmospheric transmission. One may use broader fields of view, but generally at the expense of achievable vertical resolution.

Two radiance profiles from l.r.i.r. are illustrated in figure 2. In both cases, the emission increases from small values at high altitudes, where the atmospheric paths are nearly transparent, to large values at low altitudes. The narrow CO_2 channel, in the centre of the 15 μm bands, is an example of a band that becomes opaque in the stratosphere, and for which the signals are Planck black body radiances of typical stratospheric temperatures. The radiance variations are due only to atmospheric temperature variations.

The ozone channel radiance is representative of the case in which the gaseous constituent is not opaque, but allows clouds in the upper troposphere to be seen. These give large and variable signals (black body radiances at tropospheric temperatures), but also allow some cloud height information to be derived. Whether from the gas or cloud, the large opacities at low altitudes put a lower limit on the region that may be sounded. The noise level, which determines how small a signal may be measured with useful precision, controls the upper limit of coverage. For the experiments described here, useful information has been achieved from the upper troposphere (*ca.* 11–70 km).

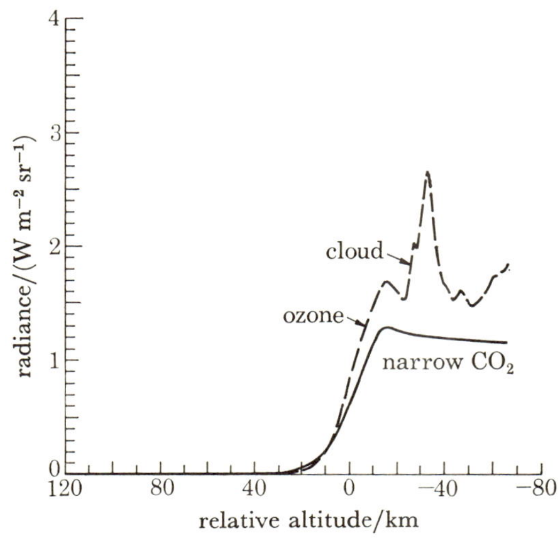

FIGURE 2. Radiance profiles in two l.r.i.r. channels. Solid line: narrow 15 μm CO channel (NCO_2). Dashed line: ozone channel.

It is also pointed out in figure 2 that radiances are measured as a function of angle, or relative altitude. It is necessary to determine the pressure (or density) for the lowest point along a ray path from the data themselves, as noted below, and thereby to obtain an absolute altitude.

In the following, § 2 contains a review of the instrumentation. The inversion of the radiance data is outlined in § 3, and some results are presented in § 4. Examples of applications of the data are given in § 5.

2. THE L.R.I.R. AND L.I.M.S. INSTRUMENTS

The optical schematic of the l.r.i.r. and l.i.m.s. is shown in figure 3.

Radiance from the Earth's limb comes through a hood and baffle assembly to the scan mirror, which directs the beam to an 18 cm off-axis parabolic primary. The radiation is chopped (940 Hz) at the focus of the primary, re-collected and directed into the detector capsule assembly.

Here the radiation passes through a series of optical elements which focus the beam through interference filters, a field of view defining mask, and on to an array of mercury–cadmium–telluride detectors.

The train from the Irtran-6 lens back to the detectors is maintained at 63 K by a cold finger from a block of solid methane, which serves as the primary cryogen. Its life is extended by a thermal shield maintained at 152 K by the solid ammonia secondary cryogen. This also maintains the temperature of the optical train from the thermal mask through the parabolas at 152 K. The ammonia and methane are contained in the solid cryogen dewar. The life of the experiment is limited by the 6.2 kg mass of methane to about 7 months.

In operation, a servo-controlled motor drives the scan mirror, so that the fields of view cross the infrared limb from above 100 km to below the hard surface at a constant rate (1° s^{-1} for l.r.i.r., 0.25° s^{-1} for l.i.m.s.). The precise relative angular position is determined from the output of an encoder on the mirror shaft, which puts out a pulse every 40 ± 2 arcsec.

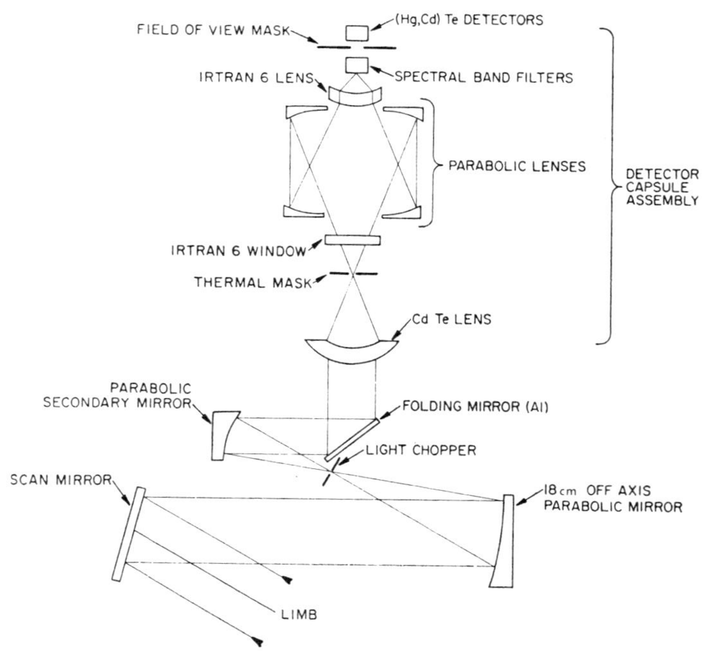

FIGURE 3. Optical schematic diagram of the l.r.i.r. and l.i.m.s.

TABLE 1. CHARACTERISTICS OF L.R.I.R. CHANNELS

channel	purpose	band/pass 50 % relative response points/cm^{-1}	field-of-view at limb/km		noise equivalent radiance/($W\ m^{-2}\ sr^{-1}$)
			vertical	horizontal	
1	NCO_2	649–672	2.0	20	0.0023
2	WCO_2	592–700	2.0	20	0.0040
3	O_3	984–1169	2.0	20	0.011
4	H_2O	412–446	2.5	25	0.008

TABLE 2. CHARACTERISTICS OF L.I.M.S. CHANNELS

channel	purpose	bandpass 50 % relative response points/cm^{-1}	field-of-view at limb/km		noise equivalent radiance/($W\ m^{-2}\ sr^{-1}$)
			vertical	horizontal	
1	NO_2	1580–1613	3.6	28	0.0006
2	H_2O	1396–1527	3.6	28	0.0021
3	O_3	947–1103	1.8	18	0.0035
4	HNO_3	859–900	1.8	18	0.0015
5	WCO_2	595–739	1.8	18	0.005
6	NCO_2	645–673	1.8	18	0.0013

The l.r.i.r. channel characteristics are displayed in table 1. The radiometer channel outputs were sampled on the encoder pulses, at a nominal sample interval of 22 ms. The data were read out in 12 bit words. In-flight calibration occurred every 32 s.

The l.i.m.s. channels are described in table 2. The major difference, apart from the additional channels and slower scan, is the clock initiated data sampling every 24 ms. A number of improvements and lesser changes were also incorporated. In-flight calibration occurs every 28 s.

The l.r.i.r. and l.i.m.s. experiments have been discussed in greater detail by, respectively, Gille *et al.* (1975) and Russell & Gille (1978).

3. Inversion of radiance profiles

The equation for the emerging radiance I within a spectral band may be written

$$I(h) = \int_{-\infty}^{\infty} B(x) \frac{\mathrm{d}\tau(h, x)}{\mathrm{d}x} \mathrm{d}x, \tag{1}$$

where h is the tangent height of the ray path, B is the Planck black body function for the band, $\tau(h, x)$ is the transmittance from the point x to the spacecraft along the ray path having tangent height h, and x is the coordinate along the ray path.

The general approach is to measure the outgoing radiance from carbon dioxide (CO_2), whose distribution is known, and for which the terms $\mathrm{d}\tau/\mathrm{d}x$ may be calculated. From these, B and, thus the temperature, are determined. For other gases, B is calculated for the appropriate band from the temperature, and used with the measured I to determine $\mathrm{d}\tau/\mathrm{d}x$ from which the required trace gas concentrations may be obtained.

A major problem as noted above is that the Nimbus attitude control and ephemeris data do not allow h for a given measurement to be determined very well. Even if this were possible it would not be enough, since it is necessary to know the absorber density at h, which is controlled by the temperature structure through the hydrostatic equation. The solution was developed in G.H.

After converting (1) to a vertical integral,

$$I(h) = \int_{h}^{\infty} B(z) \left[\left(\frac{\mathrm{d}\tau(h, z)}{\mathrm{d}x}\right)_{\mathrm{a}} + \left(\frac{\mathrm{d}\tau(h, z)}{\mathrm{d}x}\right)_{\mathrm{p}}\right] \left|\frac{\mathrm{d}x}{\mathrm{d}z}\right| \mathrm{d}z, \tag{2}$$

(where subscripts a, p refer to points on the path anterior and posterior to the tangent point), equation (2) may be written

$$I(h-z_0) = p_0 R^{-1} C_1 \int_{h-z_0}^{\infty} B(z)\, L(h, z) \left|\frac{\mathrm{d}x}{\mathrm{d}z}\right|_z \times \exp\left[-\int_{z_0}^{z} gR^{-1}T(z')^{-1}\,\mathrm{d}z'\right] T(z)^{-1}\,\mathrm{d}z = p_0 \int_{h-z_0}^{\infty} B(z)\, W(h, z)\,\mathrm{d}z, \tag{3}$$

where $L(h, z)$ contains the derivatives of the transmittance in terms of absorber amount and pressure along the path, C_1 represents the mixing ratio of CO_2, and $p_0 W$ is the weighting function mentioned above. The only reference point needed is an arbitrary level z_0, corresponding to one of the radiance samples at which the pressure is p_0. G.H. then showed that p_0 could be determined from measurements in two CO_2 channels for which the atmosphere has different opacities.

The iterative procedure described there is too slow for operational use, as well as leading to error growth when a large number of iterations is used. It has been necessary to develop a fast but accurate method for performing the retrievals. The procedure now in use (Bailey & Gille 1978) begins with a pre-inversion stage, in which an up and a down profile (taken within 4 s, or 25 km, of ground track for l.r.i.r.) are used to determine the mean spacecraft motion, then combined into a single profile and the cloud tops located from radiances in a non-opaque channel such as the ozone channel.

The data are then Fourier transformed into a spatial frequency domain, low pass filtered for noise suppression and 'boosted' to partially compensate the effects of the finite field of view (Gille & Bailey 1978) before being transformed back into physical space.

Inversion then takes place in two steps. In an initial or direct inversion (Bailey & Gille 1978), (3) is written as

$$I(p) = B[\hat{T}]\,\epsilon(p, \hat{T}), \tag{4}$$

where p is the pressure at the tangent point, $\hat{T}$ is an average temperature over the region of the atmosphere contributing to the outgoing radiance, defined by

$$\hat{T} = \int_h^H W(h, z)\, T(z)\, \mathrm{d}z \Big/ \int_h^H W(h, z)\, \mathrm{d}z, \tag{5}$$

and ϵ is an effective emittance when looking at tangent pressure p.

The procedure starts by locating a point on the more opaque (narrow) CO_2 channel where $\epsilon_N \approx 1$, and deriving $\hat{T}_N$. Substituting $\hat{T}_N$ into equation (4) for the less opaque (wide) CO_2 channel leads to ϵ_W, from which a reference pressure p can be determined. The solution proceeds up the wide and narrow channels, using values of $\hat{T}$ in the hydrostatic equation to determine the pressure at the next level. In general, p_0 must be adjusted so that $\hat{T}_N \approx \hat{T}_W$.

When the profile $\hat{T}(p)$ is determined, equation (5) is inverted to yield the 'direct' solution, $T^{(1)}(p)$.

To complete the solution, (3) is written as

$$I(z_i) = W_{ij} B(z_j). \tag{6}$$

The $T^{(1)}(p)$ is used to calculate W_{ij}, equation (6) solved for $B^{(2)}$, and the solution iterated if desired. This allows improved vertical resolution, more complete allowance for temperature and pressure dependence, and the sounding of lower levels in the atmosphere.

The constituent retrieval proceeds similarly, except now we have

$$\epsilon_i = \epsilon_i(p, \hat{C}_i, \hat{T}),$$

where C is the concentration of the ith species. After ϵ_i is determined from (4), the distribution of $\hat{C}_i$ is obtained, and inverted to give the direct solution $C_i^{(1)}$. This is then used as the initial step in an equation of the form (6). In this case, improved solutions are obtained from the W_{ij} elements.

Inversion methods will be described in detail in future publications.

4. Results of limb profile inversion

(a) *L.r.i.r. results*

We first discuss the results of temperature retrieval, since this is the first step in any inversion. Figure 4 presents a temperature profile retrieved by using the initial inversion compared to a nearby rocket measurement. There is quite reasonable agreement between these different observing techniques. We also see that the satellite retrieval does a good job of recovering fine scale structure.

There are two measures of quality to be applied to the retrievals; their repeatability (random error), and accuracy (systematic error). The former may be assessed by retrieving a sequence of radiance profiles in regions where the along-orbit gradient of atmosphere temperatures is small, a circumstance that can be checked. This is generally the case at low latitudes. Taking a series of observations, and assuming that the atmosphere is the same, the standard deviation of the temperatures at a level gives a measure of the end to end repeatability of the system of measurement plus inversion. This procedure returns values for the random error of less than 1 K from 50 to 5 mbar;† less than 2 K for 5–0.4 and $<$ 2.5K at 0.25 mbar. The results then are very stable, and comparable to the repeatability of rocket measurements.

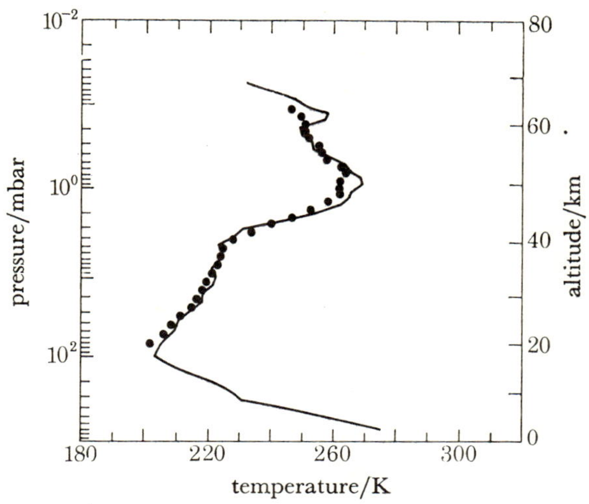

FIGURE 4. Temperature obtained from l.r.i.r. measurements, using the direct (first stage) inversion technique (points), compared to conventional rocketsonde and radiosonde measurements (solid line) at Wallops Island, Virginia (37.85° N, 75.48° W) at 1800 U.T. on 19 November 1975. The l.r.i.r. observations were obtained at 37.95° N, 73.60° W, at 1730 U.T.

The accuracy is much more difficult to assess with confidence, because of the lack of laboratory type standards for atmospheric measurements. Figure 5 presents a comparison for 47 cases using data from meteorological rockets – a familiar system which has been widely used. Comparisons include rocket results within 2° latitude and equivalent longitude, and 3 h time of l.r.i.r. results. The systematic differences are small; most are $<$ 1°, and not statistically significant. These are themselves quite good, and certainly an excellent point from which to start the more detailed inversion.

The temperature profile from the data of figure 4, after the shell inversion, is shown in figure 6. The main differences are that it has somewhat finer vertical resolution, allows elimination of errors correlated with temperature, and, perhaps best, allows an extension to higher and especially to lower altitudes.

It should be pointed out that, aside from the use of four profiles to determine an instrumental parameter not well defined by laboratory measurements, there is no tuning of the transmittances or retrievel to give these results.

† 1 mbar = 10^2 Pa.

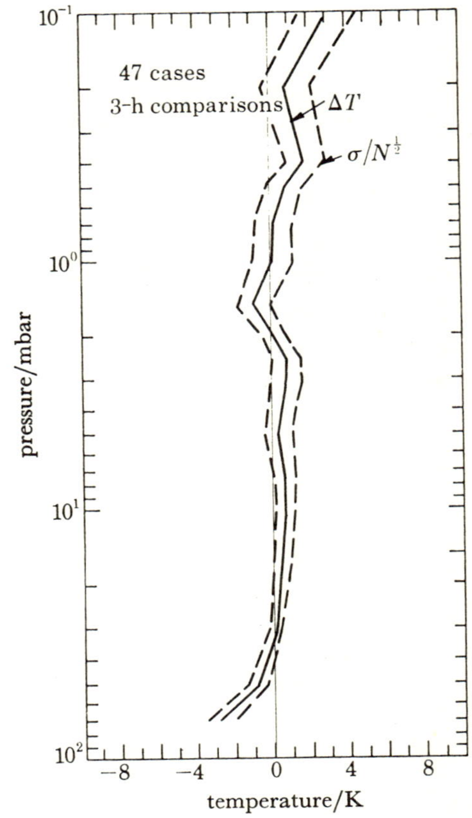

FIGURE 5. Comparison between l.r.i.r. and conventional rocketsondes and rawinsonde temperatures for 47 comparison cases (see text for criteria). Solid line displays l.r.i.r. minus conventional temperature. Broken lines bracket standard deviation of the mean difference.

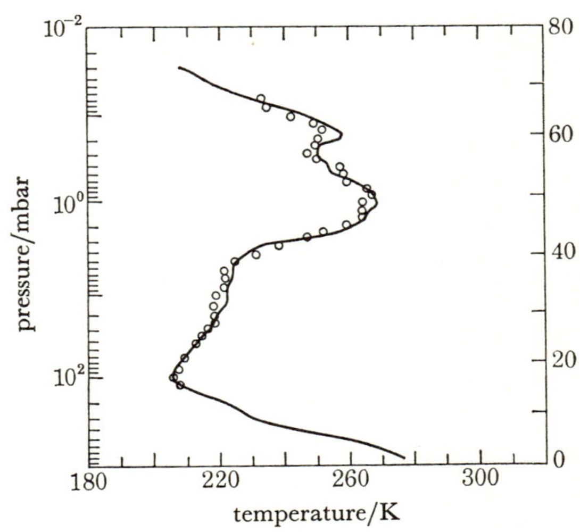

FIGURE 6. Temperature obtained from l.r.i.r. measurements, using a shell inversion to refine the first stage inversion (open circles), compared to conventional rocketsonde and radiosonde measurements (solid line). Conditions the same as in figure 4.

Figure 7 shows an l.r.i.r. ozone profile, compared to a rocket measurement by Krueger. The agreement is again good. Other noteworthy features are the resolution of the satellite retrieval, and its coverage, from 100 to 0.4 mbar, or the whole stratosphere and into the mesosphere.

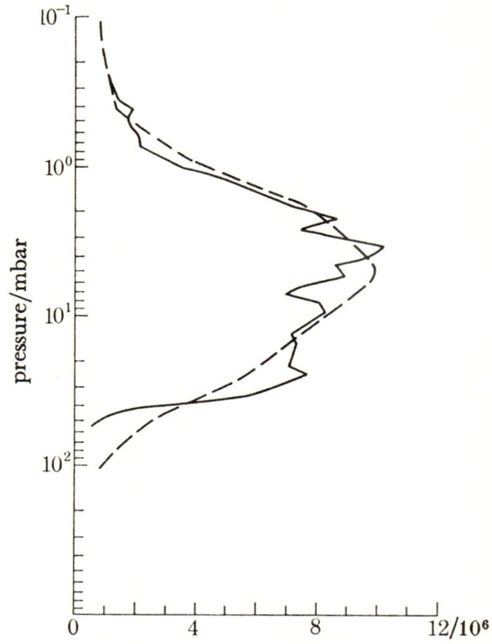

FIGURE 7. Comparison between l.r.i.r. ozone determination (dashes) and rocket ozone measurement using optical technique (solid line). Rocketsonde was obtained at Wallops Island at 1716 U.T. on 19 November 1975. The l.r.i.r. observations were obtained at 39.9° N, 75.0° W at 1730 U.T.

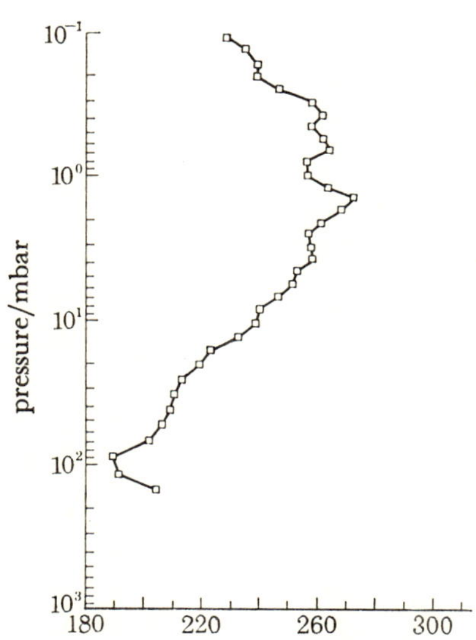

FIGURE 8. Preliminary l.i.m.s. temperature retrieval at the equator, 28 October 1978. The tropical tropopause region is clearly shown.

Again, the repeatability is less than 0.3 parts/10^6 by volume from 50 to 1 mbar, and less than 0.1 from 20 to 7 mbar, or 1 % at 10 mbar. The accuracy is difficult to assess, because of the fewer number of comparison cases and larger uncertainties of the ozone instruments, but the l.r.i.r. retrievals appear to be of comparable accuracy to the *in situ* devices.

The calibration of the H_2O channels has proved troublesome, and the signal:noise ratio is lower than expected. Results from this channel are not presented here.

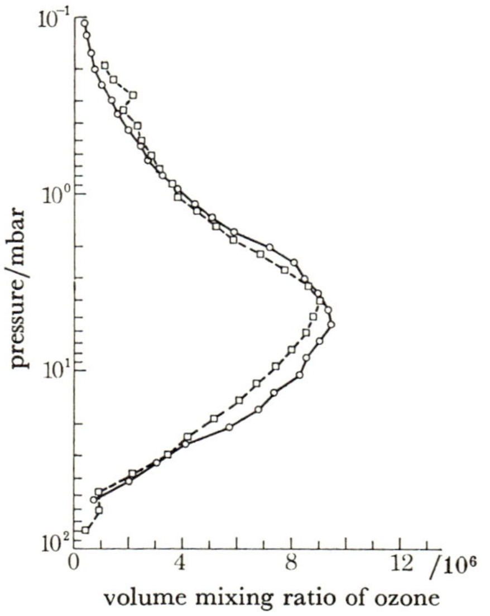

FIGURE 9. Preliminary l.i.m.s. ozone retrievals on 28 October 1978. Solid line, 109° E; dashed line, 111° W.

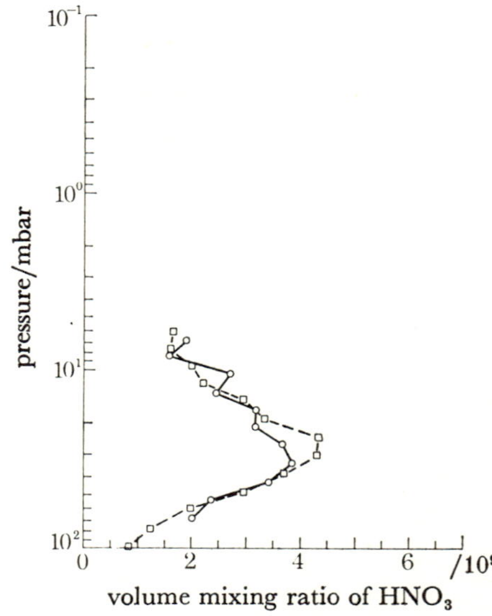

FIGURE 10. Preliminary l.i.m.s. nitric acid retrievals on 28 October 1978. Solid line, 109° E; dashed line, 111° W.

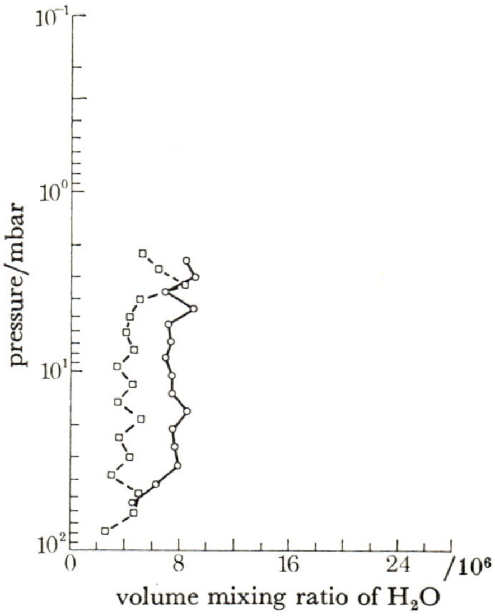

FIGURE 11. Preliminary l.i.m.s. water vapour retrievals on 28 October 1978. Solid line, 109° E; dashed line, 111° W.

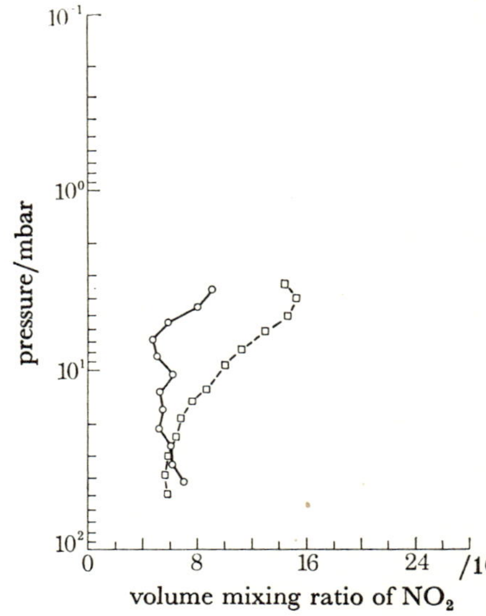

FIGURE 12. Preliminary l.i.m.s. nitrogen dioxide retrievals on 28 October 1978. Solid line, 109° E (day); dashed line, 111° W (night). A large diurnal variation is expected.

(*b*) *Preliminary l.i.m.s. results*

The l.i.m.s. results are very new at this time. All aspects of the instrument performance, atmospheric transmittances and retrieval techniques are not yet completely checked, and so the results presented here should be regarded as preliminary, but indicative.

An example of a l.i.m.s. temperature retrieval is shown in figure 8. Note that the tropical tropopause near 90 mbar is clearly seen, and the temperature retrieval extends into the upper tropopause. The lower altitude of the l.r.i.r. retrievals was limited not only by clouds, but also by the opacity of the CO_2 band itself. The l.i.m.s. channel 5 was spectrally broadened to facilitate retrieval into the upper tropopshere on cloud-free occasions.

Among the few profiles we have been able to look at to date, some go down to 240 mbar. L.i.m.s. temperature profiles, showing the lower stratosphere and tropopause region, will be supplied to the level II-*b* data sets of the global weather experiment (f.g.g.e.) to improve the depiction of the atmosphere for studies of weather forecasting.

L.i.m.s. ozone retrievals are presented in figure 9. The lower noise on l.i.m.s. allows retrievals to higher altitudes (and also lower, not shown). The results appear quite reasonable, although they have not yet been intercompared with other techniques.

Retrieved HNO_3 profiles are shown in figure 10. These are rather typical of values on this date, with peak values of 3–4 parts/10^9 by volume, peaking near 30 mbar. This appears to be typical of the Southern Hemisphere at this time. The HNO_3 profile at the equator appears to peak at higher altitude (*ca.* 10 mbar), with lower peak values (1–2 parts/10^9 by volume), with higher values at lower altitudes in the Northern Hemisphere. Initial comparison to *in situ* measurements shows reasonable agreement.

Two water vapour profiles at 30 S are presented in figure 11 (note the units are parts per million by volume: 1 part/10^6 by volume = 0.621 part/10^6 by mass). The mean concentrations are approximately 4 and 7 parts/10^6 by volume (2.5–4.4 parts/10^6 by mass) – within the range of accepted values. It is too early to say whether the indicated longitudinal variation is real or not.

Finally, two NO_2 retrievals are shown in figure 12. The daytime values are slightly lower at latitudes other than $\pm 60°$, night-time values higher. All profiles show the expected night-time increase of NO_2, by about a factor of two. Because of uncertainties in the basic spectroscopic parameters, these results are more provisional than the other l.i.m.s. results.

Note that for HNO_3, H_2O and NO_2, results are presented only to about 5 mbar because final transmittance values are not yet available. Retrievals to about 1 mbar should be possible for all these species.

5. Application of l.r.i.r. results

Observations of the type presented above allow the stratosphere to be observed globally with a detail not previously possible. We shall mention just a few of the possible uses of such data.

(*a*) *Mesospheric ozone distribution*

Riegler *et al.* (1976, 1977) have derived the night-time ozone distributions down to *ca.* 45 km from measurements of the occultation of β-Centauri obtained from the Copernicus telescope on the OAO-3 satellite. Night-time concentrations derived from l.r.i.r. measurements up to 55 km for the previous day, at the same latitude and bracketing longitudes, are a factor of 3 smaller than the Copernicus results, and in much better agreement with measurements by other techniques and theoretical models. Contrary to the Copernicus results, they indicate that photochemical theory can explain the ozone concentration in the lower mesosphere reasonably well.

(*b*) *Diurnal variation of ozone*

A Sun-synchronous orbit is not a good one for the study of diurnal changes, since a given latitude is viewed at the same two local times every day. One can measure day–night differences, however. For the l.r.i.r. scan track, there are latitudes at which a location will be observed on the day side of the Earth, and several orbits later will be observed again on the night side. In this few hours interval, and at high enough altitudes, one would expect the effects of transports to be minimal and the differences in ozone concentration to be due to photochemical effects. Preliminary results show night-time values exceeding day-time values by up to 20 % (± 10 %) in the layer from 0.4 to 0.07 mbar (56–66 km), decreasing to no detectable change at 1 mbar. These studies will be amplified and extended in the future.

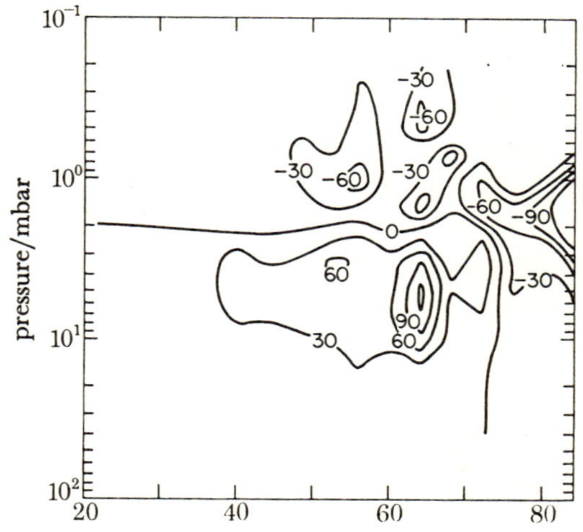

FIGURE 13. Ozone transports due to geostrophic eddies. Units are parts/10^6 (by volume) m s^{-1}.

(*c*) *Transports*

Because temperature is measured as well as constituent concentrations, geopotential heights and geostrophic winds may be derived, and transports by geostrophic eddies directly calculated. The key to this procedure is the method of analysing the spatial variation of the fields. We have used the approach suggested by Rodgers (1976), in which the observations are Kalman filtered to update the Fourier coefficient representation of the fields on a latitude circle. Analyses are produced at each 4° of latitude for T, O_3, and $\Delta\phi_{30}$, the thickness above the 30 mbar surface, at 17 pressure levels. The geopotential height is obtained by adding the thickness to the conventionally analysed geopotential at 30 mbar, to give a result of the form

$$\phi(p, \theta, t) = \phi_0(p, \theta, t) + \sum_{n=1}^{6} a_n(p, \theta, t) \sin\frac{2\pi n\lambda}{360} + \sum_{n=1}^{6} b_n(p, \theta, t) \cos\frac{2\pi n\lambda}{360},$$

where p is the pressure, θ the latitude, t the time and λ is the longitude. Calculation of the geostrophic wind, and especially of the meridional wind associated with the eddies ($v' \propto \partial\phi/\partial\lambda$)

follows immediately. Finally, when the ozone concentration is similarly expressed, the transports by wavenumber n, $\overline{v_n' O_{3n}'}$, may be immediately obtained. A cross section of eddy transports of ozone on a single day is shown in figure 13. Of course, this is only part of the transport, since there are also compensating mean meridional transports, as well as photochemical effects. With observations as detailed as these, it should be possible to derive the meridional circulation (as Hartmann (1976) did), to obtain a complete picture of the transport.

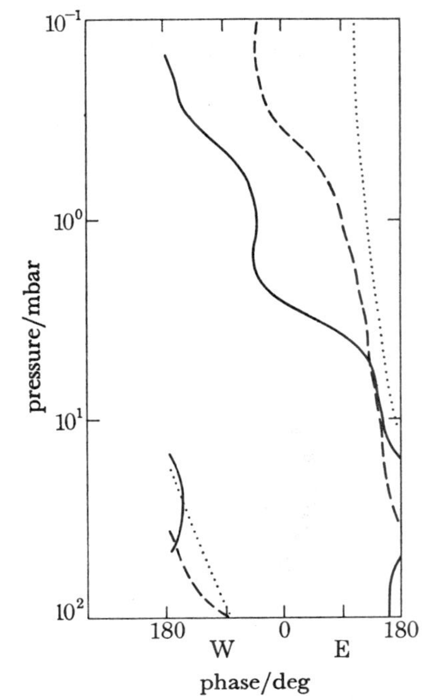

FIGURE 14. Phases of wave in temperature (dashed line), ozone mixing ratio (solid line) and geopotential height (dotted line) as a function of altitude on 1 November 1975. The in-phase relation between ozone mixing ratio and geopotential indicates dynamical control, while an exactly out of phase relation between ozone and temperature signifies photochemical control (Hartmann & Garcia (1979)).

(*d*) *Photochemistry*

Finally, these data provide information on the basic ozone photochemistry. The phase of the wavenumber 1 ($n = 1$) variation of ozone mixing ratio (μ_3), temperature and geopotential are shown for 60° N on 1 November 1975 in figure 14. Hartmann & Garcia (1979) have used a simple mechanistic model to show that when the ozone distribution is dynamically controlled, the ozone mixing ratio will be in phase with the geopotential, whereas under photochemical control it will be 180° out of phase with the temperature. This latter results from temperature dependence of the reaction rates, experimentally observed in the atmosphere by Barnett *et al.* (1975). The data shown in figure 14 thus indicate dynamical control of the ozone distribution up to 5 mbar (*ca.* 35 km), with photochemical control above 2 mbar (*ca.* 45 km). This directly confirms the behaviour predicted on the basis of the photochemical models, and the existence of the two régimes (with an interesting transition region between).

6. Conclusions

The results presented here demonstrate the ability of limb scanning radiometers to provide temperature and ozone profiles with high vertical resolution from the troposphere into the lower mesosphere that are very repeatable and agree well with rocket measurements. They also show for the first time the ability to measure constituents in the parts/10^9 range (HNO_3, NO_2) from a satellite. Work is still required to establish the accuracy of the new measurements.

Finally, the preliminary scientific results give a glimpse of the power and utility of these data in the solution of problems in the middle atmosphere.

We gratefully acknowledge the contributions of G. P. Anderson, R. Garcia, W. Kohri, S. Nolte and V. Rinehart, all of N.C.A.R., to the work presented here. This activity was partially supported by the National Aeronautics and Space Administration through Contract S40135 G and Order Number L36446 A. The National Center for Atmospheric Research is sponsored by the National Science Foundation.

References (Gille *et al.*)

Bailey, P. L. & Gille, J. C. 1978 In *Remote sensing of the atmosphere: inversion methods and applications* (ed. A. L. Fymat & V. E. Zuev), pp. 115–122. Amsterdam: Elsevier.

Barnett, J. J., Houghton, J. T. & Pyle, J. A. 1975 *Q. Jl R. met. Soc.* **101**, 245–257.

Gille, J. C., Bailey, P., House, F. B., Craig, R. A. & Thomas, J. R. 1975 In *The Nimbus* 6 *user's Guide* (ed. J. E. Sissala). Greenbelt, Maryland: National Aeronautics and Space Administration.

Gille, J. C. & House, F. B. 1971 *J. atmos. Sci.* **28**, 1427–1442.

Gille, J. C. & Bailey, P. L. 1978 In *Remote sensing of the atmosphere: inversion methods and applications* (ed. A. L. Fymat & V. E. Zuev), pp. 107–113. Amsterdam: Elsevier.

Hartmann, D. L. 1976 *J. atmos. Sci.* **33**, 1789–1802.

Hartmann, D. L. & Garcia, R. R. 1979 *J. atmos. Sci.* **36**, 350–364.

Riegler, G. R., Drake, J. F., Liu, S. C. & Cicerone, R. J. 1976 *J. geophys. Res.* **81**, 4997–5001.

Riegler, G. R., Atreya, S. K., Donaheu, T. M., Liu, S. C. & Wasser, B. 1977 *Geophys. Res. Lett.* **4**, 145–148.

Rodgers, C. D. 1976 In *Inversion methods in atmospheric remote sounding* (ed. A. Deepak), pp. 117–138. New York: Academic Press.

Russell, J. M. III & Gille, J. C. 1978 In *The Nimbus* 7 *user's Guide* (ed. C. R. Madrid), pp. 71–103. Greenbelt Maryland: National Aeronautics and Space Administration.

Phil. Trans. R. Soc. Lond. A **296**, 219–241 (1980) [219]
Printed in Great Britain

The stratospheric and mesospheric sounder on Nimbus 7

BY J. R. DRUMMOND, J. T. HOUGHTON, F.R.S., G. D. PESKETT, C. D. RODGERS, M. J. WALE, J. WHITNEY AND E. J. WILLIAMSON
Department of Atmospheric Physics, Clarendon Laboratory, Oxford, OX1 3PU

The stratospheric and mesospheric sounder (s.a.m.s.) instrument was launched on the Nimbus G satellite on 24 October 1978. It is designed to measure temperature and concentration profiles of various gases in the height range 20–100 km by detecting either their thermal emission or, in some cases, resonant scattering of sunlight.

The gases selected, CO_2, CO, CH_4, NO, N_2O and H_2O, significantly affect the upper atmosphere energy budget by their influence on the concentration of the primary sunlight absorber, ozone. This influence is disproportionate to their own concentration because of the existence of 'catalytic cycles' which destroy ozone while regenerating the catalyst.

A description of the instrument, its principles of operation and some of the methods of retrieval used is presented, together with some preliminary results from the first 3 months of operations.

1. INTRODUCTION

As has been explained earlier in this meeting by Professor Labitzke and by Dr Barnett, during the last 10 years remote sounding observations from satellites of atmospheric temperature have been extended through the stratosphere into the mesosphere. Much extra insight into atmospheric phenomena has arisen through the provision of these observations substantially continuous in time and with near global coverage. The purpose of the stratospheric and mesospheric sounder (s.a.m.s.) on Nimbus 7 is to provide similar coverage in time and space for observations of the concentration of a number of important minor constituents in the stratosphere and mesosphere, all of which are directly or indirectly involved in the chemistry of ozone (cf. Thrush, this symposium). Like previous remote sounding instruments built for Nimbus satellites in which the Oxford University Atmospheric Physics Department has been involved, s.a.m.s. observes infrared radiation emitted by the atmospheric constituents it is required to measure. In this paper, the principles of the techniques employed will first be outlined, then the different measurements being made by s.a.m.s. will be discussed together with a description of the instrument and the method of calibration. Finally, some early results will be presented.

2. PRINCIPLES OF OPERATION

2.1. *Limb-sounding*

The advantages of the limb view (see figure 7) are that emission observations can be made from as long a path through the atmosphere as possible, and that behind the atmospheric path is the cold radiation background of space. Such a limb path possesses an air mass approximately 70 times that in a vertical path above the lowest point of the path. The atmosphere's pressure profile and the geometry of the path both weight the material in the path strongly towards the tangent height giving good vertical resolution. For emission from spectral lines that are not saturated at their centres, 60% of the emission from a limb path originates from altitudes within 3 km of the lowest altitude of the path (see figure 4). By contrast, the horizontal resolution of a limb observation is poor, the length of the limb path being of the order of 400 km.

Radiometers carried by balloons or aircraft have, for a long time, observed emission from the atmosphere's limb. Results from the limb radiance inversion radiometer (l.r.i.r.) on Nimbus 6 (Gille *et al.* 1975), the first satellite radiometer to observe radiance from the atmosphere's limb, have demonstrated the vertical resolution which can be achieved in measurements of temperature and ozone (Gille, this symposium).

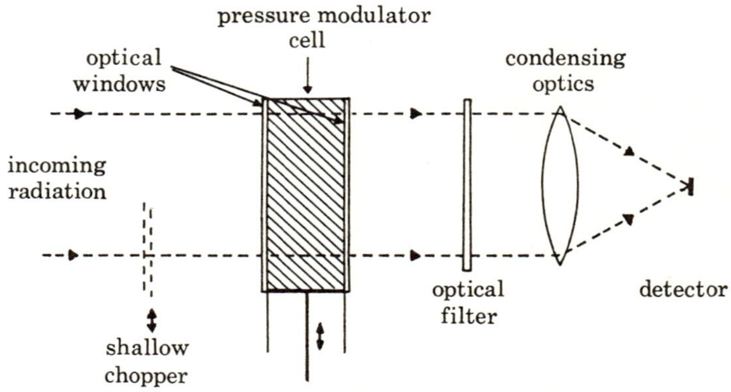

FIGURE 1. Simplified diagram of a p.m.r. optical system employing a 'shallow chopper' which only modulates part of the incoming radiation beam.

2.2. *The pressure modulator*

The pressure modulator is a device which selectively modulates the emission from a gas by using the absorption lines of the same gas as an optical filter. A cell of gas is included in the optical path of the radiometer (figure 1) and its pressure varied cyclically. The transmission of the cell is modulated only at optical frequencies that lie within the absorption lines of the gas, thus selectively modulating emission from the same gas incident on the radiometer. After detection, the signal at the frequency of the modulation may be recovered by phase-sensitive detection. A broad optical filter limits the spectral bandwidth to the appropriate emission band.

The pressure modulator technique is a development from the selective chopper technique employed for temperature sounding instruments on the Nimbus 4 and 5 satellites (Houghton & Smith 1970; Ellis *et al.* 1973). It has previously been flown in a temperature sounder on the Nimbus 6 satellite (Taylor *et al.* 1972; Curtis *et al.* 1974). The advantages of the technique for composition sounding have been described by Chaloner *et al.* (1978), who employed it in a balloon-borne radiometer for the measurement of NO, NO_2 and H_2O in the stratosphere. Further details of the technique are also given in Drummond & Jarnot (1978).

2.3. *Double chopping*

The gas pressure in the pressure modulator cell (p.m.c.) is modulated at the resonant frequency (20–40 Hz) of the mechanical system (see § 4.3). It is also useful to include a conventional chopper at a much higher frequency (*ca.* 240 Hz) in the optical system. This modulates all 'incoming' radiation with the optical filter profile and produces a 'wideband' signal at the detector. The 'wideband' signal, in the absence of emission from other species, provides information for lower in the atmosphere than the p.m.r. signal (see figure 2). The wideband signal may also be used for the elimination of signals from atmsopheric constituents whose absorption bands overlap that of the wanted species (see Drummond & Jarnot 1978), and in the determination of the atmospheric level being observed (§ 3.3). In s.a.m.s. a shallow chopper is employed (see § 4), which chops only *ca.* 2.5% of the aperture of the system. The shallow chopper is preferred over a conventional 100% chopper because it does not obstruct the beam for the pressure modulator further back in the optical chain.

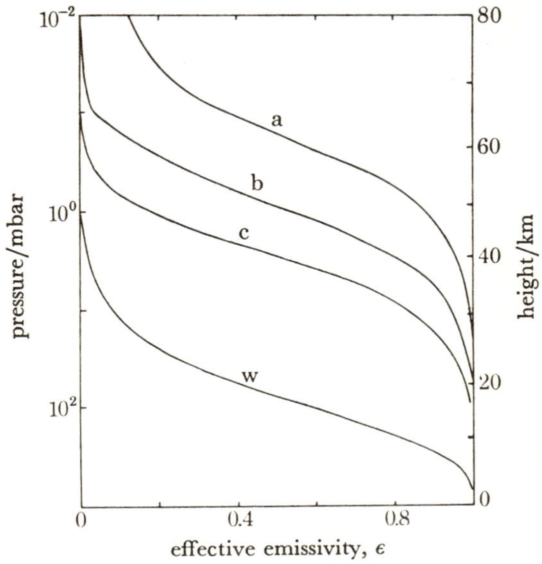

FIGURE 2. The effective emissivity, ϵ (see equation (5)) for water vapour evaluated by using a climatological equatorial temperature profile and a constant volume mixing ratio of 5×10^{-6}. Curves (a), (b) and (c) are for the s.a.m.s. B2 p.m.r. channel with mean cell pressures of 0.87 mbar, 4.48 mbar and 15.3 mbar respectively. Curve (w) is for the corresponding wideband channel and is virtually independent of modulator pressure.

3. MEASUREMENTS WITH S.A.M.S.

The radiant power, W, emitted by a limb path at frequency ν and received by one of the s.a.m.s. detectors is given by

$$W = A\Omega \int_0^\infty \int_0^\infty J_\nu(x)\, \tau_{0\nu}\, \tau_{m\nu} \frac{\mathrm{d}\tau_\nu(x)}{\mathrm{d}x}\, \mathrm{d}x\, \mathrm{d}\nu, \tag{1}$$

where A is the aperture of the telescope, Ω the field of view, $\tau_\nu(x)$ the transmission between the satellite and the position along the path described by the coordinate x, $\tau_{0\nu}$ the static component of the optical transmission in the appropriate channel, $\tau_{m\nu}$ the modulated component of the optical transmission, and $J_\nu(x)$ is the source function appropriate to the emission band in question at the location described by the coordinate x. Under conditions of local thermodynamic equilibrium (l.t.e.), the source function is equal to the Planck function $B_\nu(T)$ at the temperature T of the atmosphere at location x.

Integration is along the limb path and over the frequencies where the optical components have any significant transmission.

Calibration of the instrument is carried out by inserting a black-body at temperature T_b in the path when the radiant power W_b received is

$$W_b = A\Omega \int_0^\infty B_\nu(T_b)\, \tau_{0\nu}\tau_{m\nu}\, \mathrm{d}\nu. \tag{2}$$

The radiance, L, of an atmospheric path is then defined as

$$L = \frac{W}{W_b} \frac{\int_0^\infty B_\nu(T_b)\, \tau_{0\nu}\tau_{m\nu}\, \mathrm{d}\nu}{\int_0^\infty \tau_{0\nu}\tau_{m\nu}\, \mathrm{d}\nu}. \tag{3}$$

For channels covering a reasonably narrow frequency interval the ratio of the integrals in equation (3) is $B_{\bar{\nu}}(T_b)$, where $\bar{\nu}$ is the centre frequency of the interval.

To illustrate the radiometric measurements being made it is useful when l.t.e. applies to write a simplified expression for the radiance L measured by a particular channel, namely:

$$L = \epsilon B_{\bar{\nu}}(\overline{T}), \tag{4}$$

where

$$\epsilon = \frac{\int_0^\infty \int_0^\infty \tau_{0\nu}\tau_{m\nu} \frac{\mathrm{d}\tau_\nu(x)}{\mathrm{d}x}\, \mathrm{d}x\, \mathrm{d}\nu}{\int_0^\infty \tau_{0\nu}\tau_{m\nu}\, \mathrm{d}x} \tag{5}$$

may be defined as the *effective emissivity* of the path and where $\overline{T}$ is a mean temperature for the path.

3.1. *Temperature measurements*

If the radiance is measured from the emission band of an atmospheric constituent whose distribution is accurately known, for any given path the emissivity ϵ in equation (4) may be calculated and hence a mean temperature $\overline{T}$ for that path may be deduced. A set of radiances received from different levels as the instrument scans over the limb may be retrieved to give a temperature profile (see Rodgers (1976) for a review of retrieval methods and § 6.2). In s.a.m.s. two wideband and two pressure modulator channels are included observing emission from the 15 μm band of CO_2, a gas that is substantially uniformly mixed up to at least 90 km altitude. Above about 80 km, local thermodynamic equilibrium no longer applies for this band; the quantity that is then measured is the vibrational temperature of the band, which is of considerable importance because emission from it is the major energy sink in the upper mesosphere and lower thermosphere.

3.2. *Composition measurements*

Given that the atmosphere's temperature structure has been determined by the method of § 3.1, from radiance observations on any other band an effective emissivity ϵ may be determined from equation (4), and hence the distribution of the emitting constituent observed. In figure 2 are curves showing the variation of effective emissivity for emission from the long wave water vapour band as measured by a wide band channel and by channels containing pressure modulator cells with water vapour at different pressures, demonstrating that the water vapour distribution can be measured from the tropopause through into the upper mesosphere. Some results for H_2O from an early orbit of s.a.m.a. are given in § 6.

The method by which a detailed distribution is derived is as follows. First of all, through theoretical calculations on the basis of spectral information about the band in question, together with a programme of laboratory measurements as described in § 5, algorithms describing the transmission of any atmospheric path are derived. A 'first guess' profile for the distribution of the gas in question is then assumed and, on the basis of the temperature profile derived by the method of § 3.1, an expected set of radiances is computed. These computed radiances are then compared with the measured radiances, and by an iterative process a profile of the gas distribution is found for which computed and measured radiances agree.

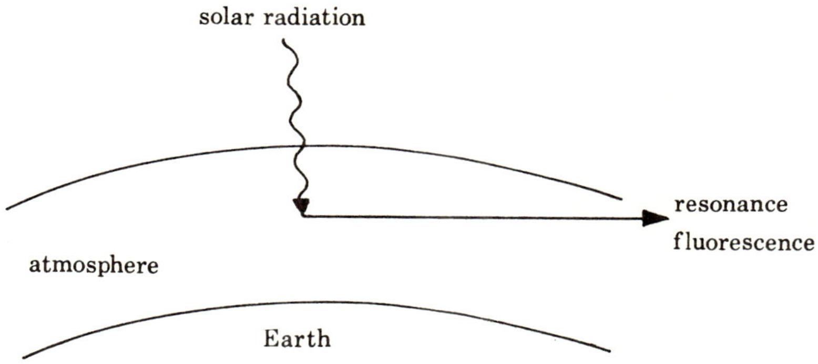

FIGURE 3. The generation of resonance fluorescence signals by scattering of sunlight within the atmosphere.

To acquire information about the distribution of some constituents in the mesosphere and lower thermosphere it is possible to use observations of resonance fluorescence of solar radiation from those molecules which have strong bands in the near infrared (figure 3) (cf. Table 1 for list of bands included in s.a.m.s.). Provided that absorption by solar radiation is the only process of excitation of the band in question, and provided that collisional quenching is negligible, the radiance observed is a function of the solar flux incident on the layer and the number of molecules along the path. In practice, an elaborate radiative transfer calculation has to be carried out for each situation because multiple scattering, excitation by other non-thermal processes (e.g. by particles), and details of quenching processes have also to be considered. Interpretation of measurements from fluorescence channels, therefore, is extremely complex.

3.3. *Reference pressure determination*

Because s.a.m.s. is a limb viewing instrument it is necessary that accurate information is available regarding the pressure of the atmosphere at the levels being viewed at any given time. Consider information such as that contained in figure 2. When, for instance, the effective emissivity of an atmospheric path is about 0.5, a change in signal corresponding to a change in emissivity of 0.01 (equivalent to a change in mixing ratio of *ca.* 6 %) could also result from a change in the level being viewed of 0.15 km, which is equivalent to a change in pressure at the tangent point of *ca.* 2 % or 0.003° in viewing direction.

The attitude of the spacecraft is controlled only to *ca.* 0.5° in all three axes; it is therefore necessary from the measurements made by s.a.m.s. itself to determine the appropriate information about the pressure at the part of the atmosphere being observed.

A method for providing this information arises from comparing the signals from a wideband and a pressure modulator channel in the 15 μm CO_2 band. Under the approximation of equation

(4) the two signals will be $\epsilon_w B(\bar{T}_w)$ and $\epsilon_p B(\bar{T}_p)$ where the subscripts w and p respectively denote the wideband and pressure modulator channels. Now it is possible to choose the mean cell pressure in the pressure modulator channel such that over a certain range of altitudes the relative contributions of different segments of the atmospheric path being observed are very similar for the two channels, even though the effective emissivity of the atmosphere for the two channels is very different (figure 4). This means that $\bar{T}_w$ and $\bar{T}_p$ will be almost equal. We then find that the ratio ϵ_p/ϵ_w is almost independent of temperature but strongly dependent on the pressure at the level of observation. Figure 5 illustrates that, for a particular cell pressure, atmospheric pressure over the range between 0.8 mbar and 20 mbar may be derived from the ratio of signals in the two channels. A first order correction for the atmospheric temperature profile can be included, enabling pressure measurement of adequate accuracy (*ca.* 2 %) to be achieved. In § 6 an example of the use of the method with s.a.m.s. data is presented.

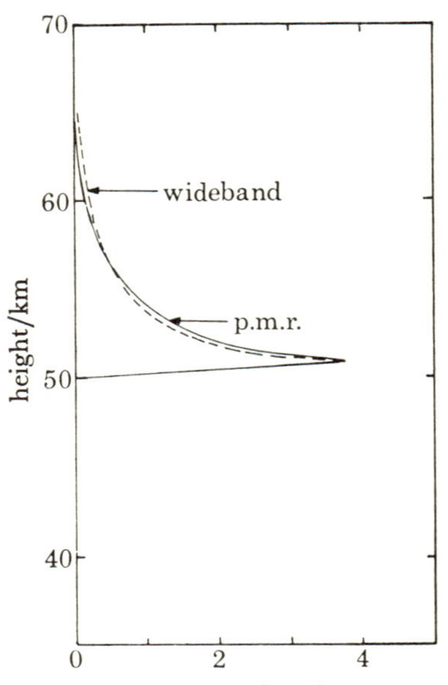

FIGURE 4. A typical pair of CO_2 15 μm weighting functions for the s.a.m.s. attitude-determining channels before the finite field-of-view is accounted for. The tangent height of the line-of-sight is 50 km.

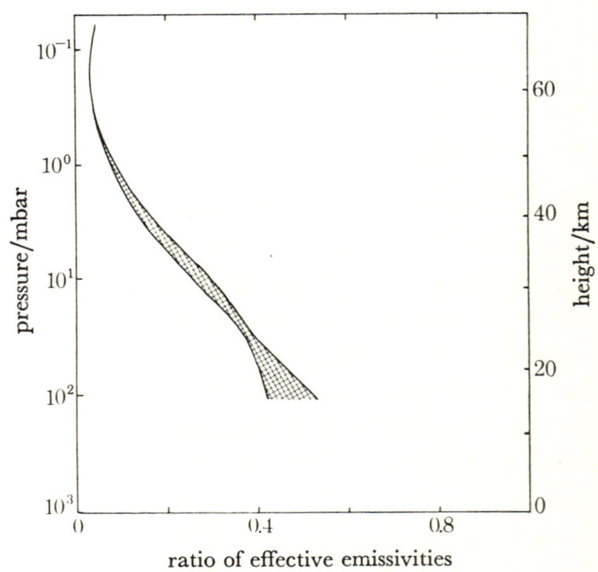

FIGURE 5. Ratio of p.m.r. to wideband emissivities for the 15 μm CO_2 channels C1 at 15 mbar mean cell pressures. The envelope shown is the extreme of a series of curves derived using climatological temperature profiles at 10° latitude intervals from Newell (1977). The effect of the finite instrument field-of-view has also been included in the calculation.

3.4. *Wind determination*

Through the incorporation in the instrument of cells containing the gases whose emission is being detected, the possibility exists of measuring atmospheric motion. If the emitting gas in the atmosphere possesses a significant velocity along the line of sight relative to the gas in the absorbing cells, the emission lines will be shifted relative to the absorbing lines. Figure 6 shows the variation of signal with azimuth angle for a particular case. The direction along which the signal is a maximum is that along which there is zero Doppler shift and hence zero relative velocity of the atmosphere with respect to the spacecraft. If the position of this maximum is determined, knowing the spacecraft velocity vector, and making allowance for the rotation of the Earth, the wind speed along the line of sight can be found. For directions nearly perpendicular to the spacecraft's velocity vector, a change in azimuth direction of 0.1° introduces *ca.* 10 ms^{-1} of the spacecraft's velocity along the line of sight.

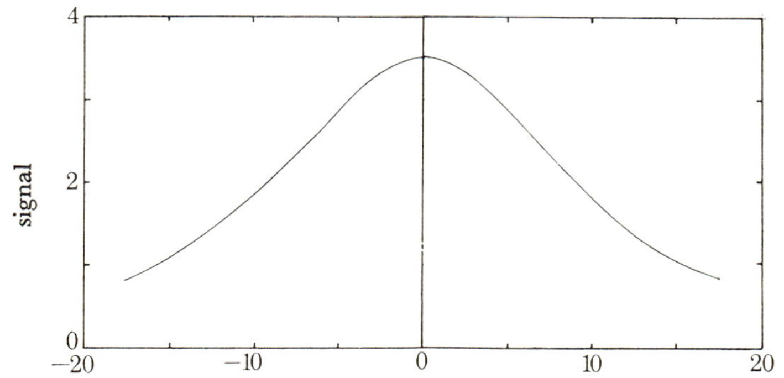

FIGURE 6. Variation of the signal from the 15 μm CO_2 p.m.r. channel with azimuth scan angle relative to the central point where there is no relative motion along the line-of-sight between the atmosphere and the instrument.

So that the influence of atmospheric variations is a minimum, the azimuth scan is carried out at such a rate that during the scan the same part of the atmosphere is kept under observation. A scan over 30° (figure 6) then takes about 240 s. The main problem in carrying out the measurement is that of the maintenance of adequate attitude stability during the period of scan so that the slope of the curve of signal against scan angle is not disturbed by attitude changes. With the Nimbus 7 spacecraft the attitude stability is not adequate for accurate wind determination, although it is hoped that the stability will sometimes be sufficient for the lower mesospheric jet at *ca.* 60 km altitude to be detected.

TABLE 1. S.A.M.S. RADIOMETRIC CHANNELS

field of view	channel	constituent (gas in modulator)	mean cell pressures / mbar	spectral band / μm	derived quantities and altitude range
A	A1	CO_2	17, 2.4, 0.69, 0.25	15	kinetic temperature 15–80 km; attitude
	A2				ν_2 vibrational temperature 80–100 km
	A3	CO	14.8, 4.5	4–5	
	A4	NO	45, 20		distribution 15–60 km
B	B1	H_2O	16, 4.5, 0.8, 0.5	2.7	distribution 80–100 km
	B2			25–100	distribution 15–100 km
C	C1	CO_2	36, 11.2, 3.25, 0.87	15	kinetic temperature 15–80 km; attitude
	C2	N_2O	24.4, 7.15	7.7	distribution 15–60 km
	C3	CH_4	47.8, 22.5		

3.5. *Instrument performance*

S.a.m.s. contains a number of channels for various measurements. Some of these channels may be operated simultaneously and some are mutually exclusive. A list of all channels is shown in table 1. Each entry in the constituent column corresponds to a modulator and each entry in the spectral band column to a detector. Since it is only possible to pass one pair of signals (p.m.r. and wideband) through a detector at any one time only one modulator signal using a particular detector may be processed at any time. Therefore, only one of the gases N_2O and CH_4 may be measured at a particular time and only one of CO, NO and 4.3 μm CO_2. A detailed description of how the channels are fitted together in the instrument is given in § 4.

TABLE 2. INSTRUMENT PERFORMANCE FOR TEMPERATURE AND COMPOSITION MEASUREMENTS FROM PRESSURE MODULATOR CHANNELS

gas	CO_2	NO	N_2O/CH_4	CO_2	CO_2	H_2O
band/μm	4.3	5.3	7.7	15	15	25–100
altitude of observation/km	120	40	40	50	100	50
approximate integration time for n.e.e. = 0.03† as measured by s.a.m.s. on Nimbus 7 /s	10000	400	10	4‡	2000	5

† n.e.e. = noise equivalent emissivity. An n.e.e. = 0.03 implies a S/N of 33 when instrument observing black body at atmospheric temperature.
‡ n.e.e. = 0.01 as this is a temperature sounding channel required to measure to ±1 K.

It can be seen from figure 2 that over a large part of the sensitive region for composition sounding a 10 % change in composition produces a change of about 3 % in the observed signal. In table 2 some typical integration times required to detect such a change are tabulated for various channels in the instrument at typical heights in the atmosphere.

4. THE S.A.M.S. INSTRUMENT

The s.a.m.s. radiometer instrument consists of two modules: the sensor housing containing the optics, mechanisms and detectors, and a separate electronics module. The overall size of the sensor (see figure 8) is 55 cm × 30 cm × 55 cm and it weighs 23.6 kg. The electronics module weighs 6.7 kg. The average power consumption is about 20 W.

There are a total of twelve radiometric channels: six pressure modulator and six wideband channels distributed between three fields-of-view denoted A, B and C. The relation between the three fields is fixed by the geometry of the instrument optics and is illustrated in figure 7. Some of the channels may be used to sense one of several constituents selected by programmable control logic within the instrument (see § 4.7). Figure 7 also shows the scanning range of the instrument for limb views and space calibration.

4.1. *Optical system*

The optical system may be conveniently subdivided into two sections:

(*a*) the front optics comprising scan mirror, telescope, calibration system and wideband chopper;

(*b*) the rear optics comprising field separator mirror, pressure modulator cells, detectors and optical relay components. The front optics is shown in figure 9. Incident radiation is first reflected from the scanning mirror (M1), which can rotate about two axes for both limb (vertical) and azimuth (horizontal) scanning (see § 4.5). It is then reflected from the main paraboloid (M2) to the primary focus forming a telescope of 177 cm^2 area and 22 cm focal length.

During in-flight calibration a small black body is rotated into the beam at the primary focus to give a known input radiance to the rest of the optical system.

After the primary focus the beam is directed by ellipsoid (M3) and plane (M4) mirrors through the fast, shallow chopper (see § 4.4) to the secondary focus at the field separation mirror

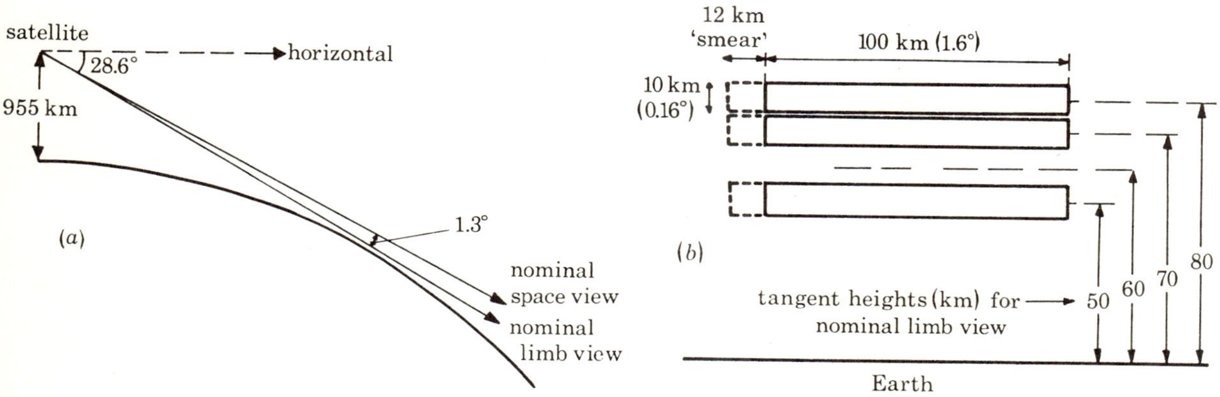

FIGURE 7. (*a*) The geometry of the s.a.m.s. viewing direction. Note that only very small angular moments are necessary to scan the entire atmosphere. As drawn the satellite velocity is into the paper. (*b*) The relative positions of the three fields-of-view A, B and C viewed along the direction of the line of sight in the 'nominal limbview' position. The horizontal smear is caused by spacecraft motion during a 2 s measurement period.

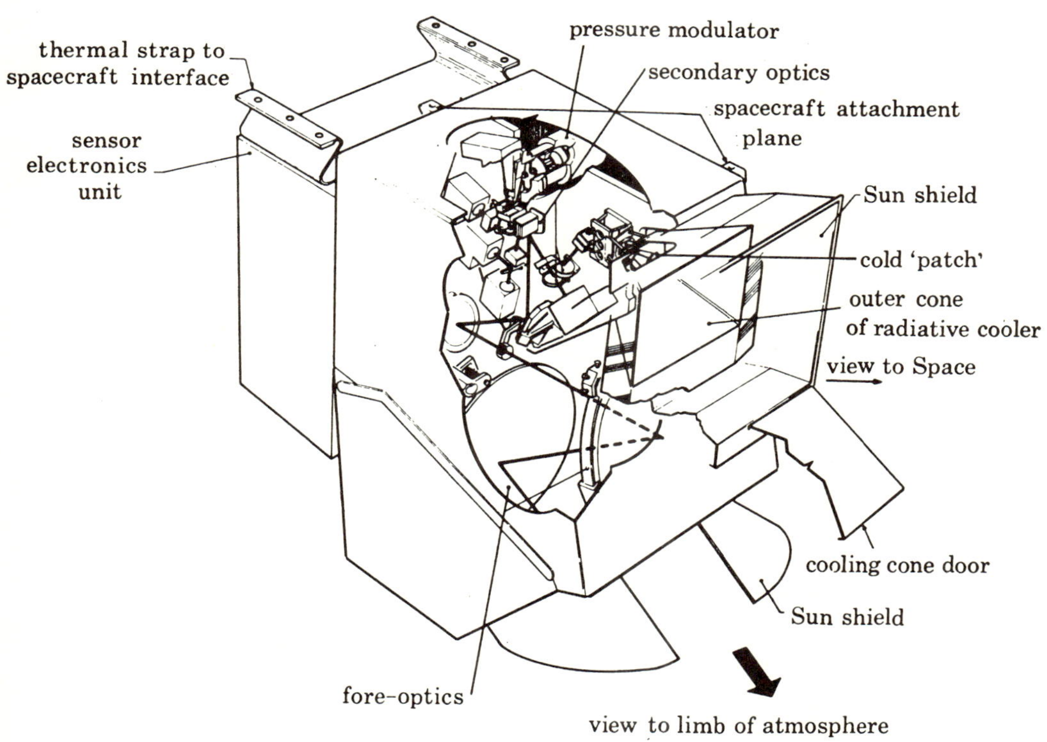

FIGURE 8. Cutaway diagram of the s.a.m.s. sensor unit showing the relative orientation of the optical components, pressure modulators and radiative cooler. The whole unit is located on the sensor ring below the satellite.

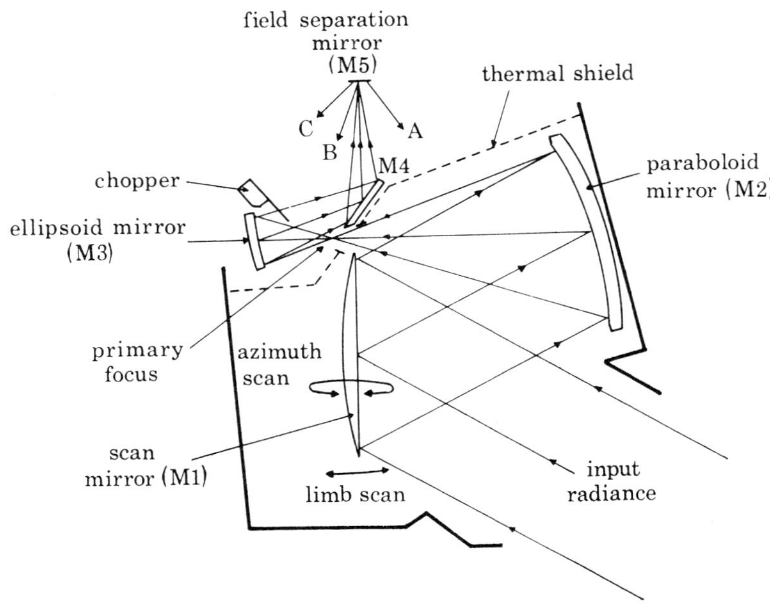

FIGURE 9. The fore-optics of s.a.m.s. in plan view. The calibration black body (not shown) is inserted into the beam at the primary focus. The thermal shield isolates the secondary optics compartment at 25 °C from the front optics compartment at 0 °C.

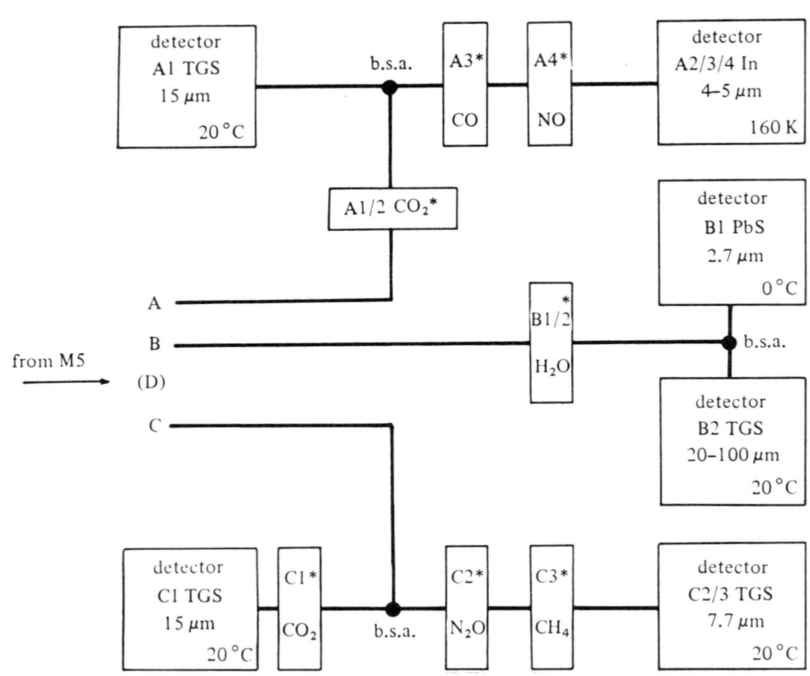

FIGURE 10. Schematic of the secondary optics of s.a.m.s. after the field-separation mirror (M5). The dichroic beamsplitters are marked b.s.A, b.s.B and b.s.C. Secondary mirrors relay lenses and other optical components have been omitted for clarity. The beam marked (D) corresponds to a facet of M5 that reflects the input beam back through the front optics and is used for optical alignment. The pressure modulator cells are identified by an asterisk (*).

(M5). The use of an ellipsoid in the M3 position cancels some of the aberrations of the off-axis paraboloid (M2).

Thermal isolation of the compartment containing the first two mirrors, i.e. those before the primary focus and calibration point, allows these mirrors to be run at a somewhat lower temperature (0 °C) than the rest of the instrument, thus reducing the thermal strays from this area and increasing the reliability of the calibration.

The rear optics are shown in the correct spatial orientation in figure 8 and in schematic form in figure 10. Field separation occurs at three of the four facets of the mirror M5 which are angled to direct parts of the beam into one of three separate optical chains, and curved to image M3 on the detector aperture stop. Each of the fields so defined is rectangular and subtends an angle in the instrument line-of-sight $0.16° \times 1.6°$, a solid angle of 7.8×10^{-5} sr (figure 7).

Broad spectral selection for the different molecular bands is achieved by dichroic beam-splitters (b.s.A, b.s.B, b.s.C) and filter packs of up to three separate components on the various detectors. The pressure modulator windows are either anti-reflexion coated or are made of low refractive index materials. In all there are about 40 individual optical components in the system after M5. High optical transmission in the system is maintained by the use of anti-reflexion coatings. As an example, the overall transmission of the C2/3 chain with 19 surfaces is still 67 % at band centre.

The rear optics, modulators and detectors are mechanically and, except for the 2.7 μm PbS and 4–5 μm InSb detectors (B1 and A2/3/4), thermally attached to the main optics plate made of aluminium 1 cm thick giving good thermal uniformity and capacity as well as mechanical stability.

4.2. *Detectors*

There are six detectors in s.a.m.s., four triglycine sulphate (TGS), one lead sulphide (PbS) and one indium antimonide (InSb). Each detector flake is imaged by its optical system on to one of the facets of M5 and, therefore, the aspect ratio of the detectors is the aspect ratio of the field of view. Each detector is 3.12 mm × 0.312 mm, giving an area of approximately 1 mm^2.

The detectors are each assembled and tested as sub-assemblies consisting of detector, pre-amplifier, condensing optical system and filters before being integrated with the rest of the instrument. A TGS sub-assembly is shown in figure 11. The condensing optical doublet, consisting of a meniscus and a plano-convex lens (see Murray 1962), is made of anti-reflexion coated germanium (except in the case of the 25–100 μm H_2O channel (B2) where silicon is used). The package consisting of the meniscus lens detector and mounting flange is evacuated to protect the detectors from moisture during ground operations.

The sub-assemblies for the cooled detectors are more complicated than those for the TGS because of the need to isolate the detector thermally while mechanically attaching it to the optics plate. The B1 and A2/3/4 detectors are cooled radiatively, the B1 to 0 °C by a radiator in the fore optics compartment and the A2/3/4 to 160 K by a separate radiative cooler (see § 4.6).

4.3. *Pressure modulator assemblies*

The pressure modulators used in s.a.m.s. are a development of those used in the Nimbus 6 p.m.r. instrument (see Curtis *et al.* 1974). The pressure of the gas in the optical cell is altered by means of a piston oscillating at between 25 and 40 Hz in a connecting volume (figure 12). A pressure variation over the cycle of the order 2:1 is achieved in this design, the exact value depending on the dimensions of the particular optical cell and the mean cell pressure. The principal

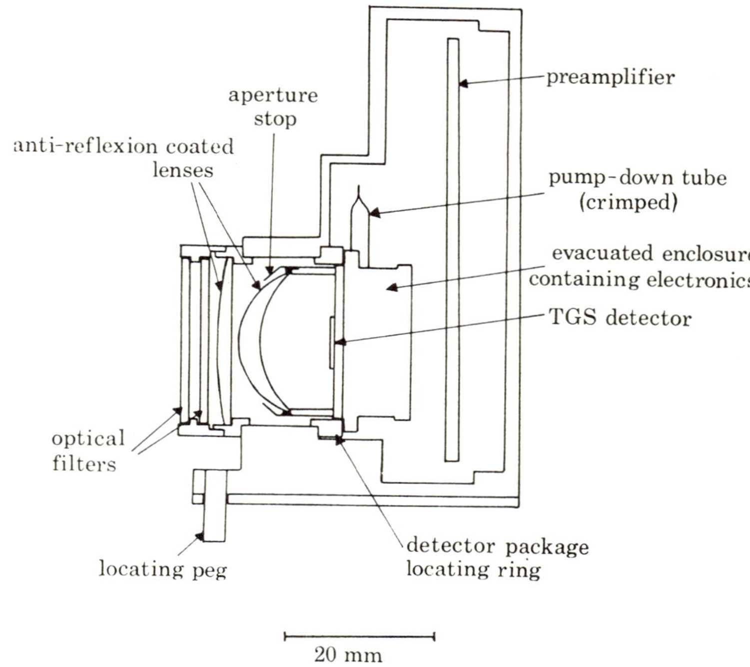

FIGURE 11. S.a.m.s. detector sub-assembly (TGS). Because the impedance of a TGS detector is very high and the signal small, part of the preamplifier is built into the back of the detector package within the evacuated enclosure. Precise optical alignment is essential, both of the package to the sub-assembly and of the sub-assembly to the secondary optics mounting plate. In the former case a close-fitting ring is used and in the latter a precision locating peg.

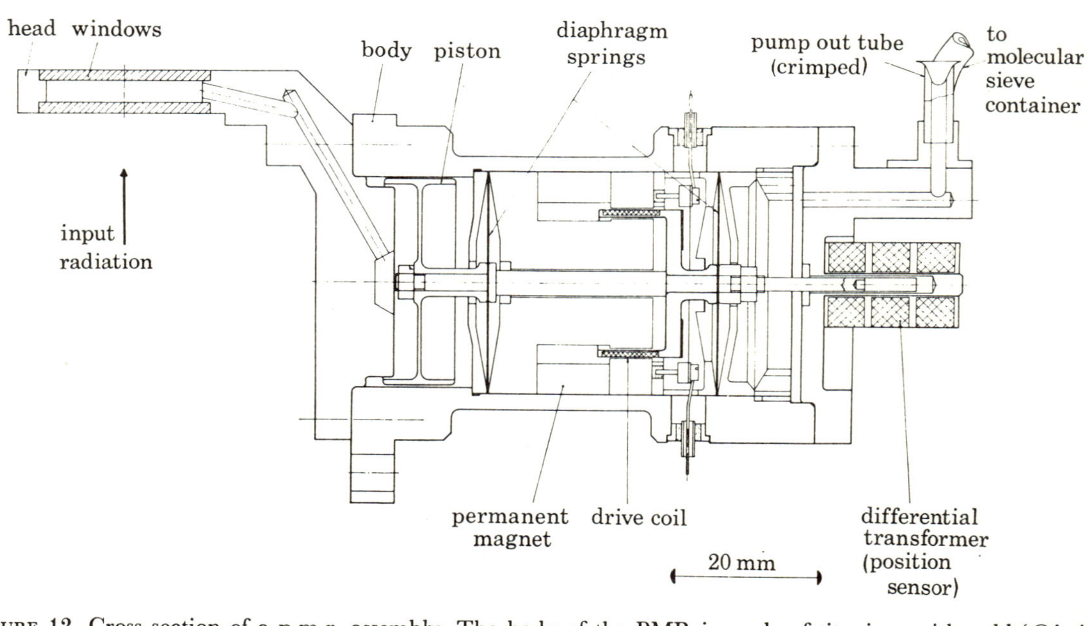

FIGURE 12. Cross section of a p.m.r. assembly. The body of the PMR is made of titanium with gold 'O' ring seals at each end. The support springs and coil lead-outs are made of beryllium–copper. The thermostatted container of molecular sieve is not shown.

changes from the devices used in the Nimbus 6 instrument are a reduction of size by a factor of approximately 2 and the use of a moving coil drive leading to better amplitude control and electrical efficiency. Other improvements have made the modulators easier to manufacture and assemble. It is necessary, as was observed in § 3.2, to alter the pressure of gas in the cell in flight, the pressure selected depending on the altitude range being scanned. This is achieved in the s.a.m.s. modulators as follows. The gas in the cell is in equilibrium with a much larger amount adsorbed on a zeolite molecular sieve material in a reservoir (figure 12). In this system the mean pressure of gas is determined by the temperature of the sieve, and provision has therefore been made for programmable temperature control of the sieve container. Each modulator is provided with two or four pressure settings, which may encompass a range of over 20:1, corresponding to a thermostat temperature range of 30–120 °C. These settings are selected by the program control logic. The adsorption and desorption of gas from the sieve is entirely reversible and the system provides good pressure stabilization. Measurement of the oscillation frequency, which changes rapidly with pressure, is used as the primary method of determining cell pressure since slow variations in gas amount, such as might be caused by a leak in a modulator, could not be detected by monitoring the sieve temperature.

4.4. *Fast chopper assembly*

The fast chopper used on s.a.m.s. only modulates 2.5 % of the beam for reasons described above (see § 2.3). It consists of two etched, black-painted copper grids, mounted one above the other. One grid is fixed to the mount but the other is free to move on the end of a flat cantilever leaf spring. Oscillation, at the mechanical resonant frequency of the spring–grid assembly, is maintained by an electronic loop using redundant pairs of piezoelectric ceramic plates as both drive and position sensors. Oscillation is at a frequency of approximately 245 Hz with a peak-to-peak amplitude of 2 mm.

4.5. *Scan mirror assembly*

The scan mirror must be rotated with high positional accuracy in two mutually perpendicular directions to produce limb and azimuth scanning. This is realized by supporting the mirror in two frames, one within the other, for the two scan directions. Each mirror drive consists of a lead-screw directly driven by a 45° stepping motor. Coupling to the mirror is by means of a recirculating-ball nut, which is lubricated by a lead coating. The axis of rotation of the mirror is defined by a pair of cross-leaf springs (flexipivots). The step sizes of the drive system, expressed in terms of motion of the line-of-sight, are 4.8 arc min in the limb direction and 7.3 arc min in azimuth.

Mirror position is independently sensed in both directions by a linear variable differential transformer (l.v.d.t.), actuated directly by the mirror, and associated electronics. The output of the limb l.v.d.t. is digitized with a resolution of 0.045 arc min and the azimuth l.v.d.t. to 1.8 arc min. It is therefore possible to monitor the limb position of the mirror to a very much higher accuracy than the step size. The limb position is under full control of the program control logic (p.c.l.) (see § 4.7), whereas the azimuth scan can only be programmed as a continuous scan or stepped.

Limit detectors, both optical and mechanical, are provided for resynchronizing the control logic and preventing damage under fault conditions.

4.6. *Radiation cooler*

The indium antimonide detector is cooled by a two stage radiation cooler, the first high-capacity stage forming a cone around the second stage as shown in figure 13. The first stage cools the front lens of the optical doublet (see § 4.2) as well as pre-cooling the rear lens and detector. In orbit, the inner and outer cooler temperatures are typically 160 K and 190 K.

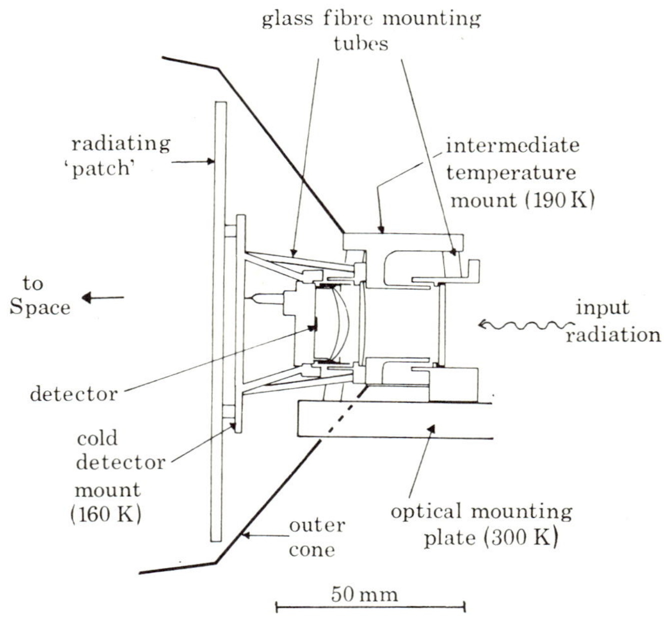

FIGURE 13. The s.a.m.s. cold detector assembly (InSb). The mount is in three parts: the detector mount at 160 K, the intermediate lens mount at 190 K and the mechanical interface at 298 K. The parts are inter-connected by 8 glass fibre tubes, some of which may be seen in the diagram. The relative sizes of the 'radiating patch', which is black-painted, and the outer cone, which is coated with vapour deposited aluminium, can be seen in figure 8.

The cooler has been designed to minimize the effects of contamination on the optics as the spacecraft outgases; however, provision has been made to heat the detector and optics electrically to remove the debris as necessary. The cooler surfaces were protected during launch and the first 3 weeks of orbit by a door which was opened on 13 November 1978. The door cannot be closed again and will remain open for the duration of the mission.

4.7. *Electronics*

The electronics of s.a.m.s. is considerably more complicated than that of previous instruments built at Oxford for three reasons:

(*a*) there are more signal channels (twelve in all) and modulator drives (seven);

(*b*) the increased number of signal channels and drives means that more care must be taken to ensure that coupling and cross-talk, both electrical and mechanical, is minimized.

(*c*) The scan and calibration control system is considerably more flexible than previous instruments allowing in-flight programming.

The major changes in construction details since previous instruments are the use of integrated circuit amplifiers in the linear electronics and CMOS in the digital electronics. A full block diagram of the system is shown in figure 14.

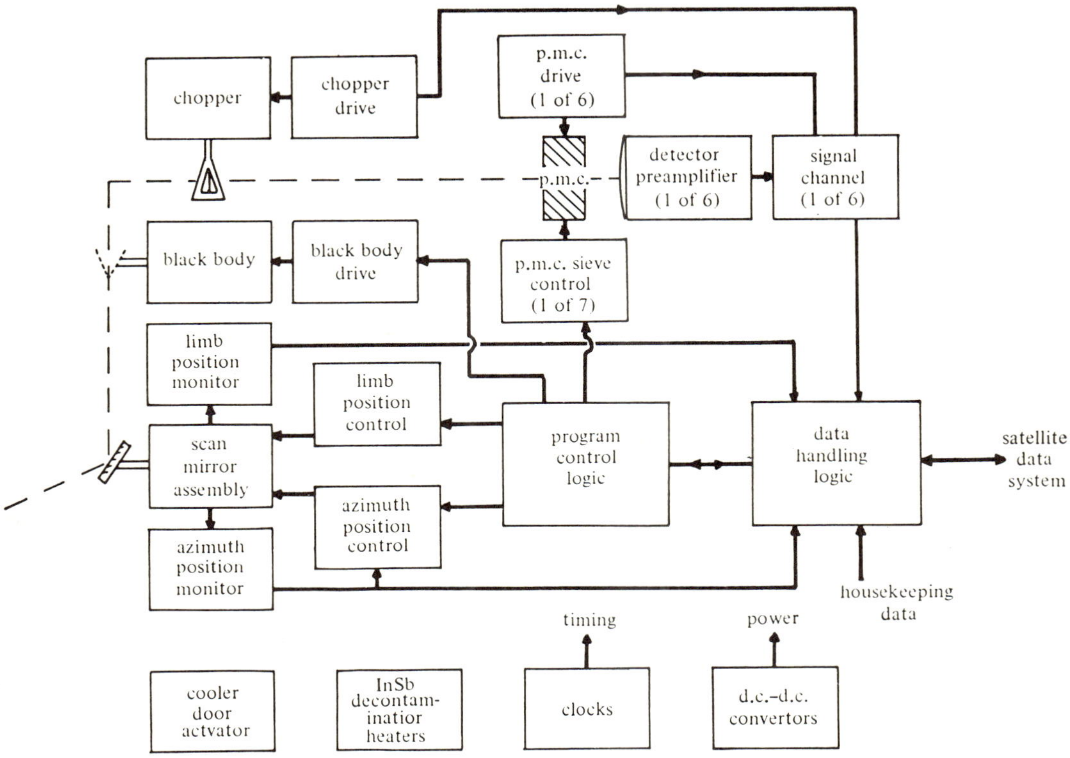

FIGURE 14. Block diagram of the s.a.m.s. electronics subsystems.

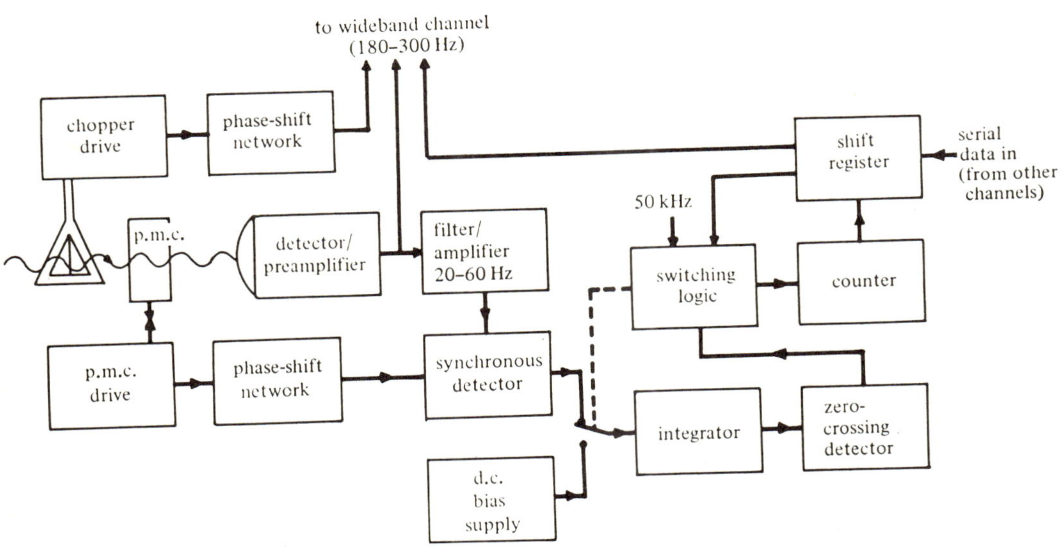

FIGURE 15. Block diagram of one half of a signal channel. Data from all channels is transferred to the data handling system in a single serial stream.

(*a*) *Signal channels*

The signal processing electronics follows the pattern of the system used for a previous balloon radiometer (see Chaloner *et al.* 1978). The high and low frequencies (wideband and pressure modulated signals respectively) present at the detector output are processed by separate channels as shown in the block diagram in figure 15. The output integrator of each channel is also made to act as a dual-slope analogue-to-digital converter. This is achieved by integrating the signal, plus a small offset to ensure that the result is positive, for 1.8 s, and then integrating a fixed negative signal until the result is zero. By timing the second integration period with the use of the 50 kHz spacecraft clock, a resolution of 1 part/10^4 is obtained in the remaining 200 ms of the 2 s period. Only 12 bits are actually telemetred to the ground but the 13th can be inferred, if necessary, by context. The 200 ms period while the signals are evaluated is also used to step the scan mirror.

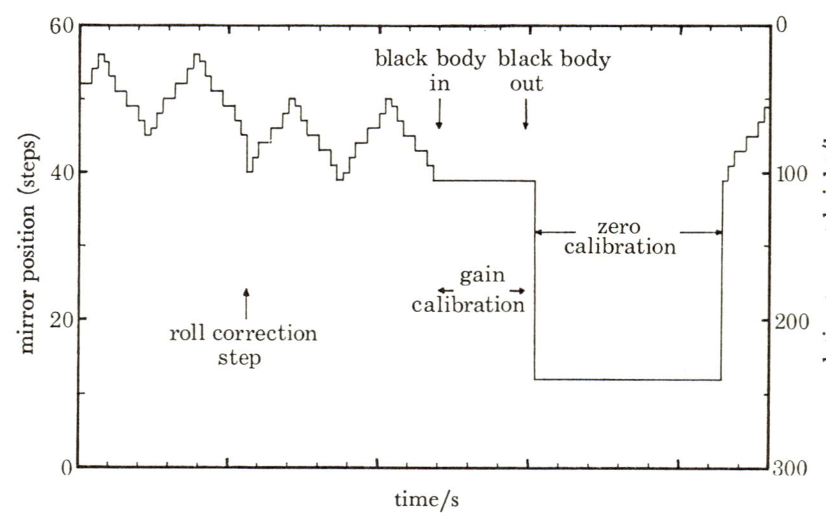

FIGURE 16. A typical scan program. There are 18 scans (up and down ramps) between calibration sequences leading to an interval between calibrations which is not synchronous with the orbital period. Roll correction is applied independently, synchronous with the orbital period.

(*b*) *Pressure modulator drives*

The drive circuits for the pressure modulators are considerably more complex than those used for the Nimbus 6 p.m.r. or the balloon instrument. The complexity is necessary because of the need to maintain a constant amplitude despite large internal pressure variations, which in turn lead to large changes in frequency and power requirements. The use of a moving coil drive, with its inherently tighter coupling than the previous moving magnet drive, assists with amplitude control and also allows active damping to be employed when a modulator is inhibited. Active damping is useful in multiple channels with only one detector (e.g. A2/3/4) since unwanted mechanical energy could otherwise be coupled into the 'off' modulators and appear as unwanted modulation in the detector output. This can be prevented by switching the drive from positive (oscillating) to negative (damping) feedback when the modulator is inhibited.

(*c*) *Program control logic* (*p.c.l.*)

The extremely large number of possible operating modes for s.a.m.s. using various combinations of modulators, scanning ranges and mean cell pressures, and the need to correct the scanning sequences for any predictable orbital roll of the spacecraft, made a programmable operating mode very desirable. The system used operates the scan mirror, modulators, molecular sieves, thermostats and calibration black body by reference to a stored 'program' in a 128 × 10 bit memory which may be reloaded from the ground. The instruction set of the p.c.l. also includes jump, wait, synchronizing and timing instructions to allow program loops to be constructed which operate in a fixed relationship to the spacecraft data system. The variations of scanning and calibration are, therefore, infinite and it is possible to adapt the scanning sequence to partially compensate for orbital roll. It is also possible to scan outside the atmosphere, for example, to locate the moon for calibration purposes.

A typical scan sequence is shown in figure 16. This program is designed to scan a height range of 55 km in 10 km steps with regular calibration sequences not related to the orbital period. A roll correction step can also be seen.

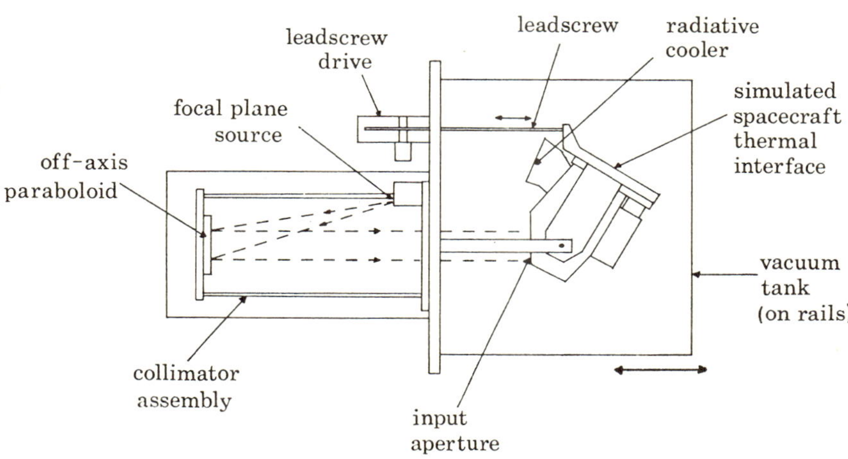

FIGURE 17. S.a.m.s. environmental test chamber at Oxford. Baffles, cooled by liquid nitrogen, surround the instrument to simulate radiative loss to space but these have been omitted for clarity. Access is provided by rolling the main vacuum tank away on rails. The collimator assembly is shown in position. The black body assembly (see text) may be substituted for the collimator when required.

5. VERIFICATION AND CALIBRATION PERFORMANCE

In addition to the tests that were necessary to prove the design of the instrument and the environmental tests specified by N.A.S.A., s.a.m.s. has undergone the following series of measurements intended to facilitate the interpretation of data when in orbit.

(*a*) The response of the instrument to known black body radiance inputs has been measured, and a calibration scheme devised to take account of variations in operating conditions.

(*b*) The response of the instrument to emission from gases in the atmosphere has been determined, using a simulated atmospheric path, in order to verify the spectroscopic calculations by which atmospheric quantities are deduced.

(*c*) The field of view profiles of all the channels and their interrelation has been measured.

(*d*) The scan mirror position telemetry (see § 4.5) has been accurately calibrated in terms of the instrument line-of-sight. A combined vacuum and thermal environment chamber at Oxford provides correct simulation of the spacecraft interface and orbital conditions. Various additional radiation sources are also available within the chamber for calibration purposes (see figure 17).

5.1. *Radiometric calibration*

The response of the instrument is determined for a range of accurately known black body radiances in order to

(*a*) verify that the instrument response is linear and has the correct gain and noise levels;

(*b*) check that the spectral selection of the optics is correct and that there are no unknown spectral 'leaks' in the overall response;

(*c*) provide a basis for the model of the instrument used in the in-flight calibration scheme;

(*d*) determine the magnitude of any stray responses, particularly those which depend on either azimuth or limb scan angle. The source used for this test is a large, baffled black-body cone, completely filling the instrument aperture, whose temperature may be held constant and uniform at any temperature between 77 K and 320 K. The input radiance is, therefore, very precisely calculable. There are in fact two sources, one on the instrument axis viewed with the azimuth scan centralized and one off-axis viewed at the limit of the azimuth scan. By using both targets, the variations of strays with azimuth scan angle may be established.

Variations with limb-scan angle are measured by tilting the whole instrument using a precision lead-screw, thus varying the relative position of the on-axis target and the instrument.

The sensor temperature is held constant during measurements of stray radiation but may be varied or cycled over a considerable range (-5 to $+40$ °C), in order to dynamically evaluate the temperature coefficients of the response. By simulating the orbital temperature variations of the spacecraft experience is gained of the likely instrument temperature variations in orbit – a verification of the sensor thermal design.

5.2. *Response of p.m. channels to atmospheric paths*

The response of the channels to paths of gas is checked by measurment of the transmission of simulated atmospheric paths in the laboratory, using either the actual sensor modulators, filters and detectors before final assembly of devices identical in design and construction. The apparatus consists of a multiple-pass absorption white cell which provides paths of up to 10 m, which may be used at temperatures down to 240 K (see Houghton & Taylor 1973). Transmissions of isothermal paths of the required absorber or absorber/nitrogen mixtures measured with this equipment as a function of pressure are then compared with the response calculated from line data, and information from both sources is combined in the transmission model used in the final attitude, temperature and composition retrievals.

5.3. *Field of view*

The attitude of the sensor line of sight is determined in orbit from the ratio of the pressure modulator and wideband signals in the CO_2 channels (see § 3.3). The line-of-sight of the other channels, therefore, needs to be measured with respect to these. The details of the field profiles, especially the shape of the lower edge, also critically affect the weighting function of the channels and must also be determined.

For field of view determination the black cones are replaced by a collimator (see figure 17). This consists of an off-axis paraboloid mirror, with a narrow slit illuminated by a hot ceramic filament in the focal plane. The lead screw system is used to rotate the s.a.m.s. sensor through the parallel beam of radiation from the collimator at a precisely known rate, and the field response is thus determined.

The collimator assembly and leadscrew are also used for various optical checks and in particular for the calibration of the l.v.d.t. system used for limb scan angle telemetry. The lead screw is driven slowly and continuously while the p.c.l. is programmed to step the s.a.m.s. scan mirror so that the edge of the field of view of a selected channel passes through the collimated beam once per step. By matching the respective points on the field response, the l.v.d.t. system is calibrated directly in terms of the line-of-sight. This test and the primary field of view determination are carried out at several temperatures over the desired operating range.

5.4. *In-flight calibration*

A two point calibration is provided in flight as follows.

(*a*) The scan mirror may be programmed to move so that the tangent height viewed in the atmosphere is sufficiently high that the atmospheric radiance is insignificant ('space' view). This is arranged in operational programs such that the minimum tangent height for the lowest field of view (the C channels) is at least 150 km.

(*b*) The black body (see § 4.1) may be placed in the optical path to provide a standard radiance at approximately 295 K. This is also under the control of the p.c.l.

In practice the estimate of the space view signal is much more important than the measurement of gain, as many of the measurements made are of small differences from the zero radiance level. Operational p.c.l. programs are therefore arranged to have a higher proportion of space view calibration than of black body view as shown in figure 16.

In flight the zero offset and radiometric gain are calculated from instrument housekeeping data by means of a model, the parameters of which have been obtained by combining the results of ground tests and in-flight calibration measurements.

6. Preliminary results

Calibrated radiance data from s.a.m.s. is processed in the following sequence. A preliminary estimate is obtained of the tangent pressure of the line-of-sight of all three fields-of-view at all times using the method described in § 3.3. Then the atmospheric temperature profile and a better estimate of the tangent pressure are derived from the preliminary estimate of tangent pressure and the CO_2 channel radiances. Finally, this information is used, together with radiances from the other channels, to retrieve constituent profiles. These three steps are described in some detail below.

6.1. *Attitude determination*

As has already been shown in § 3.3 the tangent pressure of the line-of-sight of the CO_2 channels can be determined from the ratio of the wideband to p.m.r. radiances assuming that the tangent height is within a suitable range (approximately 20–50 km). Since the space-craft roll rate is reasonably low and the orbital period long compared to the rate of sounding (1 every 2 s), smoothing along the orbit path, i.e. in time, is applied using a linear sequential estimator (Kalman–Bucy filtering, see Rodgers 1976). To achieve this the tangent pressure is converted, using a climatological temperature profile and the scan mirror position, to the tangent height of a reference scan mirror position. This tangent height estimate is then smoothed.

The estimator is stepped forwards along the orbit and for each new observation a smoothed estimate of the tangent height and its uncertainty are calculated given the observation and its error, the previous smoothed estimate and its error, the variation in scan mirror position measured by the l.v.d.t., the time between observations and a knowledge of the likely maximum roll rate. The process is repeated backwards along the orbit and the two results at each observation point are combined to derive a best estimate of the roll angle at any point. The smoothed roll-angle is then reconverted to tangent pressure for each field of view. The results for a typical orbit are shown in figure 18.

The smoothing process is capable of dealing automatically with missing data caused either by bad soundings or a tangent height out of the range where attitude determination is possible. Prolonged loss of information produces a steadily increasing uncertainty in the attitude.

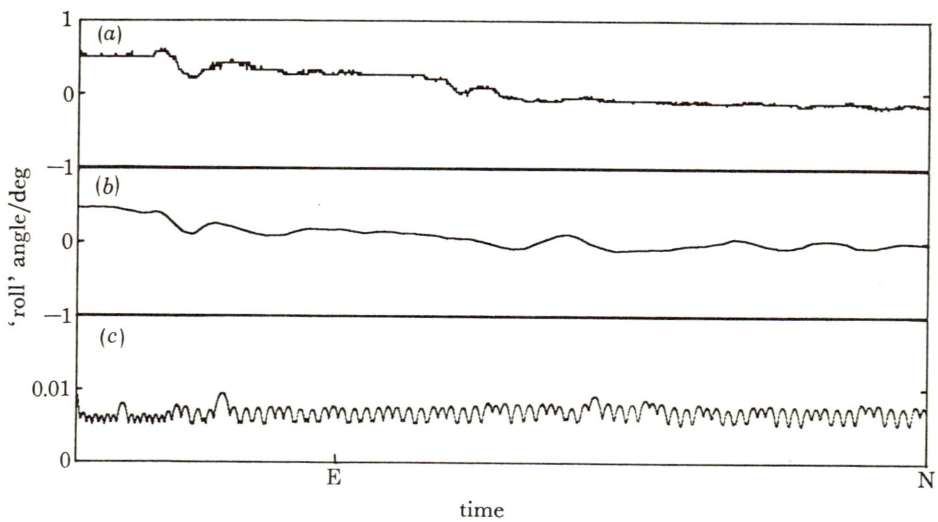

FIGURE 18. A comparison of the 'roll' angle derived as described in the text with the output of the spacecraft attitude control system (for part of orbit 135 of Nimbus 7). For reasons described in the text these are not necessarily comparable but short timescale features should be reproduced. The error in the derived roll error is in a scale magnified 100 × relative to the other two plots. E = equator, N = most northerly latitude of orbit. (*a*) Spacecraft attitude control system roll output; (*b*) s.a.m.s. derived 'roll'; (*c*) s.a.m.s. derived 'roll' error.

6.2. *Temperature retrieval*

The short wave channels are particularly sensitive to atmospheric temperature (5 % per degree for the 5 μm channel) so accurate temperature retrievals are of particular importance for making good constituent measurements. It is also possible to improve further upon the attitude determination described above which assumed climatological temperature profiles. Temperature profiles and an improved attitude are, therefore, retrieved simultaneously.

Since there are a large number of possible scan patterns in use, the retrieval method must be able to handle any pattern selected and must, therefore, be very flexible. Were the instrument quickly scanning up and down through the whole atmosphere, complete sets of radiances for a scan could be assembled and the retrieval performed by simpler methods. But there is a need to incorporate observations when and where they occur, essentially at random, and a sequential estimator is again used. For economy of computation the retrieval solution is represented by coefficients of 12 eigenfunctions, giving the deviation of source function at 15 μm from the climatological mean (interpolated linearly from monthly means at 10° latitude intervals). They are eigenfunctions of the covariance matrix

$$C_{ij} = \exp\left(-(z_i - z_j)^2/\alpha^2\right) \frac{\mathrm{d}B(z_i)}{\mathrm{d}T} \frac{\mathrm{d}B(z_i)}{\mathrm{d}T},$$

where z_i is the scale height of level i and $\mathrm{d}B(z_i)/\mathrm{d}T$ is the differential with respect to temperature at level z_i. This is the Planck function covariance resulting from a set of wave perturbations in

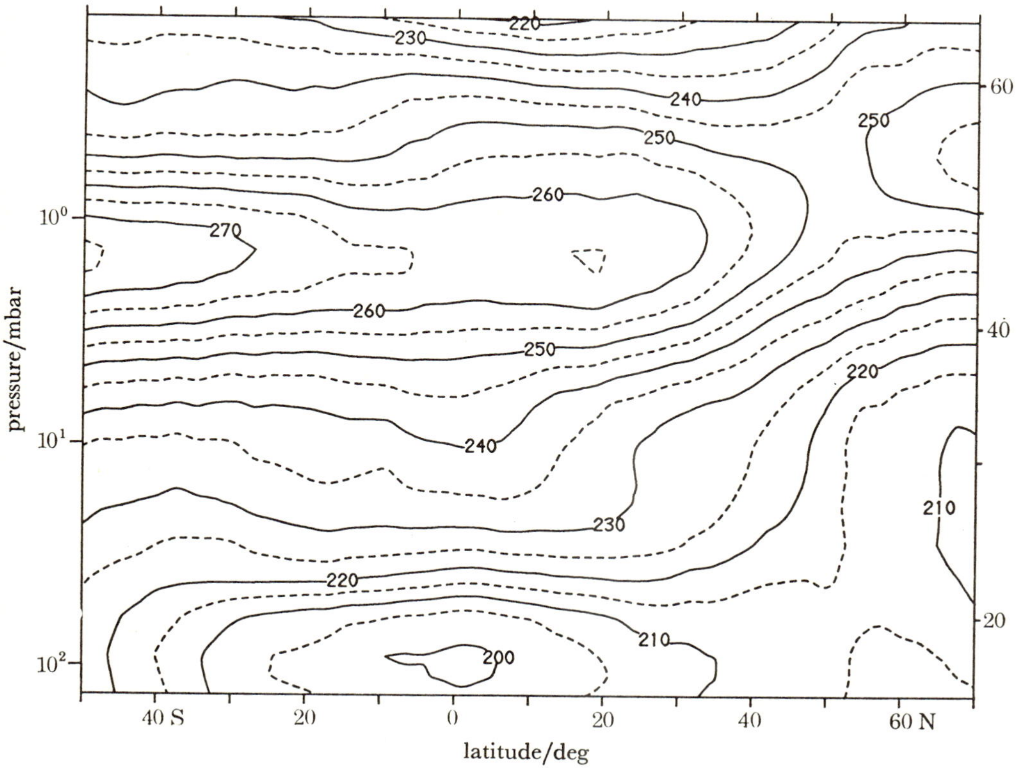

FIGURE 19. Zonal mean temperature cross section for 6 November 1978 derived entirely from Nimbus 7 s.a.m.s. data.

the vertical with a Gaussian spectrum and temperature amplitude independent of height. Using the estimate of attitude derived as above, the deviation of observed radiance from its climatological value can be represented linearly in terms of the solution which consists of the eigenfunction coefficients and the attitude. By these means a linear sequential estimator can be employed in a similar manner to the attitude determination described above. In the absence of input data the solution is made to decay exponentially with time towards zero (i.e. towards its climatological value) so that when there is no information about some or all levels, any deviations for those levels are slowly forgotten. When run forwards along the orbit the estimator is recursive and one-sided, using just past and present data to give the solution. However, an independent set of estimates is obtained by running the estimator backwards along the orbit. At $2\frac{1}{2}$° latitude intervals along the orbit both estimates are combined in accordance with their uncertainties and the result retained for derivation of temperatures and analysis on latitude–longitude grids. A typical zonal mean cross section is shown in figure 19.

6.3. *Concentration retrievals*

Using the derived tangent heights and temperatures it is now possible to obtain concentration profiles from the data. Figure 20 shows some typical results for the H_2O rotation band (channel B2) in both the p.m.r. and wideband cases.

The solid lines were calculated using the temperature data derived using the methods described above, water vapour band data from McClatchey *et al.* (1973) and Chaloner (1976) and a computer program to simulate atmospheric signals due to Eyre (1978).

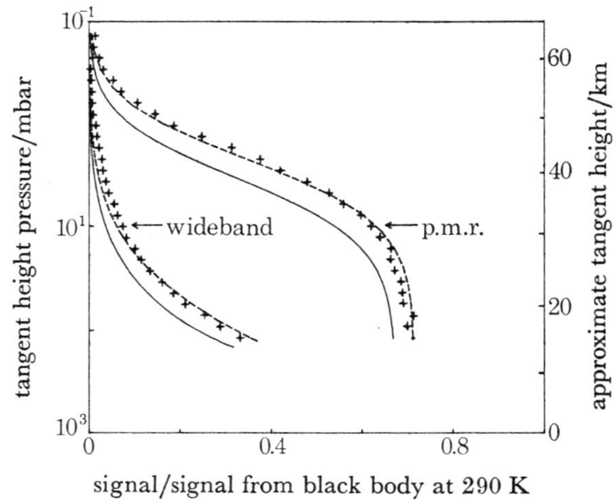

FIGURE 20. Water vapour signal from s.a.m.s. (on 6 November 1978 for latitudes between 20° N and 20° S) compared with signals calculated for constant volume mixing ratio profiles. It can be seen that the p.m.r. signals indicate a slight fall in mixing ratio from 50 to 30 km which becomes pronounced in the wideband signal between 30 and 18 km where there are uncertainties due to clouds and other effects associated with the tropopause. +, S.a.m.s. data; ---, 5 parts/10^6 by volume; —, 2 parts/10^6 by volume.

Because the data is of a preliminary nature, a profile of concentration versus altitude is not presented. However, it can be seen that the results are inconsistent with a constant mixing ratio profile but would be consistent with a slow rise starting at 20 km levelling off above 35 km. This type of behaviour would be expected if the photolytic destruction of CH_4 were a source of H_2O in the middle stratosphere (McConnell *et al.* 1971).

The project has been supported by the Science Research Council. We particularly acknowledge the invaluable work of the Project Manager, Mr K. H. Davies. Technical support at Oxford was given by Dr P. Curtis, Mr M. Clarke, Mr S. Werrett, Mr M. W. Johnson, Mr I. Colbeck and Mrs C. E. Holt. Substantial assistance in several areas was given by the Rutherford Laboratory, in particular by Mr H. Hadley, Mr J. Parker, Mr R. Wolfenden, Mr G. Stapleton, Mr T. Morgan, Mr R. Elsey and Mr P. Gear. The optical filters and windows were made by Dr J. S. Seeley and Mr R. Hunneman of Reading University. Hawker Siddeley Dynamics (now British Aerospace) at Stevenage built the flight models; we thank in particular Mr G. W. Cocks, Mr J. Bonsor, Mr C. Collins and Mr I. White. Mr W. Andiario has been the experiment representative at the General Electric Company, Valley Forge, during integration of the experiment in the spacecraft. The experiment representative with Nimbus Project, N.A.S.A., was Mr R. White. We gratefully acknowledge the generous cooperation of all members of the Nimbus Project, N.A.S.A.

REFERENCES (Drummond *et al.*)

Chaloner, C. P. 1976 D. Phil. Thesis, University of Oxford.

Chaloner, C. P., Drummond, J. R., Houghton, J. T., Jarnot, R. F. & Roscoe, H. K. 1978 *Proc. R. Soc. Lond.* A **364**, 145–159.

Curtis, P. D., Houghton, J. T., Peskett, G. D. & Rodgers, C. D. 1974 *Proc. R. Soc. Lond.* A **337**, 135–150.

Drummond, J. R. & Jarnot, R. F. 1978 *Proc. R. Soc. Lond.* A **364**, 237–254.

Ellis, P. J., Holah, G., Houghton, J. T., Jones, T. S., Peckham, G., Peskett, G. D., Pick, D. R., Rodgers, C. D., Roscoe, H. K., Sandwell, R., Smith, S. D. & Williamson, E. J. 1973 *Proc. R. Soc. Lond.* A **334**, 149–170.

Eyre, J. R. 1978 D.Phil. Thesis, University of Oxford.

Gille, J. C., Bailey, P., House, F. B., Craig, R. A. & Thomas, J. R. 1975 *The Nimbus* 6 *user's Guide*. Goddard Space Flight Center, Greenbelt, Maryland.

Houghton, J. T. & Smith, S. D. 1970 *Proc. R. Soc. Lond.* A **320**, 23–33.

Houghton, J. T. & Taylor, F. W. 1973 *Rep. Prog. Phys.* **36**, 827–919.

McClatchey, R. A., Benedict, W. S., Clough, S. A., Burch, D. E. & Calfee, R. F. 1973 *A.F.C.R.L. atmospheric absorption line parameters compilation.* AFCRL-TR-73-0096 Environmental Research.

McConnell, J. C., McElroy, M. B. & Wofsy, S. C. 1971 *Nature, Lond.* **223**, 187–188.

Murray, A. E. 1962 *Infrared Phys.* **2**, 37–47.

Rodgers, C. D. 1976 *Rev. Geophys. Space Sci.* **14**, 609–624.

Taylor, F. W., Houghton, J. T., Peskett, G. D., Rodgers, C. D. & Williamson, E. J. 1972 *Appl. Opt.* **11**, 135–141.

Phil. Trans. R. Soc. Lond. A **296**, 243–260 (1980) [243]
Printed in Great Britain

The composition of the mesosphere and lower thermosphere

By L. Thomas
Science Research Council, Appleton Laboratory, Ditton Park, Slough SL3 9JX, U.K.

The experimental information available on the neutral and ion composition of the 50–130 km height range is reviewed and, wherever possible, comparisons are made with predictions based on theoretical models.

The height distributions for various neutral constituents have been deduced from rocket-borne remote-sensing measurements and *in situ* sampling experiments, in which recent improvements in mass spectrometer techniques have been effective. The results obtained for oxygen, hydrogen, nitrogen and carbon species are described in turn. In addition, satellite-borne remote-sensing measurements have demonstrated the importance of temporal and spatial variations of molecular oxygen and nitric oxide in the lower thermosphere. *In situ* sampling with rocket-borne mass spectrometers has established that molecular oxygen and nitric oxide ions produced during or following photoionization represent the major ions down to about 85 km, with layers of metal ions also being present. Below this height, water cluster ions, $H^+.(H_2O)_n$, predominate. At lower mesospheric levels, negative ions are known to be important but the negative-ion composition has not been established.

1. Introduction

Much of the initiative for studying the ionized and neutral constituents at heights above about 50 km arose from the need to understand the formation of the ionosphere. Rocket-borne mass spectrometer observations (Narcisi 1973) showed the presence of a variety of ions, and a quantitative interpretation of these ions demanded a knowledge of the height distributions of constituents such as water vapour, atomic oxygen and nitric oxide. In the meantime, an understanding of the chemical and physical processes which govern the distributions of both neutral and ionized constituents has become a major field of study in its own right. The more general atmospheric interest in such constituents arises from their role in noctilucent clouds (Witt 1969), airglow emissions (Hunten 1967) and the thermal balance (Crutzen 1970).

The characteristics of the neutral and ionized parts of the atmosphere in the height range 50–130 km are largely determined by the absorption of solar radiations of wavelengths less than about 200 nm, except for those in the 20–80 nm range which are absorbed at greater heights. This is illustrated in figure 1, which shows the heights of unit optical depth for wavelengths up to 310 nm incident vertically, i.e. the heights at which the solar-flux intensities are reduced to e^{-1} of their free space values; these correspond to the levels of maximum absorption.

Also shown in this diagram are the wavelengths corresponding to ionization and dissociation thresholds of important constituents. The dissociation threshold of molecular nitrogen corresponds to a wavelength of 127 nm but the absorption is very weak. For ozone, photodissociation is energetically possible for wavelengths less than 1180 nm, but for the present heights of interest the spectral region below 310 nm is the most effective, producing the first electronically excited states of atomic and molecular oxygen, $O(^1D)$ and $O_2(^1\Delta_g)$. The constituents chiefly responsible for absorption in the different wavelength regions are as shown in figure 1. Absorption by molecular oxygen in the Schumann–Runge ($B^3\Sigma_u^- - X^3\Sigma_g^-$) band system between 175 nm and

205 nm is of particular interest in view of the marked variation of absorption cross section with wavelength and the pronounced changes with temperature (Ackerman *et al.* 1970). Account needs to be taken of these properties of the oxygen cross section in detailed treatments of the photodissociation of nitric oxide, carbon dioxide, and also of water in the lower mesosphere. Radiations capable of ionizing the major constituents are absorbed, in the main, above 95 km but deep windows occur between about 110 and 130 nm, one of which coincides with the strong Lyman-α line at 121.6 nm, permitting the photoionization of $O_2(^1\Delta_g)$ and NO at D-region heights.

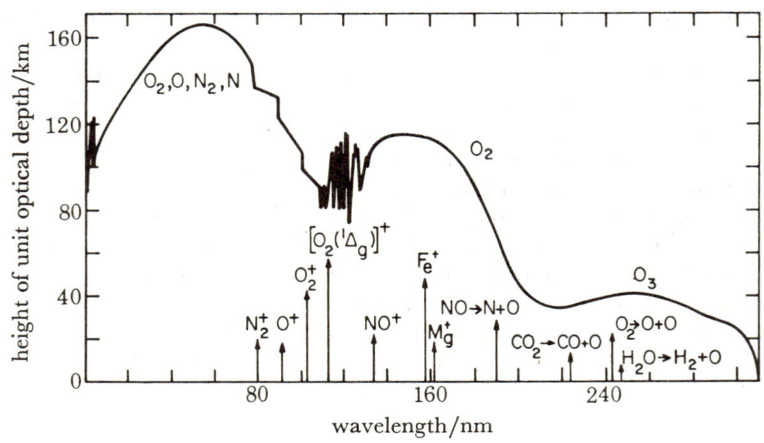

FIGURE 1. The variation with wavelength of the height of unit optical depth for solar radiations incident vertically. The wavelengths corresponding to the ionization and dissociation thresholds of certain constituents are indicated; ionization processes are represented by + sign.

Relatively few measurements of neutral composition have been made in the region between 50 and 130 km and our present knowledge and understanding are based largely on theoretical models. These involve solutions of the continuity equations for individual constituents with account being taken of photodissociation or photoionization and chemical reactions in the production and loss terms. For constituents having long lifetimes, transport processes can seriously influence the spatial variations, and increased attention has been paid in the past decade to the incorporation of such processes in theoretical models. To date, the greatest attention has been paid to vertical diffusion, the effects of both molecular and turbulent diffusion being first examined for molecular and atomc oxygen by Colegrove *et al.* (1965), but subsequently considered in connection with a number of other constituents.

The purpose of the present paper is to describe the available experimental information on the neutral and ion composition in the height range 50–130 km. For the neutral constituents it is convenient to consider in turn oxygen, hydrogen, nitrogen and carbon species. For ionized constituents a subdivision based on height of occurrence is more natural; it is found that above about 90 km the observed ions represent in the main the products of the primary photoionization processes, whereas at lower heights they arise from reactions of these products with molecules. Where possible, the experimentally determined distributions for both neutral and ionized constituents are related to corresponding theoretical predictions.

2. Neutral composition

(a) *Oxygen species*

(i) *Molecular oxygen*

As indicated in figure 1, the atmospheric penetration of solar radiations in the extreme ultraviolet (e.u.v.) wavelength region is controlled by molecular oxygen absorption. Friedman *et al.* (1951) first took advantage of this to estimate molecular oxygen concentrations in the thermosphere from rocket-borne measurements of solar e.u.v. intensities. More recently, this approach has also been used in satellite observations of occultation of the sun by the Earth's limb. The early rocket observations of ultraviolet absorption and of mass spectrometer sampling were reviewed by Offermann (1974). In the meantime, the mass spectrometer technique has been improved by the use of cryogenically cooled sources (Offerman & Tatarczyk 1973), which minimizes the formation of background gases within the source, and also eliminates the shock front ahead of the ion source; measurements with this technique can now be achieved down to 80–90 km.

Rocket-borne measurements of the molecular oxygen height distribution using a Lyman-α sensor (Weeks 1975) and a mass spectrometer (Trinks *et al.* 1978*a*) have shown the presence of wavelike structures in the height region 75–150 km. Measurements of absorption at 145 nm (Higgins & Heroux 1977), referring to the upper part of the height region of interest, indicate increases in molecular oxygen concentration at times of magnetic disturbance which are consistent with previous results derived for greater heights using satellite observations of solar ultraviolet occultation. Recent occultation measurements at more than one wavelength by Garriott *et al.* (1977) have also shown that such increases of molecular oxygen occur near 130 km but not at 100 km. These observations carried out in solar minimum years also revealed a diurnal variation of molecular oxygen near 100 km, the values deduced at sunset being larger than those at sunrise. Corresponding measurements during solar maximum years by Roble & Norton (1972) also provided evidence of a seasonal change, the summer concentrations being three times the corresponding winter values.

(ii) *Atomic oxygen*

Atomic oxygen, $O(^3P)$, is perhaps the most important single neutral constituent in the 50–130 km height range since it plays a central role in the thermal balance of the lower thermosphere, is involved in reactions responsible for a number of airglow emissions, and is an important reactant in both the positive- and negative-ion chemistry. Consequently, with molecular oxygen, it has attracted the greatest attention in neutral-composition measurements in this height range, and a total of five different rocket-borne experiments have been devised for its measurement. The most successful approaches have involved the use of improved mass spectrometers, in which the nature of the ion source has been a major consideration, and ultraviolet lamps for generating the $OI(^3P–^3S_1)$ triplet near 130 nm; the measurement of resonance scattering of this radiation from ambient atoms provides a sufficient dynamic range to cover the concentrations in the 60–140 km height range, while the simultaneous measurement of absorption over a 40 cm path furnishes the absolute calibration (Dickinson *et al.* 1979). The use of a cryogenically cooled ion source in the mass spectrometer (see, for example, Offerman & von Zahn 1971; Trinks *et al.* 1978*a*) has minimized the loss of atomic oxygen at the source walls, and the incorporation of cryogenic pumps (Philbrick *et al.* 1973; Arnold & Krankowsky 1977*a*) has enabled measurements to be extended down to 80 km; for such heights, a suitably shaped sampling cone (Arnold & Krankowsky 1975) has reduced the difficulties associated with the formation of a detached bow shock.

Even with these improvements, the atomic oxygen data available from the use of mass spectrometers are still sparse, but the repeated use of the ultraviolet lamp technique (Dickinson *et al.* 1979) has provided information on the distribution over the 60–140 km height range during a variety of conditions. The curves in figure 2 represent the results of measurements during

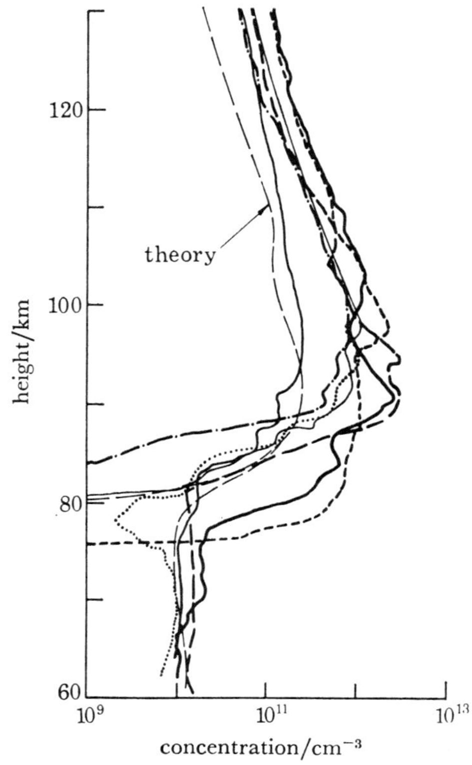

FIGURE 2. Height distributions of atomic oxygen measured at S. Uist, Scotland (Dickinson *et al.* 1979), and deduced using a theoretical model (Thomas & Bowman 1972). The distributions showing decreases to low concentrations below the maxima refer to night-time conditions, and those showing almost constant concentrations below about 77 km refer to daytime; the dotted curve represent measurements during a dawn period.

seven rocket flights at S. Uist, Scotland, during winter and equinox months in the period April 1974–February 1977. Also shown by the labelled curves are the results deduced theoretically for daytime and night-time during equinox months at mid-latitudes (Thomas & Bowman 1972). The predicted disappearance of oxygen atoms at heights below about 81 km between daytime and night-time is reproduced in the measurements, although the actual height at which this change occurs is very variable. From figure 2 it is seen that at heights near and above the maxima in the atomic oxygen distributions the observations show considerably larger concentrations than those predicted theoretically. For these heights the factors determining the atomic oxygen concentrations in the theoretical model are chiefly the solar-flux intensities, particularly in the 130–200 nm wavelength range, and the values of the coefficient representing vertical turbulent diffusion (Lettau 1951; Colegrove *et al.* 1965). The solar-flux intensities have been the subject of some controversy but the relatively low values of Parkinson & Reeves (1969) adopted in the present theoretical results have been confirmed in more recent results (Delaboudinière *et al.* 1978). A reduction in the values of eddy diffusion coefficient adopted in the theoretical treatment by a factor of 3 or 4 at all heights could be sufficient to reconcile the predicted concentrations with the observations, the reduced values still being within the range suggested by von Zahn (1967) in connection with observations of argon and molecular nitrogen concentrations.

A notable feature of the atomic oxygen distributions in figure 2 is the marked wavelike structure observed at heights above about 80 km; attention has already been drawn to the presence of such structure in corresponding molecular oxygen measurements.

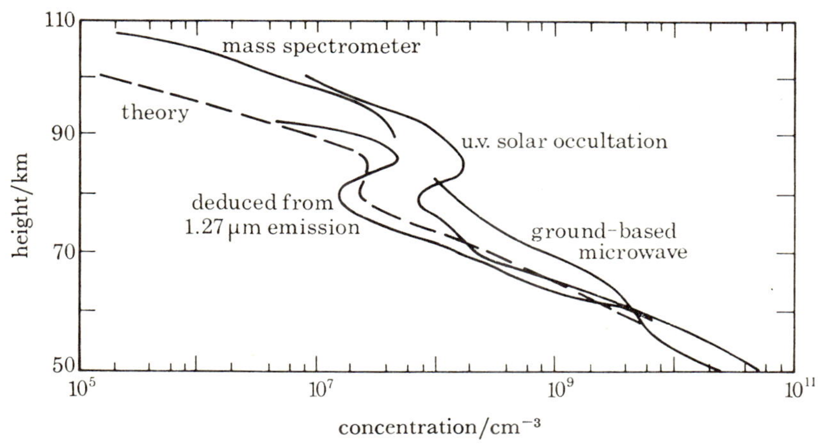

Figure 3. Daytime height distributions of ozone derived from rocket-borne observations of $O_2(a^1\Delta_g - X^3\Sigma_g^-)$ emissions at 1.27 µm (Llewellyn & Witt 1977), of solar ultraviolet occultation (Miller 1976), and of concentrations, with the use of a mass spectrometer (Weeks *et al.* 1978); results deduced from observations of thermal emission at 110.8 GHz are also shown (Penfield *et al.* 1976). A distribution obtained by using a theoretical model (Thomas & Bowman 1972) is included for comparison.

(iii) *Ozone*

The primary interest in ozone at mesospheric heights arises from its role in the excitation of infrared emissions in the airglow and in the negative-ion chemistry. Height distributions measured during daytime using four different techniques are shown in figure 3. The maximum observed near 85 km was first identified by Evans *et al.* (1968) from rocket-borne observations of the (0–0) band emission of $O_2(a\ {}^1\Delta_g - X\ {}^3\Sigma_g^-)$ at 1.27 µm in the dayglow, on the basis that the excited molecules are produced by the photolysis of ozone by absorption of solar radiation in the Hartley continuum (Vallance Jones & Gattinger 1963). The corresponding profile in figure 3 was derived for daytime by Llewellyn & Witt (1977) from measurements during twilight at Kiruna, Sweden, during March 1975. The solar ultraviolet occultation method takes advantage of the dominant influence of ozone on absorption of radiations of wavelengths above about 220 nm, the distribution shown in figure 3 being measured by Miller (1976). Observation from a satellite of the attenuation of ultraviolet radiations from stellar sources (Hays & Roble 1973) have also shown the presence of the mesospheric maximum in the ozone height distribution. Theoretical studies reproduce this maximum in the distribution, as illustrated by the results of Thomas & Bowman (1972) represented by the broken curve in figure 3; these studies have demonstrated that the distinction and magnitude of this feature depend on the values of the eddy diffusion coefficient, and for large values of this coefficient it almost disappears.

The mass spectrometer results of Weeks *et al.* (1978) extend to the greatest heights in figure 3 and the microwave measurements from about 82 km downwards refer to measurements of the emission of the 100.8 GHz rotational line by Penfield *et al.* (1976). The latter measurements showed a day-to-night increase in concentration by a factor of about 2 over the 55–80 km height range. Some evidence for such a diurnal change has also been found by Noxon (1975) from an analysis of the difference between the morning and evening twilight enhancements of the (0–1) band of $O_2(b\ ^1\Sigma_g–X\ ^3\Sigma_g^-)$ emission at 864 nm, assuming that excitation occurs by energy transfer from $O(^1D)$ atoms produced by ozone photolysis. Such a day-to-night change in ozone concentration is predicted by theoretical models (Thomas & Bowman 1972; Moreels *et al.* 1977).

(iv) $O_2(a\ ^1\Delta_g)$

Apart from their role as the source of the infrared atmospheric system in airglow emissions, $O_2(a\ ^1\Delta_g)$ molecules represent an important ionizable constituent in the D-region and are also involved in the negative-ion chemistry. The height distributions of these excited molecules during daytime and twilight, as deduced from airglow observations, can be understood on the basis of current photochemical models. However, the observed emission intensities at 1.27 μm during night-time are larger than are predicted by these models. It has been usually considered that the recombination of oxygen atoms in the presence of a third body is responsible for these excited molecules at night-time, but the discrepancy between theory and experiment implies that some other mechanism is involved (Moreels *et al.* 1977).

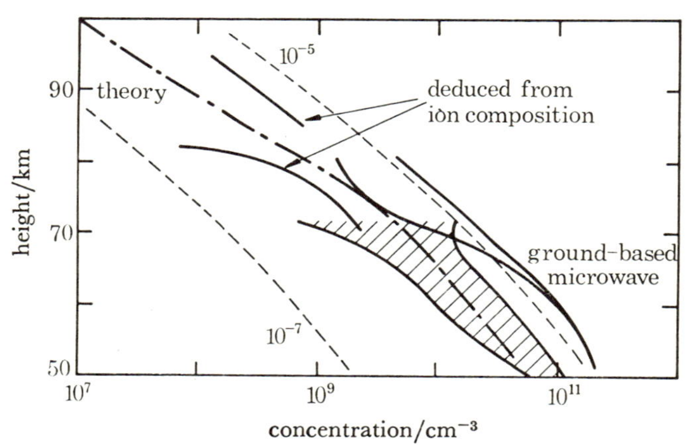

FIGURE 4. Height distributions of water vapour derived from rocket-borne observations of thermal emission in the wing of the 6.3 μm band (Rogers *et al.* 1977) and of ion composition (Swider & Narcisi 1975; Arnold & Krankowsky 1977*b*), and from ground-based observations of thermal emission at 22.2 GHz (Radford *et al.* 1977). Also shown is the distribution deduced by using a theoretical model (Thomas & Bowman 1972), and those corresponding to constant mixing ratios of 10^{-5} and 10^{-7}. Shaded region, rocket-borne infrared.

(*b*) *Hydrogen species*

(i) *Water*

The distribution of water vapour in the mesosphere is of interest in studies of noctilucent clouds, in the quantitative interpretation of complex positive ions observed below 80 km (see § 3*b*), and in considerations of the escape of hydrogen atoms from the upper thermosphere.

The limited information available on the water concentration at mesospheric heights is illustrated in figure 4.

The shaded region refers to estimates derived by Rogers *et al.* (1977) from rocket-borne observations of thermal emission in the long-wavelength wing of the 6.3 μm band, using a liquid helium cooled spectrometer; the range of values shown correspond to the 3σ limits on the radiance measurements. Since collisional deactivation of the $H_2O(010)$ molecule is more rapid than spontaneous emission at heights below about 85 km, the interpretation of the data required information on mesospheric temperatures and concentrations of molecular oxygen, nitrogen and carbon dioxide. From this interpretation Rogers *et al.* suggested that the measurements, which were carried out at Chatanika, Alaska, following an auroral glow, were consistent with a mixing ratio of 3.5 ± 2 parts/10^6 between 49 and 70 km. The second direct estimates shown in figure 4 were derived by Radford *et al.* (1977) from ground-based observation of emission of the line centred at 22.2 GHz, with the use of the 37 m radio telescope at Haystack Observatory, Massachusetts. The height range of the measurements was limited by receiver bandwidth at the lower heights and by the inability to resolve the Doppler-broadened hyperfine components of the line at the upper heights. Radford *et al.* reported errors of ± 40 % in the derived concentrations. It is to be noted that the water concentrations derived were rather larger than those normally expected; the largest value corresponded to a mixing ratio of 15 parts/10^6 at 60 km.

The two other height distributions included in figure 4 were deduced from ion mass spectrometer measurements, on the basis of the known ion chemistry. That shown for the 70–82 km height range is based on data obtained at Fort Churchill, Canada, during a solar proton event (Swider & Narcisi 1975), and that for the 85–95 km range refers to measurements at high-latitude sites in Scandinavia (Arnold & Krankowsky 1977*b*). The two sets of results could have been influenced by fragmentation of weakly bound ambient ions during sampling. It is interesting to note that these two distributions imply quite different conditions of vertical turbulent transport; that of Swider & Narcisi shows little evidence of transport of water molecules from below to replace those destroyed by photodissociation, whereas that of Arnold & Krankowsky indicates a slope similar to those of the lines representing constant mixing of 10^{-5} (10 parts/10^6) and 10^{-7} (0.1 parts/10^6) and, therefore, implies very efficient vertical mixing. The two distributions can be compared in figure 4 with the theoretical results derived by Thomas & Bowman (1972) for intermediate values of eddy diffusion coefficient and a mixing ratio of 3 parts/10^6 at 60 km.

(ii) *Water dissociation products*

According to present ideas, water represents the principal form of hydrogen in the mesosphere, the remainder being present as molecular hydrogen with small amounts of OH, H and HO_2, arising directly or indirectly from water dissociation, and also as methane. At thermospheric heights atomic and, perhaps, molecular hydrogen represent the only significant forms. The only direct measurement of these constituents in the height range of present interest is that of OH by Anderson (1971*a*, *b*) from rocket-borne observations of resonant scattering of the (0–0) band of OH($A\,^2\Sigma$–$X\,^2\Pi$) at 306.4 nm during evening twilight conditions. The local concentrations deduced: 4.4×10^6 cm^{-3} at 50 km, 5.5×10^6 cm^{-3} at 60 km, and 3.5×10^6 cm^{-3} at 70 km are reasonably consistent with values predicted theoretically (Thomas & Bowman 1972; Moreels *et al.* 1977).

Estimates of hydrogen atom concentrations have been made in the interpretation of airglow emission. Evans & Llewellyn (1973) analysed simultaneous measurements of the oxygen infrared atmospheric band and of hydroxyl emissions to deduce, first, the height distribution of ozone and subsequently that of atomic hydrogen, on the basis that the reaction of these two constituents gives rise to the vibrationally excited hydroxyl radicals and, hence, the hydroxyl emissions. These results showed an increase in atomic hydrogen concentration from 10^6 cm^{-3} near 60 km to 3×10^7 cm^{-3} between 80 and 90 km, with values of 10^7 cm^{-3} near 100 km. Noxon (1975) also suggested an upper limit of 8×10^6 cm^{-3} for the atomic hydrogen concentration near 100 km during night-time, to permit the increase of ozone required for the dawn enhancement of the (0–1) band of the atmospheric system of molecular oxygen (see § 2*a* (iii)). These two sets of results imply water mixing ratios much smaller than the 3 parts/10^6 assumed at 60 km in the theoretical curve of figure 4.

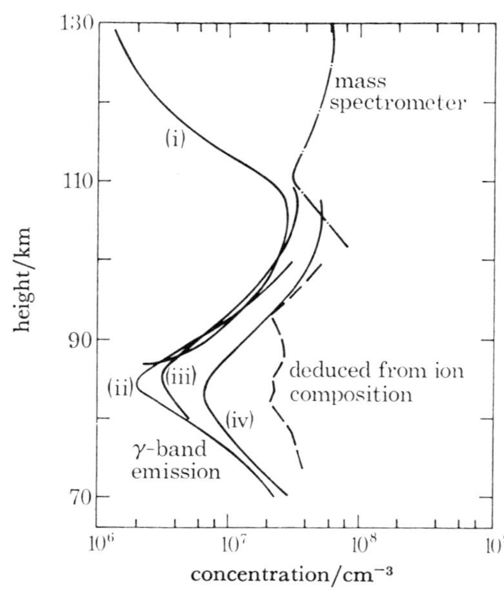

FIGURE 5. Height distributions of nitric oxide derived from rocket-borne observations of the γ-band fluorescence in the dayglow: (i) Tohmatsu & Iwagami (1976), (ii) Baker *et al.* (1977), (iii) Tohmatsu & Iwagami (1975), (iv) Meira (1971); of concentrations with the use of a mass spectrometer (Trinks *et al.* 1978*b*); and of ion composition during a disturbed winter day (Arnold & Krankowsky 1977*a*).

(*c*) *Nitrogen species*

Interest in the nitrogen constituents in the height range of interest has been inspired in part by the importance of nitric oxide in the formation of the ionospheric D-region and as a charge-transfer agent in the E-region. Studies of the other important form of nitrogen, N(^{4}S), at these heights of interest have been limited to theoretical models, although a concentration of 1.9×10^7 cm^{-3} has been deduced by Feldman & Takacs (1974) from the interpretation of twilight airglow observations. Almost all measurements of the nitric oxide distribution in the mesosphere and lower thermosphere have been based on observations of the γ-band fluorescence in the dayglow, after Barth (1964); the (1–0) and (2–0) bands at 214.9 nm and 204.7 nm have generally been used. At mesospheric heights, the need to remove background radiation due to Rayleigh scattering is a major complication. This component was previously deduced from atmospheric density measurements and the resultant small γ-band column emission rates showed large uncertainties; differentiation of the column rates provided the volume emission rates and, hence, the nitric oxide concentrations. In a development introduced by Tohmatsu & Iwagami (1975), an ultraviolet photometer coupled with a nitric oxide gas-filled absorption cell enables more accurate corrections for the Rayleigh-scattered background radiation to be made and thereby furnishes improved measurements of the dayglow emission; the cell is placed cyclically in the optical path and serves as a rejection filter for the gamma bands.

Figure 5 shows four height distributions of nitric oxide deduced from γ-band observations. Curves (i), (ii), and (iii) were derived from measurements using a nitric oxide filled cell; curve (i) extending from 87 to 130 km is based on measurements of Tohmatsu & Iwagami (1976) in early morning at equatorial latitudes; curve (ii) represents the results of Baker *et al.* (1977) near sunset at mid-latitudes; and curve (iii) was derived for sunset conditions at mid-latitudes by Tohmatsu & Iwagami (1975). The fourth set of results based on γ-band emission measurements, curve (iv) in figure 5, represents the earlier measurement of Meira (1971) near noon at mid-latitudes, and refers to a disturbed winter day (Geller & Sechrist 1971). In all these distributions deduced from γ-band emissions, account has been taken of the recent revision of the emission rate factor (Cravens 1977). Even larger increases in nitric oxide concentrations at mid-latitudes during disturbed winter conditions than those implied by curve (iv) are illustrated by the distribution represented by the broken curve in figure 5. This was deduced from ion-composition measurements carried out during an anomalous winter day in Spain by Arnold & Krankowsky (1977*a*). It is believed that such enhancements in nitric oxide concentrations contribute to the increased electron concentrations which give rise to anomalously high radio-wave absorption on such winter days.

The large concentrations in the lower thermosphere indicated by the mass-spectrometer measurements of Trinks *et al.* (1978*b*) for mid-afternoon at mid-latitudes, figure 5, are supported by simultaneous measurements of the (1–0) band. On this basis, it appears that the differences between the two distributions in the height range 110–130 km in figure 5 cannot be attributed to differences in the experimental techniques adopted. It seems likely that this difference arises in part out of the different local times of measurement, as expected from the theoretical results of Ogawa & Kondo (1977). These workers draw attention to the influence at the heights of interest of nitric oxide transported from upper thermospheric levels where the production from reaction of $N(^4S)$ atoms with oxygen molecules is very dependent on temperature and, hence, on local time. For heights below 100 km the main uncertainty in the theoretical understanding is in the vertical transport but the quantum yield of $N(^2D)$ atoms from various reactions is also relevant since at these heights nitric oxide is chiefly produced by the reaction of these excited atoms with oxygen molecules.

The variability of the nitric oxide concentrations indicated in figure 5 has been examined for lower thermospheric heights by satellite-borne measurements using the γ-band fluorescence technique; these measurements refer to heights above 90 km where the background due to Rayleigh scattering is small. The nadir-viewing experiment on OGO 4 satellite provided measurements of the vertical column density of nitric oxide above the shadow line at twilight; with this experiment Rusch & Barth (1975) found highly variable quantities of nitric oxide at latitudes above 60° N and Gérard & Barth (1977) observed a factor of 4 enhancement at high latitudes during a magnetic storm, the concentration at 110 km reaching a value of 4×10^8 cm^{-3}. More recently, the limb-scanning experiment on Atmospheric Explorer C satellite has provided information on the global morphology of nitric oxide at 105 km, including the effects of magnetic activity (Cravens & Stewart 1978), and has also demonstrated diurnal and seasonal variations at low latitudes (Stewart & Cravens 1978). The results show concentrations of $4 \times 10^7\ cm^{-3}$ to $6 \times 10^7\ cm^{-3}$ at high latitudes, up to 68° N, with considerable variability; near the equator the concentrations are nearer $2 \times 10^7\ cm^{-3}$ and show little change with longitude or magnetic activity. Also, at low latitudes, the concentrations show an increase from about $10^7\ cm^{-3}$ at sunrise to about $2 \times 10^7\ cm^{-3}$ during the afternoon, followed by a reduction towards sunset; an asymmetry is observed in the nitric oxide distribution about the equator, except during equinox periods, the concentrations being larger on the summer than winter side. No detailed comparison with theory is possible but Stewart & Cravens (1978) deduce that the diurnal changes are consistent with a chemical loss of nitric oxide by reaction with ambient nitrogen atoms, a daytime atomic concentration of $10^7\ cm^{-3}$ being required, which can be compared with the value of $1.9 \times 10^7\ cm^{-3}$ reported previously by Feldman & Takacs (1974).

(*d*) *The carbon species*

The main carbon constituent at 50–130 km is carbon dioxide, although on the basis of theoretical models the monoxide is expected to show comparable concentrations near the upper boundary of this height range (Wofsy *et al.* 1972; Shimazaki & Cadle 1973). The carbon dioxide is of interest primarily because of its role in the thermal balance of the mesosphere and lower thermosphere, and because of its involvement in both the positive and negative-ion chemistry of the D-region.

The measurement of the height distribution of carbon dioxide by mass spectrometer techniques was facilitated by the inclusion of cryogenically cooled ion sources (Offermann & Tatarczyk 1973); these serve to suppress reactions involving oxygen, whose products obscured the carbon dioxide within earlier forms of ion sources. Night-time measurements by Scholz & Offermann (1974) and others showed complete mixing of the constituent up to the turbopause, with a mixing ratio similar to that in the troposphere, i.e. 3×10^{-4}. At greater heights the distribution corresponded to that of diffusive equilibrium. Recently, a similar mass spectrometer experiment successfully measured the carbon dioxide concentrations over the height range 90–150 km during daytime (Trinks & Fricke 1978). The results again showed that below the turbopause the mixing ratio corresponded to that at tropospheric heights but at greater heights the decrease with height was more rapid than that expected for diffusive equilibrium. This is attributed mainly to photolysis up to about 115–120 km, with the production of carbon monoxide, and a reaction with O^+ ions at greater heights with unspecified products. Such photolysis of carbon dioxide by the Schumann–Runge bands is believed to be the main source of carbon monoxide in the mesosphere and upper stratosphere, the dissociation by the Schumann–Runge continuum being more important in the lower thermosphere. Theoretical models incorporating eddy diffusion have predicted the height distribution of carbon monoxide in the mesosphere (Hays & Olivero 1970; Wofsy *et al.* 1972), and these have been considered by Waters *et al.* (1976) in their interpretation of ground-based measurements of the absorption of the carbon monoxide rotational line at 115.3 GHz in the solar spectrum to yield the height distribution of this constituent. The results obtained were consistent with a mixing ratio of about 0.1 parts/10^6 near 50 km increasing to about 10 parts/10^6 near 78 km, and a constant value of about 40 parts/10^6 above 100 km.

3. Ion composition

(a) *Upper mesosphere and lower thermosphere*

Measurements of ion composition at heights between about 90 km and 130 km have been chiefly obtained by rocket-borne magnetic and quadrupole mass spectrometers. The results of several measurements (Narcisi 1973) have shown that NO^+ and O_2^+ are the dominant ions during daytime, their concentrations being comparable. At night the relative proportion of NO^+ increases, its concentration being about an order of magnitude greater than that of O_2^+.

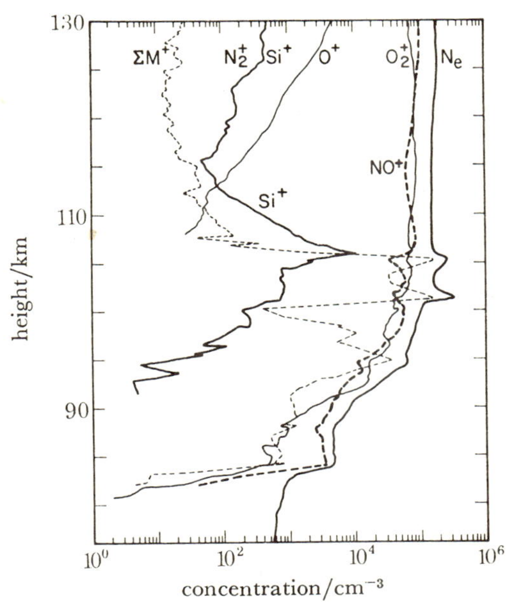

Figure 6. The positive-ion composition and height distributions of total metal-ion and electron concentrations, as derived from rocket-borne mass spectrometer, probe, and radio-propagation measurements (Herrmann *et al.* 1978) at Wallops Island in daytime during August 1976.

Typical daytime results for these ions are shown in figure 6 and are derived from measurements with a magnetic type mass spectrometer by Herrmann *et al.* (1978). Also shown are the height variations of total metal-ion concentrations, ΣM^+. The electron concentrations, N_e, were deduced from probe and propagation experiments. The observations of metal ions, with Mg^+ and Fe^+ being predominant, in the region 85–120 km is a common finding of mass spectrometer observations; the layer with a peak near 94–95 km is a permanent feature of the lower ionospheres, but the upper layer is more intermittent.

It is usually considered that the occurrence of O_2^+ and NO^+ ions is a well understood characteristic of the E-region. Only a few chemical reactions control the relative proportions of NO^+, O_2^+, O^+ and N_2^+ ions but these reactions involve the minor neutral constituents O, NO, N, as well as the major constituents O_2 and N_2, and a knowledge of the neutral composition is, therefore, essential. In view of the marked temperature gradient in this height region, data on the temperature dependence of chemical rate coefficients is also an important requirement. Attention has been paid to this temperature dependence in laboratory measurements over the past decade and information is also being obtained from analyses of Atmospheric Explorer satellite data for greater heights (Torr & Torr 1978). For night-time conditions, a detailed understanding of the ion-composition measurements requires information on the production of ionization by scattered Lyman-α and Lyman-β radiations. Furthermore, with the longer lifetimes for ions resulting from the smaller ionization densities, account needs to be taken of the movements of ionization in the upper part of the height region under consideration (Strobel *et al.* 1974).

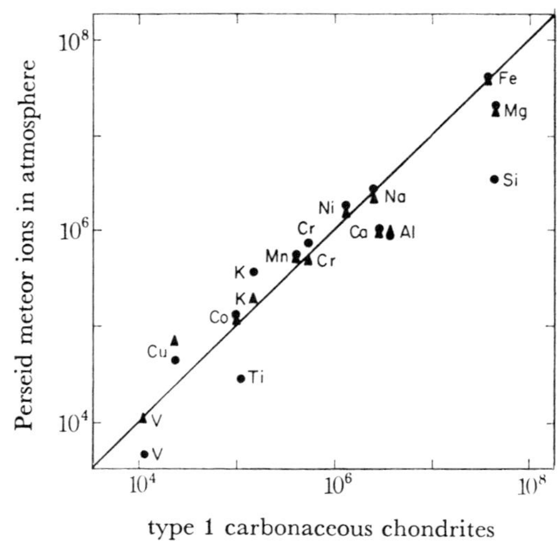

FIGURE 7. A comparison of the relative concentrations of metal ions observed between 91 and 107.5 km and of metal atoms in carbonaceous chondrites, normalized to the values for iron (Herrmann *et al.* 1978). ●, 99–107.5 km; ▲, 91–99 km.

The observations that the metal-ion concentrations are enhanced during periods of meteoric activity (Narcisi 1968; Goldberg & Aikin 1973) have suggested a meteoric source for the corresponding metal atoms. Support for this suggestion has been found in the agreement between the relative abundance of observed ions and of metal atoms in chondrite meteorites, first noted by Goldberg & Aikin (1973). In this connection, the relative concentrations of metal ions observed during the measurements of figure 6, which were carried out 12 h after the Perseid meteor shower, have been compared by Herrmann *et al.* (1978) with the metal-atom concentrations in carbonaceous chondrites, as shown in figure 7; here the values were normalized to those of iron. It is seen that for elements such as Ni, Cr, Co, Mn and Na the agreement is very good, but marked discrepancies can be seen for Ti and Si. The importance of a meteor source for alkali metals has also been demonstrated in some measurements of the column densities of sodium and potassium atoms in the height ranges 80–100 km from the resonance scattering of suitably tuned ground-based laser beams (Megie *et al.* 1978). However, it is suggested that during winter months the relative concentrations of these elements are no longer consistent with a meteoric source and a possible contribution from sea spray is invoked.

Consideration of the cross sections for collisions at high energies has suggested that the production of metal ions during meteor ablation is relatively unimportant. Instead, a large proportion of the observed ions is believed to arise from photoionization or charge-transfer processes. The ionization thresholds for iron and magnesium correspond to wavelengths of 157.5 nm and 162.2 nm, respectively, and, as seen from figure 1, these do not penetrate below 100 km. However, because of the relatively low energies corresponding to these thresholds, and still smaller values for sodium and others atoms, charge transfer can occur even with NO^+ ions. The direct loss of the metal ions by radiative recombination is slow and it seems likely that transport to lower levels followed by three-body processes are chiefly responsible for their disappearance.

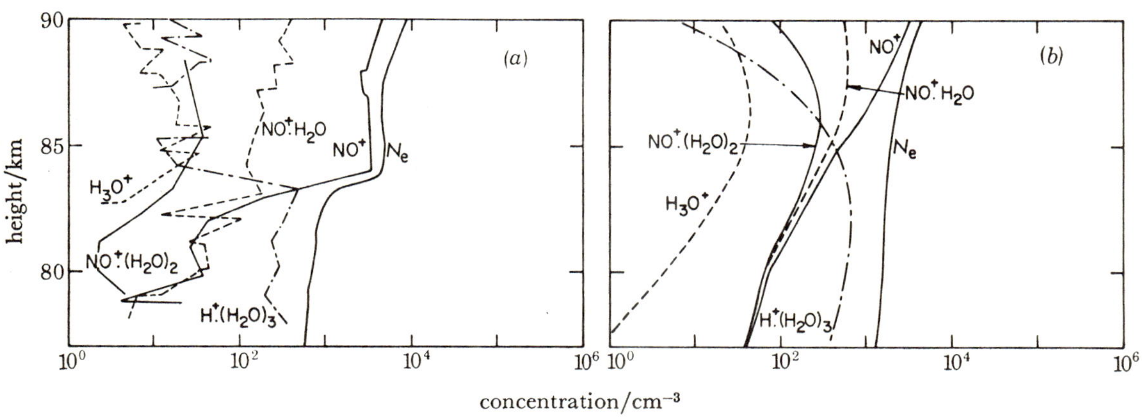

FIGURE 8. A comparison of (*a*) a simplified version of the ion composition measured in daytime using a rocket-borne mass spectrometer at Wallops Island during August 1976 (Kopp *et al.* 1978) and (*b*) the ion composition predicted for summer daytime conditions by using a theoretical model in which NO^+ represents the precursor of water cluster ions (Thomas 1976).

(*b*) *Middle and lower mesosphere*

(i) *Positive ions*

Rocket-borne measurements with quadrupole and magnetic mass spectrometers, incorporating suitable pumping arrangements, carried out since 1963 at a number of sites have shown that under normal conditions ions of the type $H^+.(H^2O)_n$ dominate the positive-ion composition below about 83 km in daytime, and below about 86 km during twilight or night-time. The majority of early measurements showed $H^+.(H_2O)_2$ as the dominant cluster ion with significant concentrations of H_3O^+ and $H^+.(H_2O)_3$ also being present. There is, however, some doubt whether these measurements represented ambient conditions since fragmentation of complex weakly bound ions could occur through thermodynamic break-up because of increased temperatures at the shock layer associated with the supersonic velocity of the rocket, or through collisions resulting from the draw-in electric fields. To overcome these effects, modified geometries of the sampling region have been introduced and lower attracting potentials (Arnold & Krankowsky 1975; Kopp *et al.* 1978). With these improvements, higher-order cluster ions have been customarily observed.

The main features of the results during a daytime flight of the magnetic mass spectrometer are shown in figure 8*a*. This represents a simplified version of the observations presented by Kopp *et al.* (1978); thus the O_2^+ ion distribution is not shown and that of $H^+.(H_2O)_4$, which is almost identical with that of $H^+.(H_2O)_3$, is also omitted. Similar results have been found on a number of occasions using a quadrupole mass spectrometer experiment (Arnold & Krankowsky 1977*a*).

The early measurement of these water cluster ions by Narcisi & Bailey (1965) was quite unexpected on the basis of the prevailing ideas concerning the formation of the D-region (Nicolet & Aikin 1960); NO^+ and O_2^+, produced both by photoionization and by ion–molecule reactions were predicted as the dominant ions. Although a reaction scheme for the formation of cluster ions from O_2^+ ions was formulated on the basis of laboratory measurements, the supply of O_2^+ ions from the photoionization of $O_2(^1\Delta_g)$ (Huffman *et al.* 1971) is believed to be too small to be significant. Instead, NO^+ is now believed to be the precursor of water cluster ions under normal conditions. Reactions involving the successive hydration by three-body attachment of water molecules to produce $NO^+.(H_2O)_3$ followed by the formation of $H^+.(H_2O)_3$, were first proposed by Fehsenfeld & Ferguson (1969). The inclusion of more rapid hydration processes involving weakly-bound intermediate ions, such as $NO^+.CO_2$ and $NO^+.N_2$, has helped to formulate a reaction scheme for the formation of water cluster ions from NO^+ ions. One aspect of this scheme, which arises from the temperature dependence of the three-body processes giving rise to these intermediate ions and of the collisional dissociation of these ions, is a marked temperature dependence expected for the ion composition of the ionospheric D-region. Recent calculations based on this scheme have shown that the main features of the ion composition are reproduced. This is illustrated in figure 8*b*, which shows results of calculations for summer conditions, as described by Thomas (1976). It is seen that the relative concentrations of different ion species are similar to those in the experimental results, presented in figure 8*a*. By suitable choice of the height distribution of nitric oxide concentration, and of temperature, it would be possible to reproduce the general forms of the ion distributions. A notable difference in figure 8*a*, *b* is that the theoretically predicted concentrations of $NO^+.(H_2O)_2$, and to a lesser extent of $NO^+.H_2O$, are substantially larger than those actually observed. Reid (1977) has suggested that this discrepancy between theory and observation arises because of dissociation of the NO^+ hydrates during mass spectrometer sampling. Alternatively, Charkrabarty *et al.* (1978) have invoked collisional dissociation of the first and second hydrates of NO^+, and have adopted values of rate coefficients to ensure agreement between theoretical results and observations. However, there is no supporting evidence for such values from laboratory investigations.

(ii) *Negative ions*

It is well known that the relatively large atmospheric pressures encountered at D-region heights permit the formation of negative ions by the three-body attachment of electrons to neutral atoms and molecules. Early laboratory measurements showed that such attachment occurs rapidly for molecular oxygen, but more recent studies have revealed that the O_2^- ions take part in a complicated sequence of reactions involving the neutral constituents O_3, CO_2, NO, NO_2, and H_2O.

The information on negative-ion composition, derived from two sets of mass spectrometer measurements, has shown serious discrepancies, as illustrated in Figure 9*a*, *b*. The results of figures 9*a* refer to measurement by Arnold *et al.* (1971) during weak auroral conditions at high latitudes, and those of figure 9*b* represent the measurements of Narcisi *et al.* (1972) carried out at middle latitudes near totality of a solar eclipse. Figure 9*a* indicates that CO_3^- was the dominant ion present, significant concentrations of HCO_3^- and NO_3^- were present between about 70 and 80 km, and substantial concentrations of Cl^- ions, of isotopic masses 35^- and 37^-, were also observed. In figure 9*b* the negative-ion composition is dominated by cluster ions of masses corresponding to $NO_3^-.(H_2O)_n$, with possible admixtures of $CO_3^-.(H_2O)_n$, where n = 0–5, and a layer of ions of masses in excess of 150^- was also observed near 88 km.

Neither of these sets of mass spectrometer observations are consistent with what is predicted from theoretical models based on laboratory measurements of negative-ion reactions. However, the relative concentrations of negative and positive ions, measured during the total-ion mode of operation of mass spectrometers and with electrostatic probes for total positive-ion concentrations, show good agreement with theoretical predictions. This is illustrated in figure 10, which compares the results of daytime measurements during winter at two mid-latitude sites (Arnold & Krankowsky 1977*a*, 1978) with results derived from a theoretical model (Thomas *et al.* 1973). These height variations indicate that free electrons are the dominant negatively charged species at heights above about 70 km.

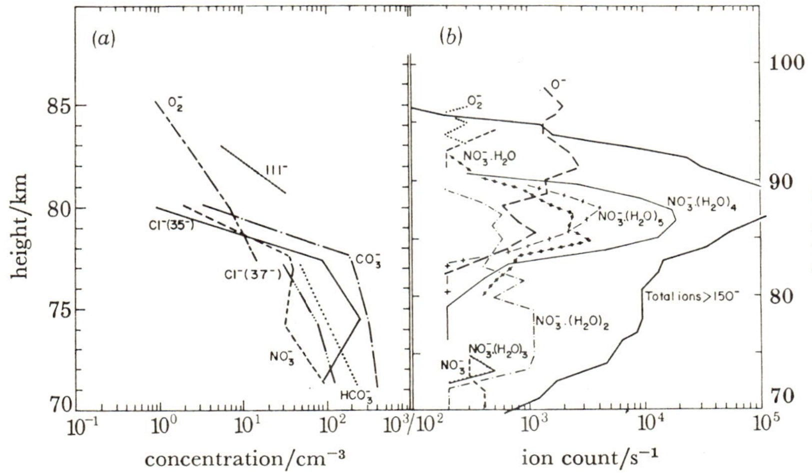

FIGURE 9. The negative-ion composition of the D-region as measured by rocket-borne mass spectrometer: (*a*) Arnold *et al.* (1971); (*b*) Narcisi *et al.* (1972).

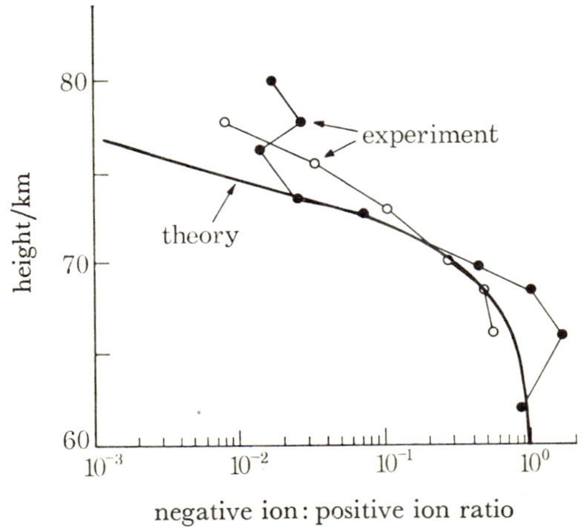

FIGURE 10. A comparison of the ratios of negative-ion to positive-ion concentrations, as deduced from rocket-borne measurements with mass spectrometers and probes (Arnold & Krankowsky 1977*a*, 1978), with those derived using a theoretical model (Thomas *et al.* 1973).

4. CONCLUSIONS

Measurements of neutral and ionized constituents in the mesosphere and lower thermosphere have been limited by the fact that *in situ* sampling can, in general, be achieved only from rockets. Furthermore, remote-sensing measurements of minor neutral constituents based on optical techniques have encountered serious difficulties, especially from Rayleigh scattering at mesospheric heights. As a consequence, most investigations involving a knowledge of the neutral or ionized composition at these heights have relied heavily on theoretical models. In this connection, the agreement found between the main features of the theoretical predictions and the limited experimental data acquired in recent years (see §§ 3 and 4) is encouraging.

The height distributions measured for certain constituents, such as those for atomic oxygen and nitric oxide in figures 2 and 5, respectively, have indicated considerable variability. To understand such variability, measurements with a particular technique need to be extended in time and in geographic coverage, as illustrated by the satellite-borne measurements in the lower thermosphere of molecular oxygen from observations of solar e.u.v. occultation (see § 2*a*(i)) and of nitric oxide from observations of the γ-band fluorescence in the dayglow (see § 2*c*). Similar observations of thermal emissions and solar occultation in the infrared and microwave spectral regions hold considerable promise of measuring several minor neutral constituents in the heights of present interest on a worldwide basis.

Improved information on the distributions of minor neutral constituents, such as atomic oxygen and nitric oxide, will help resolve some of the uncertainties in the current understanding of the ionospheric D- and E-regions. It has become clear that the neutral and ionized chemistries cannot be realistically separated. For instance, nitric oxide represents a critical constituent in the ion chemistry of the D- and E-regions but its production is itself dependent on the formation of $N(^2D)$ atoms, largely during processes involving NO^+ and N_2^+ ions. The need for coordinated measurements of neutral and ion composition is, therefore, indicated. Such measurements of neutral and positive-ion composition will benefit from recent developments in mass spectrometer techniques but, to date, these have not been successful in establishing the ambient negative-ion composition of the lower D-region.

References (Thomas)

Ackerman, M., Biaume, F. & Kockarts, G. 1970 *Planet. Space Sci.* **18**, 1639–1651.

Anderson, J. G. 1971*a* *J. geophys. Res.* **76**, 4634–4652.

Anderson, J. G. 1971*b* *J. geophys. Res.* **76**, 7820–7824.

Arnold, F., Kissel, J., Krankowsky, D., Wieder, H. & Zahringer, J. 1971 *J. atmos. terr. Phys.* **33**, 1169–1175.

Arnold, F. & Krankowsky, D. 1975 *Proc. Int. Symp. on Solar Terrestrial Physics*, Sao Paulo, Brazil, vol. 3, pp. 30–50. May 1974.

Arnold, F. & Krankowsky, D. 1977*a* *Dynamical and chemical coupling between the neutral and ionized atmosphere* (ed. B. Grandal & J. A. Holtet), pp. 93–127. Dordrecht: D. Reidel.

Arnold, F. & Krankowsky, D. 1977*b* *Nature, Lond.* **268**, 218–219.

Arnold, F. & Krankowsky, D. 1978 *Report of Max-Planck-Institute for Nuclear Physics*, Heidelberg, F.R. Germany, V9.

Baker, K. D., Nagy, A. F., Olsen, R. O., Oran, E. S., Randhawa, J., Strobel, D. F. & Tohmatsu, T. 1977 *J. geophys. Res.* **82**, 3281–3286.

Barth, C. A. 1964 *J. geophys. Res.* **69**, 3301–3303.

Chakrabarty, D. K., Chakrabarty, P. & Witt, G. 1978 *J. atmos. terr. Phys.* **40**, 437–442.

Colegrove, F. D., Hanson, W. B. & Johnson, F. S. 1965 *J. geophys. Res.* **70**, 4931–4941.

Cravens, T. E. 1977 *Planet. Space Sci.* **25**, 369–372.

Cravens, T. E. & Stewart, A. I. 1978 *J. geophys. Res.* **83**, 2446–2452.

Crutzen, P. 1970 *Q. Jl R. met. Soc.* **96**, 767–769.

Delaboudinière, J. P., Donnelly, R. F., Hinteregger, A. E., Schmidtke, G. & Simon, P. C. 1978 *Intercomparison/compilation of relevant solar flux data related to aeronomy*, *COSPAR, Technique Manual* 7.

Dickinson, P. H. G., Bain, W. C., Thomas, L., Williams, E. R., Jenkins, D. B. & Twiddy, N. D. 1979 *Proc. R. Soc. Lond.* A (In the press).

Evans, W. F. J., Hunten, D. M., Llewellyn, E. J. & Vallance Jones, A. 1968 *J. geophys. Res.* **73**, 2885–2896.

Evans, W. F. J. & Llewellyn, E. J. 1973 *J. geophys. Res.* **78**, 323–326.

Fehsenfeld, F. C. & Ferguson, E. E. 1969 *J. geophys. Res.* **74**, 2217–2222.

Feldman, P. D. & Takacs, P. Z. 1974 *Geophys. Res. Lett.* **1**, 169–171.
Friedman, H., Lichtman, S. W. & Byram, E. T. 1951 *Phys. Rev.* **83**, 1025–1030.
Garriott, O. K., Norton, R. B. & Timothy, T. G. 1977 *J. geophys. Res.* **82**, 4973–4982.
Geller, M. A. & Sechrist, C. F. 1971 *J. atmos. terr. Phys.* **33**, 1027–1040.
Gérard, J. C. & Barth, C. A. 1977 *J. geophys. Res.* **82**, 674–680.
Goldberg, R. A. & Aikin, A. C. 1973 *Science, N.Y.* **180**, 294–296.
Hays, P. B. & Olivero, J. J. 1970 *Planet. Space Sci.* **18**, 1729–1733.
Hays, P. B. & Roble, R. G. 1973 *Planet. Space Sci.* **21**, 273–279.
Herrmann, U., Eberhardt, P., Hidalgo, M. A., Kopp, E. & Smith, L. G. 1978 *Space research* XVIII (ed. M. J. Rycroft), pp. 249–252. Oxford: Pergamon.
Higgins, J. E. & Heroux, L. 1977 *J. geophys. Res.* **82**, 3295–3298.
Huffman, R. E., Paulsen, D. E., Larrabee, J. C. & Cairns, R. B. 1971 *J. geophys. Res.* **76**, 1028–1038.
Hunten, D. M. 1967 *Space Sci. Rev.* **6**, 493–573.
Kopp, E., Eberhardt, P. & Herrmann, U. 1978 *Space research* XVIII (ed. M. J. Rycroft), pp. 245–248. Oxford: Pergamon.
Lettau, H. 1951 *Compendium of meteorology* (ed. T. F. Malone), pp. 320–333. Boston, Mass.: American Meteorological Society.
Llewellyn, E. J. & Witt, G. 1977 *Planet. Space Sci.* **25**, 165–172.
Megie, G., Bos, F., Blamont, J. E. & Chanin, M. L. 1978 *Planet. Space Sci.* **26**, 27–35.
Meira, L. G. 1971 *J. geophys.* **76**, 202–212.
Miller, D. E. 1976 Reported by Llewellyn & Witt (1977).
Moreels, G., Megie, G., Vallence Jones, A. & Gattinger, R. L. 1977 *J. atmos. terr. Phys.* **39**, 551–570.
Narcisi, R. S. 1968 *Space research* VIII (ed. A. P. Mitra, L. G. Jacchia & W. S. Newman), pp. 360–369. Amsterdam: North-Holland.
Narcisi, R. S. 1973 *Physics and chemistry of upper atmosphere* (ed. B. M. McCormac), pp. 171–183. Dordrecht: D. Reidel.
Narcisi, R. S. & Bailey, A. D. 1965 *J. geophys. Res.* **70**, 3687–3700.
Narcisi, R. S., Bailey, A. D., Wlodyka, L. E. & Philbrick, C. R. 1972 *J. atmos. terr. Phys.* **34**, 647–658.
Nicolet, M. & Aikin, A. C. 1960 *J. geophys. Res.* **65**, 1469–1483.
Noxon, J. F. 1975 *J. geophys. Res.* **80**, 1370–1373.
Offermann, D. 1974 *J. geophys. Res.* **79**, 4281–4293.
Offerman, D. & Tatarczyk, H. 1973 *Rev. scient. Instrum.* **44**, 1569–1572.
Offerman, D. & von Zahn, U. 1971 *J. geophys. Res.* **76**, 2520–2522.
Ogawa, T. & Kondo, Y. 1977 *Planet. Space Sci.* **25**, 735–742.
Parkinson, W. H. & Reeves, E. M. 1969 *Solar Phys.* **10**, 342–347.
Penfield, H., Litvak, M. M., Gottlieb, C. A. & Lilley, A. E. 1976 *J. geophys. Res.* **81**, 6115–6120.
Philbrick, C. R., Faucher, G. A. & Trzcinski, E. 1973 *Space research* XIII (ed. M. J. Rycroft & S. K. Runcorn), pp. 255–160. Berlin: Akademie-Verlag.
Radford, H. E., Litvak, M. M., Gottlieb, C. A., Gottlieb, E. W., Rosenthal, S. K. & Lilley, A. E. 1977 *J. geophys. Res.* **82**, 472–478.
Reid, G. C. 1977 *Planet Space Sci.* **25**, 275–290.
Roble, R. G. & Norton, R. B. 1972 *J. geophys. Res.* **77**, 3524–3533.
Rogers, J. W., Stair, A. T., Degges, T. C., Wyatt, C. L. & Baker, D. J. 1977 *Geophys. Res. Lett.* **4**, 366–368.
Rusch, D. W. & Barth, C. A. 1975 *J. geophys. Res.* **80**, 3719–3721.
Scholz, T. G. & Offermann, D. 1974 *J. geophys. Res.* **79**, 307–310.
Shimazaki, T. & Cadle, R. D. 1973 *J. geophys. Res.* **78**, 5352–5361.
Stewart, A. I. & Cravens, T. E. 1978 *J. geophys. Res.* **83**, 2453–2456.
Strobel, D. F., Young, T. R., Meier, R. R., Coffey, T. P. & Ali, A. W. 1974 *J. geophys. Res.* **79**, 3171–3178.
Swider, W. & Narcisi, R. S. 1975 *J. geophys. Res.* **80**, 655–664.
Thomas, L. 1976 *J. atmos. terr. Phys.* **34**, 1345–1350.
Thomas, L. & Bowman, M. R. 1972 *J. atmos. terr. Phys.* **34**, 1843–1858.
Thomas, L., Gondhalekar, P. M. & Bowman, M. R. 1973 *J. atmos. terr. Phys.* **35**, 397–404.
Tohmatsu, T. & Iwagami, N. 1975 *Space research* XV (ed. M. J. Rycroft), pp. 241–245. Berlin: Akademie-Verlag.
Tohmatsu, T. & Iwagami, N. 1976 *J. Geomagn. Geoelect., Kyoto* **28**, 343–358.
Torr, T. G. & Torr, M. R. 1978 *Rev. Geophys. Space Phys.* **16**, 327–340.
Trinks, H. & Fricke, K. H. 1978 *J. geophys. Res.* **83**, 3883–3886.
Trinks, H., Offermann, D., von Zahn, U. & Steinhauer, C. 1978*a* *J. geophys. Res.* **83**, 2169–2176.
Trinks, H., von Zahn, U., Barth, C. A. & Kelly, K. K. 1978*b* *J. geophys. Res.* **83**, 203–206.
Vallance Jones, A. & Gattinger, R. L. 1963 *Planet. Space Sci.* **11**, 961–974.
Waters, J. W., Wilson, W. J. & Shimabukuro, F. I. 1976 *Science, N.Y.* **191**, 1174–1175.
Weeks, L. H. 1975 *J. geophys. Res.* **80**, 3655–3660.
Weeks, L. H., Good, R. E., Randhawa, J. S. & Trinks, H. 1978 *J. geophys. Res.* **83**, 978–982.

Witt, G. 1969 *Space research* IX (ed. K. S. W. Champion, P. A. Smith & R. L. Smith-Rose), pp. 157–169. Amsterdam: North-Holland.
Wofsy, S. C., McConnell, J. C. & McElroy, M. B. 1972 *J. geophys. Res.* **77**, 4477–4493.
von Zahn, U. 1967 *J. geophys. Res.* **72**, 5933–5937.

Phil. Trans. R. Soc. Lond. A **296**, 261–268 (1980) [261]
Printed in Great Britain

A winter anomaly campaign in Western Europe

By D. Offermann
University of Wuppertal, 56 Wuppertal 1, Postfach 10 01 27, Federal Republic of Germany

An integrated campaign of ground-based, balloon- and rocket-borne measurements was performed in Western Europe during winter 1975–76. The campaign was aimed at the study of the atmospheric–ionospheric coupling during pronounced disturbances of ionospheric D-region electron density (winter anomaly). Various atmospheric and ionospheric parameters were measured in the altitude regime 30–120 km. A large set of middle atmosphere temperature data as obtained from satellites is also available for the time period of the campaign. Some major aspects of the campaign are described and respective results presented as they are available at the present state of data evaluation.

Introduction

The D-region winter anomaly was detected many years ago by radio propagation measurements at medium latitudes. Radio-waves of suitable frequency passing through the ionospheric daytime D-region and being reflected from some higher ionospheric layers are attenuated (absorbed) in the D-region (75–95 km). The degree of absorption is dependent on the electron density and collision frequency in the D-region. The electron density is assumed to be mostly produced by photoionization of NO by solar Ly-α radiation. Therefore, a seasonal variation of the D-region absorption with a minimum in winter is to be expected. This is found, indeed, but superimposed on it there are frequent and sometimes very strong enhancements of absorption with a duration of one to several days. They are called 'winter anomaly events'. The intensity and distribution of these events has been studied by ground based radio propagation measurements for quite a number of years and at many places.

The events appear to occur in 'patches' with horizontal extensions of several hundreds of kilometers, and sometimes up to 2000 km. Latitudinal variations have been found as well as correlations with solar activity during the solar cycle. Short period variations frequently occur as wavelike structures with periods between 5 and 25 days. Coupling to other atmospheric or ionospheric layers are not infrequent, and there is also some influence of geomagnetic disturbances. When separating the geomagnetic and solar flare induced effects from the rest of the phenomena, one remains with a class of events which are called the 'meteorological type' of winter anomaly. Only this type will be considered here. (For a compilation of recent winter anomaly literature see Offermann (1979*a*)).

Campaign description

On the face of it the winter anomaly phenomenon appears to be a rather simple electron density effect. It turns out, however, to be very complicated as soon as one considers the reasons for the density variations. Many atmospheric and ionospheric parameters appear to influence them, and it is this fact that makes the winter anomaly so interesting for the study of atmospheric and ionospheric processes and of their interactions. The Western European Winter Anomaly Campaign described here therefore contained a large number of different experiments. They can be roughly grouped into five different categories:

1. definition and verification of meteorological type winter anomaly events;
2. determination of large scale structures (as winter anomaly patches or wind systems);
3. determination of the local thermodynamic structure (vertical wind, temperature, and density profiles);
4. measurements of the neutral atmosphere composition including trace constituents;
5. measurements of plasma parameters as electron density, ion composition, ion mobility and conductivity.

For a detailed description see Offermann (1977, 1979*b*). Each of these five groups contained ground-based, balloon-borne, and rocket-borne experiments. A number of remote sensing satellite experiments were also in operation during the campaign. The balloons and rockets were launched from the site 'El Arenosillo' in Southern Spain (37°6′ N; 6° 44′ W). The ground stations included several radio-wave propagation experiments as well as a number of meteor wind and incoherent scatter radars. They were located at various places in Spain, France, Great Britain, Italy, Austria and Germany. The campaign lasted from the end of December 1975 until mid-February 1976. A total of 22 scientific groups participated in it. They performed 38 different types of experiments on 47 rockets, 23 balloons, and in 12 ground-based stations. Quite a few of these experiments were performed not to study the D-region itself, but to explore the atmospheric layers below and above it to look for mutual couplings. The campaign yielded a large amount of data, which in part are still being evaluated. Only a few selected results are therefore summarized here.

Electron densities

Seventeen *in situ* electron density determinations were performed during the campaign (Friedrich *et al.* 1979). Good correlation with ground based radio-wave absorption measurements was found. The data show that there were density enhancements during winter anomaly events up to one order of magnitude, and that they essentially occurred in the altitude régime 75–95 km.

The reason for enhanced electron densities may be increased ion production rates as well as decreased recombination rates. Enhancements of the production rate may be caused by temperature changes, which affect the chemistry of minor atmospheric constituents. Other causes might be horizontal and/or vertical transports of minor constituents with low ionization potential (like NO). Decreases of the electron recombination rate, too, can be effected by temperature changes, for instance by the temperature sensitivity of the water cluster ion chemistry. Similar effects could be imagined to be produced by transports (horizontal and/or vertical) of suitable minor constituents (such as H_2O, etc.). Therefore the question arises whether the production rate was increased, or whether the recombination rate was decreased during the winter anomaly events found.

Ion production rates

Ion production rates were calculated by Thrane *et al.* (1979) for 2 days of the campaign (4 and 21 January, 1976) from *in situ* solar flux and particle precipitation measurements. They found that the two major production sources in the D-region were ionization of NO (by Ly-α) and of $O_2(^1\Delta_g)$. To determine whether the ion production rate from NO was normal or enhanced, one has to consider the solar Ly-α flux and the NO density. The Ly-α flux was found to have been normal. To make a similar statement about the NO density, a NO reference profile for undisturbed conditions is required. Such a profile does not exist. To obtain a rough idea of how it might look, figure 1 shows a collection of NO profiles inferred from ion spectrometer measurements under quiet conditions (Swider 1978). The data show considerable fluctuations.

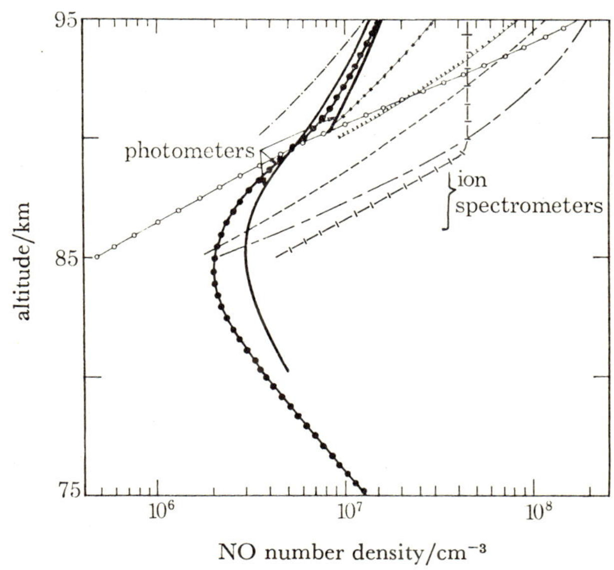

FIGURE 1. NO densities in the D-region during undisturbed conditions. The two heavy lines represent photometer measurements by Tohmatsu (1977), the heavy dotted profile is a photometer experiment of Baker *et al.* (1977), and the remaining seven profiles (thin lines) represent densities inferred from ion spectrometer measurements (Swider 1978).

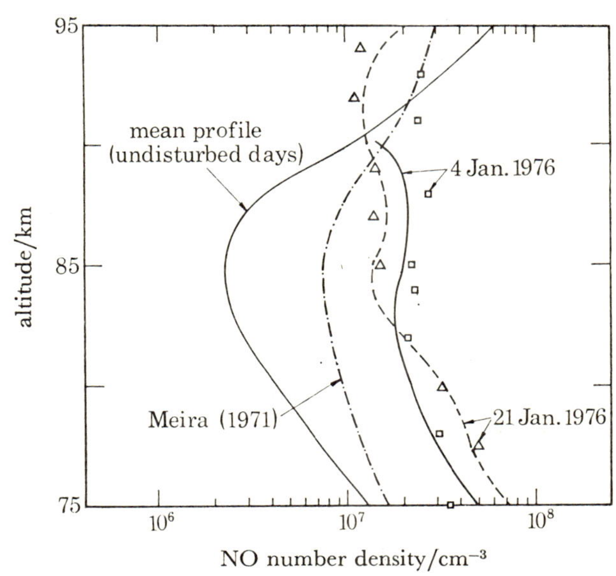

FIGURE 2. NO densities measured during the winter anomaly as compared to a mean quiet time profile derived from figure 1. The squares and triangles give densities obtained from ion spectrometers. Other data shown were measured by photometers. The profile of Meira (1971) is given for comparison.

A number of quiet time NO measurements by photometers are also shown in figure 1 (Tohmatsu 1977; Baker *et al.* 1977). Despite the scatter of the data and although the different techniques may conceal different systematic errors, a mean curve was calculated from the data given in figure 1, and is shown in figure 2. This is, of course, a very crude method, and it is used only to obtain an approximate idea of what an undisturbed NO density profile might be. In figure 2 this mean profile is compared to NO densities as obtained on winter anomalous days (4 and 21 January 1976) during the campaign described here. Results of two photometers and two ion spectrometer measurements are shown. Meira's (1971) profile is also given for comparison. Figure 2 shows that in the important régime 80–90 km the NO densities were considerably enhanced on both days. This means that also the ion production rates calculated by Thrane *et al.* (1979) represent enhanced values.

$O_2(^1\Delta_g)$ densities were obtained from a 1.27 μm infrared experiment, and were also inferred from two ion spectrometer measurements during the campaign (Beran & Bangert 1979; Arnold & Krankowsky 1979). Ion production rates calculated on the basis of these densities are about a factor of 3 higher than one would normally expect for an undisturbed atmosphere. They are comparable to those obtained from NO (Thrane *et al.* 1979).

The conclusion, therefore, is that the ion production rate appears to have been enhanced during winter anomalous days of the campaign. It should be noted in this context that all NO measurements during winter anomaly days which are in the literature so far, do show considerable density enhancements in the D-region (Offermann 1979*a*). On the contrary respective $O_2(^1\Delta)$ enhancements appear to be present only occasionally. This makes it difficult to answer the question how these two types of enhancement come about. Combined action of two atmospheric parameters (as, for instance, temperature plus transport) may be required.

Electron loss rates

As mentioned above, another reason for enhanced D-region electron density may be a decrease of the electron loss rate. Effective loss rates were calculated by Thrane *et al.* (1979) from the ion production rates and the measured electron densities. In the régime 80–90 km the loss rate is found to be low, indeed. In the lowest part of the D-region (around 78 km) it shows, however, a steep increase by more than an order of magnitude on both days considered (4 and 21 January 1976). This is attributed to an important change in ion composition occurring at that altitude: ion spectrometer measurements show that at this level the number density of NO^+ ions became equal to the sum of the densities of the water cluster ions (Arnold & Krankowsky 1979). This height level is called the 'transition height' (below it the water cluster ions dominate). The transition height occurs at 83–85 km under normal conditions. During the winter anomaly it is found to be appreciably lowered on all occasions for which it was reported in the literature, sometimes even below the 78 km level reported here (Offermann 1979*a*).

The low transition height means that the water cluster ions essentially disappear from the D-region during winter anomaly. This indicates an explanation of the low loss rates reported here, because the recombination rate of the cluster ions is much larger than that of the NO^+ ions. Explanations of ion composition changes of this kind could be found in temperature changes or in transports of suitable minor constituents. As in the case of the production rate also here one is referred to the thermodynamic structure of the atmosphere for a more detailed understanding of the winter anomaly.

Thermodynamic structure of the atmosphere

Temperatures and winds measured during the campaign were analysed by Offermann *et al.* (1979), Rees *et al.* (1979) and Labitzke *et al.* (1979). As concerns the temperatures, strong wavelike structures were found in the vertical temperature profiles in the altitude régime 30–110 km above the rocket range on the 2 days considered above (4 and 21 January 1976, see Offermann *et al.* (1979)). These structures were very similar on those 2 days. An analysis of more rocket measurements and quite a number of satellite data at four different altitude levels between 31 and 80 km shows that these wavelike structures persisted in the altitude régime covered for at least 1 month. They contained short, medium, and long period components (periods of roughly 6–7 days, 12–15 days, and of the order of a month), which appeared to have fixed phase relations at the various altitude levels. This indicates that the thermodynamic state of the atmosphere was governed by pronounced wavelike disturbances during the entire campaign. In the 80 km temperature field these structures were horizontally extended to distances of at least 2000 km.

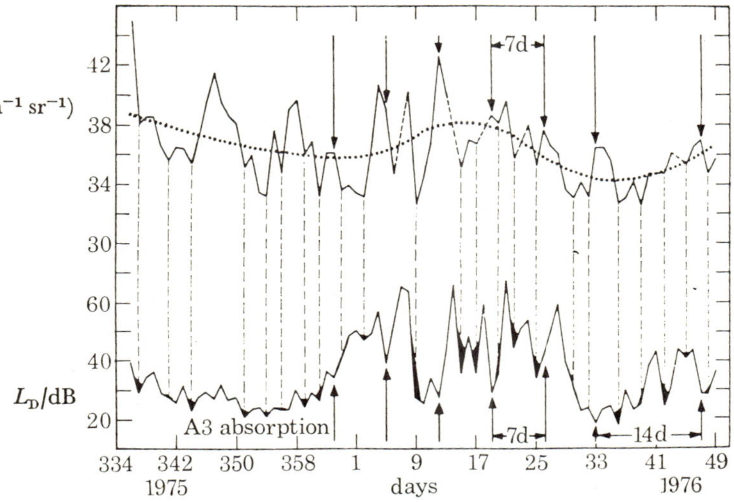

FIGURE 3. Comparison of winter anomaly activity and temperatures at 80 km. Strength of winter anomaly is represented by daily means L_D of A3 radio wave absorption. Infrared fluxes as measured by the p.m.r. experiment (channel 3000) of the Nimbus 6 satellite are taken as a measure of 80 km temperatures. The long term temperature trend is indicated by a dotted line; it roughly parallels the mean intensity of the winter anomaly activity. Good correspondence of short term fluctuations is found when connecting the minima in the two curves by vertical dashed lines. The respective correspondence of maxima is much less pronounced, and is definitely missing on a number of days, for instance, those indicated by arrows. (For a detailed discussion see Offermann *et al.* 1979).

Disturbances of this kind could influence the D-region electron densities. The decrease of water cluster ion density discussed above would, for instance, be compatible with a temperature enhancement in the same altitude régime. Such enhancements in the D-region are found indeed in the vertical temperature profiles measured on 4 and 21 January 1976. Furthermore, the satellite data indicated generally increased temperatures around 80 km (long term enhancements) during those time periods of winter 1975–6 when the winter anomaly activity was strong. This is shown in figure 3. A more detailed correlation between variations of 80 km temperatures and winter anomaly activity (as represented by the strength of absorption of radiowaves L_D), i.e. a correlation of short term fluctuations was sought. As shown in figure 3, there appears to be a correspondence of low values of temperature and winter anomaly. High values, however, do not correlate regularly but only occasionally. From this, one may conclude that high temperature is a necessary condition for winter anomaly to occur, but not a sufficient one. Something else (a 'cofactor') is necessary in addition to the high temperature to produce winter anomaly events on this timescale (Offermann *et al.* 1979). There is some indication that this cofactor might be controlled by atmospheric dynamics. (Note the 7-day intervals between the days marked by arrows in figure 3.)

As regards the nature of the cofactor one could obviously think of some horizontal and/or vertical transport of a suitable atmospheric trace constituent. NO and $O_2(^1\Delta_g)$ (or related species) might be good candidates as they were found enhanced during winter anomaly. Horizontal NO transports from the auroral zones to medium latitudes by large scale wind systems have been suggested in the literature as an explanation of the winter anomaly. For the time of the campaign therefore a circulation analysis for the Northern Hemisphere was performed on the basis of temperature and wind data obtained from radiosondes, rockets, satellites, and meteor wind radars (Labitzke *et al.* 1979). Circulation charts were computed for 1 day per week, and a total of 6 weeks. The circulation was generally found to be highly variable. If one compares the wind speeds and directions obtained to the winter anomaly data discussed here one does not arrive at convincing evidence that horizontal transport alone is a sufficient means of explaining the winter anomaly events. In addition, the magnetic activity was generally low during the campaign (Offermann 1979*b*). Hence only moderate NO production in the auroral zone is expected for that time period, if one may assume that the results of Cravens & Stewart (1978) at 105 km are also indicative for the D-region.

Labitzke *et al.* (1979) also calculated mean vertical motions, which could transport, for instance, NO from the E-region reservoir into the D-region. The authors feel that combination of both horizontal and vertical motions may explain some winter anomaly events. It should be mentioned, however, that the analysis made by these authors refers to radio-wave absorption measurements taken at a different place and in a somewhat higher altitude régime than the data discussed here.

Vertical transport other than by mean motions could be effected by atmospheric turbulence. NO or O might be transported downward from the 100 km level by enhanced eddy diffusion. Enhanced O densities in the D-region could lead to enhanced ozone production and thereby to the increased $O_2(^1\Delta)$ densities mentioned above. A transport mechanism of this kind would have to be compatible with a number of constraints imposed by the following measurements:

(1) Enhancements of NO density appear to occur regularly during winter anomaly events, whereas $O_2(^1\Delta)$ density increases are not always found (see above). Furthermore, there are measurements of enhanced NO densities that were accompanied by high O densities, as well as other measurements that found high NO densities with decreased O densities (Krankowsky *et al.* 1979; Dickinson 1977). As NO and O are necessarily transported at the same time, it appears difficult to reconcile these different findings by a mechanism with only one free parameter (transport).

(2) Turbulence must be strong enough to effect strong D-region modifications sometimes in as short a time as half a day (see figure 3). When estimating the eddy coefficients, values of the order of 10^7 cm^2/s are obtained which must have been present in a large altitude régime during the campaign described here. The more serious problem, however, appears to be the removal – at the end of a winter anomaly event – of the large amounts of NO from the D-region, where

it is rather stable against photodissociation. The mechanism of removal has to be as fast as the build-up process, as is seen from figure 3. Turbulent downward transport into the stratosphere appears to be questionable, because the Richardson numbers calculated for the lower mesosphere (50–70 km) indicate that the region was rather stable during the campaign. This assumes, however, that turbulent conditions were similar above Spain and some surroundings. This assumption is not trivial, because wind speeds at D-region heights were large (Labitzke *et al.* 1979), and, therefore, vertical transports a few thousand kilometres away from Spain might have influenced the measurements above the rocket range. The ongoing data analysis will therefore have to check on this assumption. Vertical eddy diffusion also would have to be modulated, as only part of the winter anomaly variability shown in figure 3 can be attributed to temperature variations (see Offermann *et al.* 1979). Such modulation should be detectable by analysing the Richardson numbers measured in the D-region during the campaign. First results of this analysis indicate that turbulence in the D-region was strong throughout the campaign. It does, however, not appear to be modulated. (The same caveat concerning horizontal extension of turbulent conditions applies as above.)

In consequence, a vertical eddy transport model also appears to have its difficulties. The future data analysis will therefore have to carefully consider the interplay of horizontal wind systems, vertical mean motions, vertical eddy transport, and temperature changes. It will then hopefully improve the understanding of the role of minor constituents and of the cofactor during winter anomaly events.

Conclusions

During the Western European Winter Anomaly Campaign 1975–6, an indication was found that winter anomaly events (i.e. electron density enhancements in the D-region) were accompanied by enhancements of ion production rates as well as by decreases of electron loss rates in the D-region. Ion production rate increases were due to density enhancements of NO and $O_2(^1\Delta_g)$. Decreases of loss rates could be attributed to decreased densities of water cluster ions. Destruction of water cluster ions might have been caused, at least in part, by considerable temperature enhancements. Density variations of the minor constituents mentioned (NO, $O_2(^1\Delta_g)$, and related species) could be brought about by combined action of horizontal and vertical transports. This latter effect will be analysed in more detail during the future data evaluation.

The Western European Winter Anomaly Campaign 1975–6 was supported by Bundesministerium für Forschung und Technologie, Bonn; Norwegian Defence Research Establishment, Kjeller; Instituto Nacional de Técnica Aerospacial, Madrid; Austrian Council of Scientific Research, Wien; U.K. Science Research Council, London.

References (Offermann)

Arnold, F. & Krankowsky, D. 1979 *J. atmos. terr. Phys.* (submitted).
Baker, K. D., Nagy, A. F., Olsen, R. O., Oran, E. S., Randhawa, J., Strobel, D. F. & Tohmatsu, T. 1977 *J. geophys. Res.* **82**, 3281–3286.
Beran, D. & Bangert, W. 1979 *J. atmos. terr. Phys.* (submitted).
Cravens, T. E. & Stewart, A. I. 1978 *J. geophys. Res.* **83**, 2446–2452.
Dickinson, P. H. G. 1977 *Geophys. J. R. astr. Soc.* **49**, 295.

Friedrich, M., Torkar, K. M., Spenner, K., Rose, G. & Widdel, H. U. 1979 *J. atmos. terr. Phys.* (submitted).
Krankowsky, D., Arnold, F., Friedrich, V. H. & Offermann, D. 1979 *J. atmos. terr. Phys.* (submitted).
Labitzke, K., Petzoldt, K. & Schwentek, H. 1979 *J. atmos. terr. Phys.* (submitted).
Meira, L. G. 1971 *J. geophys. Res.* **76**, 202–212.
Offermann, D. 1977 *J. Geophys.* **44**, 1–13.
Offermann, D. 1979*a* *J. atmos. terr. Phys.* **41**, (In the press.)
Offermann, D. 1979*b* *J. atmos. terr. Phys.* **41**, (In the press.)
Offermann, D., Curtis, P., Cisneros, J. M., Lauche, H., Petzoldt, K., Rose, G. & Satrustegui, J. 1979 *J. atmos. terr. Phys.* **41**, (In the press.)
Rees, D., Scott, A. F. D., Cisneros, J. M., Satrustegui, J., Widdel, H. U. & Rose, G. 1979 *J. atmos. terr. Phys.* **41**, (In the press.)
Swider, W. 1978 *J. geophys. Res.* **83**, 4407–4410.
Thrane, E. V., Bangert, W., Beran, D., Friedrich, M., Grandal, B., Hagen, O., Loidl, A., Spenner, K., Schwentek, H., Torkar, K. M. & Ugletveit, F. 1979 *J. atmos. terr. Phys.* **41**, (In the press.)
Tohmatsu, T. 1977 *Space Res.* **17**, 247–251.